Windows Assembly Language & Systems Programming

16- and 32-bit low-level programming for the PC and Windows

2nd edition

Barry Kauler

Lecturer, Edith Cowan University
M.Sc.(EE), C.Eng.

CRC Press
Taylor & Francis Group
Boca Raton London New York

CRC Press is an imprint of the
Taylor & Francis Group, an **informa** business

CRC Press
Taylor & Francis Group
6000 Broken Sound Parkway NW, Suite 300
Boca Raton, FL 33487-2742

First issued in hardback 2017

© 1997, Barry Kauler. Published by Taylor & Francis.
CRC Press is an imprint of Taylor & Francis Group, an Informa business

No claim to original U.S. Government works

ISBN 13: 978-1-138-41253-8 (hbk)
ISBN 13: 978-0-87930-474-4 (pbk)

**Visit the Taylor & Francis Web site at
http://www.taylorandfrancis.com**

**and the CRC Press Web site at
http://www.crcpress.com**

Contents

Preface

The *foreward*, or *preface*, often has the misfortune of being written last, or "tacked on" to finish the book, and tends to have much the same status as the *index* in the Author's mind.

I am no exception, though I determined that the index for this book would be thorough, and hope you find it to be so.

The preface however does not surrender itself to the simple mechanical process involved in producing an index. Instead I have to try and nicely summarise the entire book and categorize it in such a way that the browser will want to buy it, or, having bought it, will want to obtain a quick idea of what the book is all about.

My problem is that I can't neatly pigeon-hole this book. Perhaps it would be more correct to cut it into pieces and stuff it into several pigeon-holes. The problem behind my problem is its uniqueness. There is no other book in existence, at least at the time of writing, that specialises in **Windows assembly language**.

Let me identify those several pigeon-holes, and you can decide which ones are of interest to you ...

1. Windows Assembly Coding

First is the issue of learning to program in assembly language for Windows. The majority of programmers wouldn't have a clue where to start, due to an almost complete lack of published information. Many believe that Windows assembly language programming is very difficult, a viewpoint gained by comments made by some journalists or after examining the documentation provided in the SDK (Software Development Kit).

This book teaches Windows assembly language, assuming virtually no prior knowledge of assembly language, and shows Windows programming to be very approachable and in fact just as easy as programming in C or some other high-level language.

The first pigeon-hole is that this is a textbook that introduces and demystifies Windows assembly language.

2. Introduction to Windows

Approaching Windows assembly language first requires an understanding of what Windows programming is all about. Quite apart from the assembly language aspect, this book is a means to learn about Windows architecture and programming for someone who knows only about DOS. Thus the second pigeon-hole is that this book is an introduction to generic Windows architecture and programming, not just to Windows assembly language.

3. Advanced Architecture

The "nuts and bolts" approach means that you are learning from a level that gives you a solid foundation on which to build later knowledge. The closeness to the architectural features of the PC and CPU builds an intimate understanding of how the hardware interacts with the operating system and application software. The advanced features of the 286/386/486/Pentium, in particular the Protected modes, pose incredible new challenges to the programmer and engineer. Thus another pigeon-hole is that this is a hardware architecture course on the modern CPUs and PCs.

4. OOP Demystified

Always we are being inundated in trade journals with the words "object oriented programming" and with a seemingly endless convocation of object oriented products and tools hitting the market. Yet for many programmers, engineers, and enthusiasts, the topic remains a mystery. Even those proud new owners of Borland or Microsoft C++, or something similar, still aren't sure what it is they've got or how to fully utilize it.

OOP textbooks can be baffling in the extreme, but I have taken a very simple approach that completely demystifies OOP. Thus another pigeon-hole is that this book is also an introduction to advanced **software** architecture, especially OOP.

5. Assembly Language OOP

Not only does this book introduce OOP, but it also introduces assembly language OOP, and shows how complete Windows programs can be written in fewer than a dozen lines. Nobody anywhere has done anything like this, and it's pretty exciting stuff. Not only do I show how to write stand-alone assembly language OO programs, but I also show how they interface with object oriented high-level languages, particularly C++. Thus, yet another pigeon-hole is that this book introduces, and is an in-depth treatment of, assembly language object oriented programming for Windows.

6. Systems Programming

Windows programmers are fed a carefully regulated diet of information from Microsoft in the form of the SDK and various languages and support documentation. Those wanting deeper knowledge of the operating system and its hooks into the architecture have to hunt around. Information is to be found in obscure places, or by paying a lot of money, or by having a special status with Microsoft (such as being an approved ISV (Independent Software Vendor)). I have brought together extremely useful systems-level service information, such as the DPMI (DOS Protected Mode Interrupt) services, and other little-publicized functions and interrupts. Thus another pigeon-hole is that this book is an introduction to Windows systems programming and a reference manual for low-level unpublicized services.

7. MASM & TASM

I developed code for this book using both Microsoft MASM and Borland TASM, and I have included discussion and practical code using the new features of MASM version 6.0+ and TASM version 5.0+. These new features can be difficult to grasp from the manuals, so yet another pigeon-hole is that this book is a supplement to, or clearer explanation of, the material presented in the official manuals. Large sections of the TASM manual in particular may be incomprehensible due to its OO orientation, a situation that this book remedies.

8. Legacy Code

This book takes you right through, from 16-bit Windows 3.x to 32-bit Windows 95 programming, from ring three to ring zero, from DLLs to VxDs, from the 8086 to the Pentium. This book is a *foundation*, the "base of the triangle", if you wish, and I have not jumped into high-level work without the underlying theory and historical background. This is very important. In the workplace you may have to handle old PC systems, port DOS applications, maintain 16-bit Windows applications, upgrade applications, or communicate between old and new applications. See the "ladder of learning" in Figure 15.1, page 343.

Summing Up ...

There you go, eight solid reasons for buying this book!

Getting behind the scenes to do the tasks that other Windows programmers wouldn't even know how to start, such as **direct access to memory and I/O** explained in Chapter 9, **synchronising with real-time events** in Chapter 10, or **interaction between Protected and Real modes** explained in Chapter 11, is the kind of information that Windows programmers have been crying out for.

However, this book is targeted not just at professional programmers. Educators should seriously ask themselves how many of the concepts covered in this book are in their current courses. If your institution has units on assembly language, or systems programming, or microcomputer architecture, are those units really "with it"? It may be that you are teaching yesterday's technology and using yesterday's textbooks.

I also envisage a great deal of interest from hobbyists and enthusiasts: especially "bit fiddlers" and those who want to make Windows do what no one thought it could.

It does not really matter what language you use: if your Windows compiler is capable of in-line assembly, you're in business. It is even possible to make do with compilers that have their own mechanisms for emulating assembly language instructions. The opposite end of the spectrum is to have a stand-alone assembler, such as Borland's Turbo Assembler (TASM).

I would like to make one last, but certainly not least, comment that programming Windows at this level is **fun** as well as practical. Dare I add that I foresee a whole new era of Windows "hackers" (using the word in its original, legal, sense).

Acknowledgements

Thank you to my family, for the encouragement.

Thank you to colleagues at Edith Cowan University, for the advice and support.

And thank you to my computer, for surviving the tortures I put you through.

Barry Kauler, 1997

1

CPU Architecture

Preamble

Structure of the first two chapters

This chapter starts off from when a PC is first turned on, and I have assumed only a basic familiarity with computer principles. The focus is on the architecture of the CPU, that is, how the processor itself works.

Chapter 2 takes the next step by introducing the instruction set, the machine instructions that the processor understands.

A study method

I have structured the material like a ladder from a very basic level, so feel free to jump over any parts that you are already familiar with.

I recommend going through it with the objective of picking up the overall ideas, not worrying too much about nitty gritty details. A practical plan of action is to surge forward until you get to the chapters with some hands-on examples, then when you need to know some of the fine details, refer back as needed. You'll find the index to be comprehensive, with this in mind.

x86 family compatibility

In keeping with the ladder concept, I have covered the entire x86 family of processors, from the humble 8088 to the Pentium.

It is very important to note that these later CPUs are downward compatible, meaning they will run software from an earlier CPU, though the reverse is not necessarily true.

In this chapter I point out some of the major differences between the CPUs of the Intel family.

A "grassroots" approach

As the book develops, we get into areas of programming that will require you to have a knowledge of the architectural concepts presented in this chapter. This book is about Windows programming and can even be used as an introductory text for Windows, but the emphasis is at a more fundamental level than found in other Windows programming books.

Having such a fundamental knowledge will make it easier for you to do all kinds of "tricks" with Windows, such as direct keyboard input, direct video output, and signalling via interrupts.

Power-up the PC

It is a nice place to start: from when we sit down and turn on the computer. What happens behind all that whirring of the hard drive and text and graphics flashing on the screen?

Power-on

Load bootstrap program from BIOS ROM

Load another bootstrap program from the "Boot Record" on the disk

... which loads the system files from the disk

Finally COMMAND.COM executes and the DOS prompt appears, OR WIN.COM executes and Windows loads.

IO.SYS, MSDOS.SYS, COMMAND.- COM, WIN.COM

You must have a *system* disk in either drive C: or drive A:, that is, a disk that contains the files IO.SYS, MSDOS.SYS, and COMMAND.COM if DOS is to run or WIN.COM and the rest of the Windows files if Windows is to run. Note that the first two are "hidden", that is, you can't see them with the normal DOS DIR command; however, they are there in the root directory. Note also, that on IBM PCs, these two hidden files are named IBMIO.COM and IBMDOS.COM.

The boot sequence

After first turning on your PC, or after pressing the key combination <alt-ctrl-del>, it will execute a bootstrap program that is permanently stored in the PC's *ROM* (Read Only Memory). When this bootstrap program executes, it will look at drive C: to see if the system files are on it. If not (or if drive C: does not exist), it will then look at drive A:. This second choice is where you have an opportunity to "boot" from a floppy disk — if there is a floppy disk inserted that has the system files on it.

DOS prompt or Windows

The end result of the above sequence is that COMMAND.COM is loaded and executed, at which point in time you will see the DOS prompt, which usually shows the current drive, followed by a ">" character. For example: A:>

Or, WIN.COM executes which loads the rest of Windows.

In the case of Windows 3.x, COMMAND.COM loads first, optionally followed by WIN.COM. However, Windows 95 bypasses COMMAND.COM.

The System Files

Boot Record

To boot DOS or Windows requires a *Boot Record* on the system disk. The boot program in ROM on the PC's motherboard looks for the presence of the Boot Record, part of which is a program that is then loaded into RAM and executed. When loaded and executed, the Boot Record checks to see if the system files are stored on the disk. It looks for and loads into RAM the files IO.SYS and MSDOS.SYS.

During the loading process, the files CONFIG.SYS and AUTOEXEC.BAT are looked for and referenced if they exist. They help to configure the system and create a personalised environment for the user:

IO.SYS

IO.SYS contains extensions to the ROM-BIOS. These extensions may be changes or additions to the basic I/O operations and often include corrections to the existing ROM-BIOS, new routines for new equipment, or customised changes to the standard ROM-BIOS routines.

This file contains the DOS version number — yes, Windows 95 still has DOS, so there is still a version number.

MSDOS.SYS

MSDOS.SYS contains all the DOS service routines. The MSDOS.SYS routines are more sophisticated, and we can think of them as the next level up from the BIOS routines.

That is, the file contains code. However, with Windows 95 the functionality of MSDOS.SYS is changed. It is merely a small text file.

Booting to DOS or Windows

You might like to investigate this for yourself if you are currently using Windows 95. Make sure that Windows Explorer is setup to show all filename extensions and hidden files. Then, view the C: drive root directory. Double-click on MSDOS.SYS and select NOTEPAD.EXE to view the file.

Be careful about making changes, but do note the following very interesting lines:

```
[Options]
BootGUI=1
```

If you change the entry to "BootGUI=0", the next time you boot the PC it will load COMMAND.COM, not WIN.COM, so you will be in plain old DOS!

Of course, the DOS service routines are elsewhere (in the case of Windows 95), not in MSDOS.SYS, but the boot process still knows where they are and loads what is required.

BIOS & DOS service routines

Note that the BIOS and DOS service routines are there for us to use when writing programs. There is a simple method for us to call any one of these "subroutines" from our program. Basically, these routines enable us to interface with the hardware of the computer, such as the keyboard, screen, printer, disk drives, and serial port.

These service routines existed before Windows was conceived of, so are primarily designed for use with DOS. They still work under Windows, but there are many "ifs" and "buts" here. Complete books have been written around this issue.

COMMAND.-COM

DOS itself is really the COMMAND.COM program. In the case of Windows 3.x, DOS is started first, i.e., COMMAND.COM then WIN.COM are executed.

COMMAND.COM is the keyboard interpreter. It reads what you type at the keyboard and obeys your command. If you tell it to load another program, such as WIN.COM, it will do, even though it means starting another operating system (Windows) on top of DOS.

Windows 95 simply eliminates the COMMAND.COM step, but COMMAND.COM is loaded if you start a "DOS box" inside Windows or choose to exit to DOS from Windows.

Internal & external DOS commands

COMMAND.COM contains the routines that interpret the commands from the keyboard when we are in the DOS command mode. Note that there are two classes of commands: internal and external.

The internal commands are contained within COMMAND.COM, while the external commands are kept on disk. FORMAT.COM, for example, is the program for the FORMAT command and is external. DIR is internal.

The reason that some of DOS's commands are kept as separate programs on disk is due to space constraints in RAM. Obviously there is limited RAM, so it makes sense to keep the less-used portions of DOS on the disk, bringing them in as needed.

CONFIG.SYS "System" files have an extension of .SYS and may be programs or text files. I have already mentioned above that MSDOS.SYS is a code file in early versions of DOS and a text file in Windows 95 systems. A major group of .SYS files are what is known as device drivers: these are programs that load and become semi-permanently resident in memory.

CONFIG.SYS is a system text file that is automatically read from disk during the PC's startup procedure. CONFIG.SYS can be created by any text editor and consists of a number of commands. Here is an example of a CONFIG.SYS file:

```
FILES=          40
BUFFERS=        40
DEVICE=         ANSI.SYS
DEVICE=         GMOUSE.SYS *21
COUNTRY=        061
```

Real mode Refer to your *DOS User's Manual* for more details. An important
device drivers point to note here is that "DEVICE=" is a command that allows you to load more device drivers into the system. GMOUSE.SYS, for example, is driver software for a mouse, and loading this driver will allow any program that can utilise a mouse to do so. But note that this will be what is called a Real mode driver designed to work with DOS.

Windows applications can use Real mode drivers, but there is a performance penalty. Therefore, Windows has its own drivers, that are not specified in CONFIG.SYS (instead, they are specified in another file, SYSTEM.INI, located in C:\WINDOWS\SYSTEM directory).

AUTOEXEC.- After DOS has loaded CONFIG.SYS, it then looks on the disk for
BAT AUTOEXEC.BAT. Any file with an extension of .BAT is known as a "batch" file, and AUTOEXEC.BAT is a special batch file that DOS looks for at power up. Here is an example of an AUTOEXEC.BAT file:

```
@echo off
PATH=C:\;C:\SYSTEM\DOS;C:\GALAXY
PROMPT $p$g
WIN
```

In a nutshell, a batch file is created by any text editor and contains DOS commands, as well as special batch commands, that enable you to automate the operation of DOS. Instead of having to type in the same DOS commands every time you start the computer, by putting them into the AUTOEXEC.BAT file, DOS will execute them automatically for you every time.

Successive versions of Windows have made less and less use of CONFIG.SYS and AUTOEXEC.BAT. However, even Windows 95 will still obey whatever you put in these files.

Power-on self test (POST)

The power-up sequence of the PC is quite involved, and many references are made to it throughout this book. Of particular interest is the configuration RAM that the BIOS uses during the Power-On Self Test (POST) sequence.

The configuration CMOS RAM is a part of the real-time clock chip.

Number Systems

Well, maybe I shouldn't assume too much knowledge on the part of my reader! I'm already throwing around words like "RAM", "ROM", and "boot". Perhaps some discussion of the mathematics is in order before I throw more words at you, like "hex", "byte", "ascii", and "BCD".

You also need to understand the concepts of "address" and "data".

RAM and ROM

The computer has memory, called either RAM or ROM, in which information is stored. Floppy and hard disks also store information. This information can be either data, such as documents that have been typed in, or programs, such as a word processor.

Byte-addressed memory

All information is stored in the computer as binary values, that is, as 1's and 0's. The computer's memory, whether RAM, ROM, floppy disk, tape, or hard disk, records information in groups of 8 bits. That is, each memory *location* contains 8 binary bits.

Furthermore, every memory location is *addressable*, which is logical, since the computer must be able to store and retrieve the information from each location. So, there is an address, which is a binary number, and it references a location, which is an 8-bit binary code. This is shown pictorially in Figure 1.1.

So, what does "00110100" mean? "00110100" (in Figure 1.1) is just a string of binary bits stored in memory, but the PC will interpret it in some meaningful way. There is a pictorial answer in Figure 1.2.

Figure 1.1: A memory location.

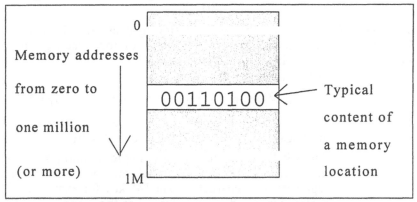

Figure 1.2: Interpretations of a memory content.

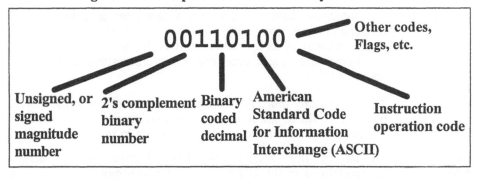

Straight binary Considering each of the above, 00110100 could be treated as a magnitude only, "straight" binary number, or the most significant bit (MSB) could be a sign, leaving 7 bits for the magnitude: this is not the same as 2's complement.

Considering it as a straight binary number:

00110100

$= 0*2^7 + 0*2^6 + 1*2^5 + 1*2^4 + 0*2^3 + 1*2^2 + 0*2^1 + 0*2^0$

$= 52$ decimal.

That is, 52 decimal is represented in memory by 00110100 binary.

2's complement Positive numbers in 2's complement look just like unsigned straight binary numbers. They are distinguished as +ve by the MSB, i.e., the left-most bit, being 0.

A 2's complement negative number is distinguished by the MSB being 1.

The value 00110100 is 52 decimal — so, how does -52 look? The rule is, invert all bits to the left of the first-bit-from-the-right-set-to-1. Thus: 11001100.

Binary coded decimal (BCD)

BCD is another way of storing decimal values in the computer. The bits are grouped into lots of four, each group converted to decimal:

0011 and 0100 become 3 and 4

That is, 34 decimal is represented in memory by 00110100.

ASCII

The code does not represent a number at all, but a character. From an ASCII table, 00110100 represents ASCII character 4.

An ASCII "character" is any single symbol, such as a numeric digit, letter, punctuation symbol, etc. Each key on the keyboard represents one character. The "4" key for example, when pressed, is stored inside the computer not as the binary value 00000100 but as the ASCII code 00110100.

Instruction code

The bit-string does not represent a character or a value, but an instruction operation code. Machine language instructions are stored in memory as op-codes followed by zero or more operands, depending upon the instruction: the interpretation of this code is up to the CPU.

There are many more possibilities: for example, a value stored in memory could be an address. This may at first seem confusing, but you could store address pointers to data in memory.

Base or radix

So far in this section, we have looked at the different interpretations of 00110100, but it is also important to know that there are different bases, or radices, in which the number can be represented.

We saw above that 00110100 is 52 in decimal. That is, 00110100 is the representation in binary, with radix = 2, and 52 is the representation in decimal, with radix = 10. Note that 52 is just a "digit-string", like 00110100, and interpretation as a number is our choice.

Actually, numbers can be represented in any base. Most important for us, apart from binary and decimal, is the *hexadecimal* representation:

Hexadecimal

The next section will start to throw "hex" numbers at you, so now is the time to be clear on what they are.

Hexadecimal numbers are base-16, i.e., are based upon a number system with 16 digits, rather than the 10 in decimal or the 2 in binary. They are:

```
Hex:
0    1    2    3    4    5    6    7    8    9    A    B    C    D    E    F
 Decimal:
0    1    2    3    4    5    6    7    8    9    10   11   12   13   14   15
 Binary:
0000 0001 0010 0011 0100 0101 0110 0111 1000 1001 1010 1011 1100 1101 1110 1111
```

The first row shows hex digits, followed by decimal, then binary: note that hex numbers are just a shorthand notation for binary, which is why they're used. Each hex digit represents 4 binary bits, and FFFF0 hex is the same as 1111 1111 1111 1111 0000 binary. Note that hex is not quite the same as BCD, as described above.

Registers and Memory

We haven't finished covering the really basic stuff.

x86 CPU initialisation
When powering up the PC, you are also powering up the 8088, 8086, 80286, 80386, 80486, or Pentium CPU, whichever your particular computer has. The CPU has internal registers that are initialised to certain values at power up. Two of them, the *Code Segment* (CS) register and the *Instruction Pointer* (IP), are initialised in a very special way.

Registers
Hey, before we go on about *registers*, just what are they? Figure 1.3 introduces the registers of the PC, but do note that registers can be in any integrated circuit, such as the video adaptor card.

20-bit address
To return to the Code Segment register, CS, and the Instruction Pointer, IP: at power on, or after resetting, the CPU combines these two in a certain way to produce an *address* on the address bus.

Note that this address is 20 bits in size, because the 8088 and 8086 have a 20-bit physical address bus. The 286, 386, 486, and Pentium have larger physical address buses but only use 20 bits in startup "Real mode". That is, even the powerful Pentium starts up behaving just like its ancestor — this is a very important point.

Major components of a PC
Figure 1.3 also shows the major components of a PC: note that they are all connected by something called the *bus*. To access memory contents, addresses must be sent out from the CPU (microprocessor), and the very first address is a combination of CS and IP.

Figure 1.3: What is a register?

WHAT IS A REGISTER?

Just as the PC has **memory** chips, called

RAM   (changeable, contents lost when turned off) or
ROM   (permanent program storage),

which sit somewhere in the *memory map*, so too the **CPU** chip has internal RAM memory, called **registers**.

Registers are 8-, 16-, or 32-bit memory locations but are not addressed like external memory. Instead, each of these registers has an explicit unique name that can be used in the machine language instructions.

The 8086 has these registers:

AX, BX, CX, DX, SI, DI, SP, BP, CS, DS, SS, ES

Some are general purpose and some have special purposes: this is something that is learnt with time, as you practise with the machine instructions.

These registers are 16 bits in size, so representing the values in hexadecimal notation requires four hex digits.   The address is 20 bits, and so requires five hex digits.   At power on, the CPU initialises CS and IP, as Figure 1.3 shows.

Power-on address calculation

CS = FFFF

IP = 0000

Address produced

= FFFF0

At power on, the CPU will put this address onto the address bus and fetch the first instruction from this address.

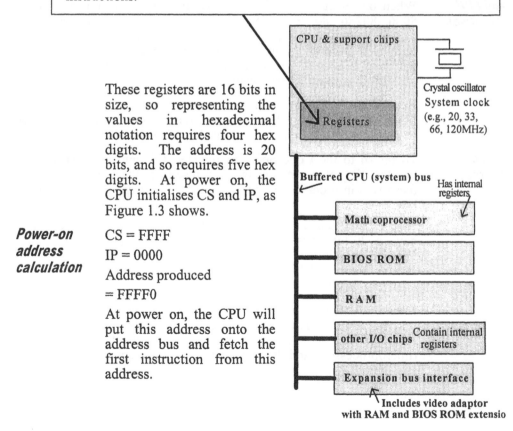

BIOS ROM   Thus, the program that takes control when the computer is switched on must start at address FFFF0. Furthermore, it must be in *BIOS ROM* so that it is there at power on (see Figure 1.3 alongside).

This program has a special name — it is part of the BIOS — Basic Input/Output System.

The BIOS routine with the start address of FFFF0 takes control of the boot procedure. This routine looks for the Boot Record on the system diskette in drive A: or C:. The Boot Record is a small program, with certain important system information, that is then loaded from disk and executed. The Boot Record has already been introduced earlier in this chapter.

We will do a number of passes through the same topics as the chapter progresses, going a little deeper each time. So, to find out more about the registers, memory, etc., read ahead.

Figure 1.4: Memory map of the PC.

ADDRESS	CONTENTS
0000h →	Interrupt Vector Table
00400h	BIOS data area
00500h	DOS data area
	Resident part of DOS, device drivers, & TSRs
	Free memory
640K(dec)	Transient part of DOS
A0000h	Color display RAM
B0000h	Monochrome display RAM
B8000h	Color display RAM
C0000h	VGA BIOS extensions
C8000h	Hard disk BIOS extensions
F0000h	ROM extensions
F4000h	User ROM space
F6000h	ROM BASIC (maybe)
FE000h	ROM BIOS
FFFFFh → (1M)	64K higher memory area

This is a memory map of the PC. Any of the x86 CPUs running in Real mode, which is the default at power on, only utilise a 20-bit address bus, so only address up to 1M. This is referred to as the conventional memory, while that above 1M is extended memory.

The memory map has RAM and ROM hardwired into fixed addresses, and in some locations there may be nothing.

Extended memory

Memory Map of the PC

***Size of the
address bus***

The 8088 and 8086 have a 20-bit address bus, which means that they can address 2^{20} - 1 = FFFFF (hex) = 1 megabytes (decimal). The 80286 has 24 bits, and the 80386 has 32 bits.

A good starting point is at the beginning. For the 8088/6 the memory organisation on the PC looks like Figure 1.4.

***The famous
640K limit***

You know the 640K RAM specification given for the PC — this RAM exists in the memory map from address 00000 up to the 640K shown in Figure 1.4. You can see that the first 140K or thereabouts is occupied by various things, which leaves about 500K free for user programs. Of course this free memory is a very variable thing, depending on a number of factors.

For example, if the CONFIG.SYS file specified some device drivers, they would be loaded into memory and kept there. If DOS is to run, COMMAND.COM will have to load, and for Windows it will be WIN.COM. Same for any resident "pop-up" programs such as Sidekick — though, it is rare that anyone uses these today. All of these will reduce your free memory.

Figure 1.4 is simplified, and in practise there will be a lot more functions occupying the memory space. Do be clear on one thing: not all of the address space is necessarily occupied. The address range from 00000h down to 9FFFFh (640K) is occupied by contiguous RAM, and the region marked as "free RAM" is available for a program to load into from disk.

At the end of the 1M region is the BIOS ROM, and maybe other ROMs before it. Basically, ROM and RAM in this middle region are provided by plug-in expansion cards. Examples of video and hard drive cards are shown, though the addresses are in some cases adjustable.

***Scan for
extra ROM***

It is useful to note at this point that during the power-on sequence, the region C0000h to F4000h is scanned to see if any programs are present in ROM, and if they are, subject to certain identification, they are immediately executed. Thus, these programs are able to modify the system to suit themselves.

The CPU & Support Chips

One step at a time: we are focusing for now on the 8086, as that is the mode that all of the x86 family power up into. It is the CPU used in the first IBM PC — no, strictly speaking, it was the 8088. The only difference between the two is that the 8088 had an 8-bit data bus, while the 8086 had a 16-bit data bus. The first PC was thus able to use the cheap and readily available interface chips designed for an 8-bit data bus.

Embedded systems

The 8086 is still used, not just in PCs, but in a host of dedicated (embedded) controller applications, though usually in the latter case it is some derivative of the 8086. So, the 8086 is not dead, and its presence is to be found even in the very latest Pentium CPUs. Why? — because of the requirement for backward compatibility.

In fact, you may be very surprised. I don't have the exact figures, but processors for embedded systems far outweigh annual sales for PCs. Quite literally, billions of processors are manufactured annually for embedded systems. This would include humble home appliances, such as your washing machine and video player. The 8088 would be considered too powerful for most of these applications! But you will find that the 8088 derivatives are selling very strongly today, possibly in larger quantities than the Pentium.

Book on design of embedded systems

I'm very much into the design of the lower end of embedded systems, using the more humble 8- and 16-bit processors, as covered in my book:

Flow Design for Embedded Systems, R&D Books/Miller Freeman, USA, 1997. For more information see:

`http://www.rdbooks.com/`

Figure 1.5: Three parts of a computer bus.

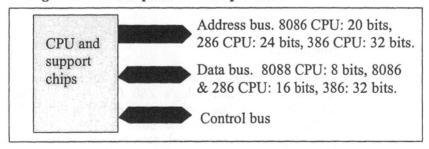

Three parts of a computer bus

The 8088 is the *Central Processing Unit* (CPU) of the PC. It executes instructions contained in RAM or ROM.

The CPU and a few support chips produce various signals, known as the bus, which can be broken down as shown in Figure 1.5.

The lines are physical wires going to and from the CPU and support chips. The bus goes to all the memory and I/O (input/output) chips in the computer and is the means by which everything communicates.

Conventional and Extended Memory

Extended, conventional, high

Extended memory is that above the 1 megabyte (M) address limit, while *conventional* memory is below 1M. Expanded memory is bank-switched memory that can be mapped into the conventional memory area. The first 64K of extended memory is sometimes referred to as *high* memory.

Memory map of the PC

The map in Figure 1.4 is a rough indication of how everything looks.

When Windows has loaded, however, the processor will be in what is called *Protected mode*.

Protected mode

Chapter 12 goes a lot more into the particular complications of Protected mode, in which the basic memory map in Figure 1.4 can no longer be considered as residing at the actual physical address range zero to 1M. The 386 is capable of creating *virtual machines*, each with an apparent 1M address space. Note that the addresses in these virtual machines are called *virtual addresses*.

Unfortunately, the PC is a mess. It started out life in 1980 with a text-only screen, cassette mass-storage (no hard disk), no real-time clock, only 64K of RAM, and an 8088 CPU. Features got tacked on over the years, and the operating system and hardware grew and grew.

One of the most fundamental problems inherited from the 8088, and something that causes headaches for programmers now, is — *segments*.

Segments

Popular desktop PCs prior to IBM's PC used 8-bit processors, such as Intel's 8008 and Zilog's Z80 (Figure 1.6). They have an 8-bit-wide data bus, while the 8088 introduced the 16-bit architecture. Although the 8088 has only 16-bit data paths internally, with an external data bus of 8 bits. The 8086 is identical to the 8088, except it has a full external 16-bit data bus.

History of Intel CPUs

The earlier 8-bit CPUs had 16-bit address buses. Now, if you are up to some binary calculation, this means that the possible range of addresses is from zero to $(2^{16} - 1)$. In binary, that is an upper limit of 1111111111111111, or in hexadecimal FFFF, or in decimal $(64*1024) - 1 = 65,535$. We normally refer to this memory capacity as 64K, where K represents "times 1024" (note that a megabyte is 1024*1024, so $1M = 1*1024*1024 = 1,048,576$ bytes).

Figure 1.6: 16-bit addressing prior to the x86 family.

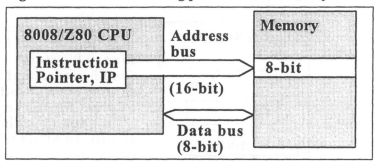

The upgrade from 16- to 20-bit address bus

The Intel engineers thought that with the 8086 family they would increase the memory capacity to something more than plenty, so they gave the 8088 and 8086 a 20-bit address bus. $2^{20} = 1M$. A million or so bytes of memory seemed like an enormous amount at the time, but this has turned out to be a serious limit and constitutes half of our headache.

The other half of the headache is how the engineers designed the chip to address that 1M.

16-bit IP matches a 16-bit address bus

They wanted to make it easy to port software from the 8-bit CPUs. Internally, those earlier CPUs have an Instruction Pointer, IP, which is a register that marks the address of the next instruction to be executed. It is 16 bits, to match the external address bus.

Byte addressing

Note that each memory location, as addressed by a unique address, is 8 bits (1 byte), and this is retained even for the latest Intel x86 CPUs with 32-bit or more data buses.

They decided to keep the 16-bit instruction pointer, but it only addresses 64K and so is incompatible with a 20-bit external address bus.

How segments started

Enter the segment. The designers introduced registers, called *segment registers*, to map the 64K region addressed by IP to anywhere in the 1M range. Code could still think it was in a 64K space, but transparently would be mapped to wherever the segment registers specified. One tick for compatibility, and one also for complexity. Figure 1.7 shows how it is done.

64K segment limit

It is absolutely vital that you understand this process. The 16-bit IP is added to the 20-bit starting address, to give the 20-bit address from where the CPU will fetch the next instruction. Thus, the IP is an offset within the segment; **therefore the segment can have a maximum size of only 64K.**

Figure 1.7: Concept of segmentation.

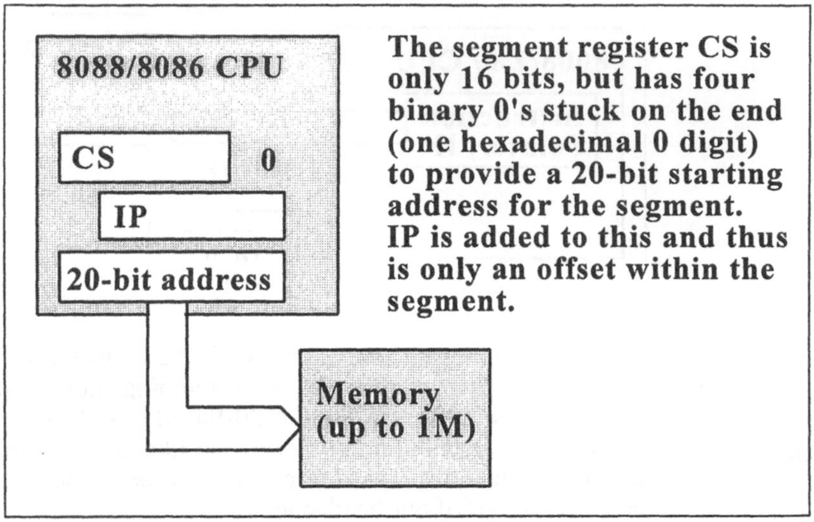

8088/8086 CPU

CS   0

IP

20-bit address

Memory
(up to 1M)

The segment register CS is only 16 bits, but has four binary 0's stuck on the end (one hexadecimal 0 digit) to provide a 20-bit starting address for the segment. IP is added to this and thus is only an offset within the segment.

.COM executable file format

Executable files of .COM format are restricted to 64K maximum, as they were born back in the 8-bit-CPU days; however, the engineers realized this to be a problem and "solved" it by introducing three more segment registers: *DS* (Data Segment), *SS* (Stack Segment), and *ES* (Extra Segment). To support these registers, the designers introduced the .EXE executable file structure that allows code to be stored in the segment pointed to by CS, data to be in another segment pointed to by DS, the stack to be in yet another segment pointed to by SS, and ES to be a segment that can be used by the application programmer. Figure 1.8 shows a pictorial representation of how these registers might be laid out in memory.

.EXE executable file format

Although segments are still only 64K maximum, it is possible to have multiple segments of code and data for large programs.

One aspect of the headache associated with segments is this 64K limit. Obviously large code or data could exceed this, and problems arise. Another aspect is that this segmentation scheme of addressing has carried over to the 286/386, etc., again, for compatibility reasons.

Figure 1.8: .EXE program segments.

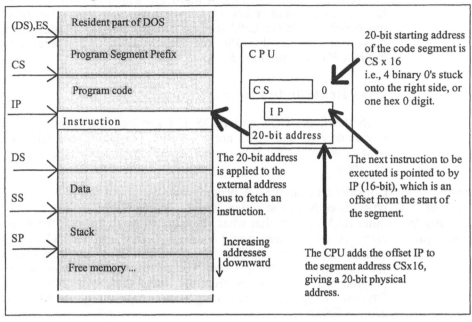

Real Mode

The 8088/8086 operate in what we now call *Real mode*. This means that the segment registers hold real addresses, in accordance with Figures 1.7 and 1.8. The 286/386, etc., chips also run in Real mode when first turned on and employ the same 20-bit segmented addressing mode. For the sake of compatibility, the more advanced CPUs can only address 1M, and the extra address lines are inactivated.

High memory

The fundamental problem with DOS and DOS applications is the 1M limit. There is, however, a qualification to this: Real mode addressing can address over the 1M limit, by an extra 64K, known as the *high memory segment*. A quick look at Figure 1.8 will show why. If we put maximum values in them, that is, CS = FFFF and IP = FFFF, the computed 20-bit address is 10FFEF. The 1M limit is FFFFF.

The 21st address bit

The 20-bit upper limit is FFFFF hex (1M - 1), but the offset IP allows, in theory at least, addressing just over this. Physically this would require a 21st address bit, which the 8088/8086 don't have, and it's disabled on the 286, etc. chips also. But the PC can be instructed to turn on the 21st address bit on the 286/386 chips, thus allowing them access to that 64K above the 1M.

An extra 64K is peanuts. The 286 has 24 address lines and can theoretically have 2^{24} bytes of memory, which is 16M. The 386 has 32 address lines, allowing 4.3 gigabytes (G). But all of this is inaccessible with the CPU in Real mode. Or is it? Later in this book you will see that 386 and later processors *can* access the entire 4.3G from Real mode.

DOS Real Mode Programming

DOS itself, the DOS and BIOS I/O services, DOS applications, device drivers, and TSRs are all designed to work in Real mode. This is because they rely on real addresses being in the segment registers.

Using segment registers in a program

Consider an example. A programmer could use the ES register to write directly to the video RAM. Video RAM is just like any other RAM, except that what you store there appears on the screen also. A programmer could load B800 hex into ES, which will address the CGA video RAM (i.e., the full 20-bit starting address for the segment will be B8000 hex). In assembly language, the programmer writes an instruction like "MOV ES:[DI],AL" or "STOSB", to store a value from general-purpose register AL to the address ES:DI.

Note that the terminology ES:DI refers to the address in the form of segment:offset. ES is the segment, and DI (or some other 16-bit register) has the offset. DI and AL would both need to be loaded beforehand, of course.

The point here is that the program loads an actual segment address into ES. This reliance upon real values being in the segment registers means that DOS programs cannot work above 1M.

Selectors

The 286/386/etc. CPUs can be switched into what is known as *Protected mode*, which allows memory access up to the 16M or beyond limit, but the real addresses have to be dumped from the segment registers. Instead, they contain *selectors* or indexes into tables, and the tables have the real addresses of the segments.

DOS does have some mechanisms for switching into and out of Protected mode, and a couple of early DOS services are introduced here.

DOS Protected Mode Programming

Oh, what a can of worms! I didn't quite know where to start, as there are so many considerations. I have written about the problems, at least some of them, of running DOS applications in Protected mode, and DOS's own failure in this regard.

INT-15h

However, Microsoft gradually extended DOS, and one of the first services they added was the functions invoked via INT-15h. So I'll start with these.

The idea was to provide some means of switching from Real mode to Protected mode and back, to transfer code to and from conventional and extended memory, and to transfer execution from a Real mode program to a Protected mode program.

INT-15h,
AH = 88h,
AH = 87h

INT-15h, AH = 88h, will tell us how much extended memory there is, though not how it is being used. It simply returns with a value in the AX register. There is a picture of the CPUs registers in Figure 1.11.

INT-15h, AH = 87h, moves a block of data between conventional and extended memory. A use that immediately comes to mind with this is a TSR manager that could keep them all out of the way and bring one back as needed (there are such managers available).

Local and
global
descriptor
tables

A problem with Protected mode is all the housekeeping required, that is, the various tables required for addressing. The segment registers no longer have the actual addresses: they are kept in the *Local Descriptor Tables* (LDTs) or the *Global Descriptor Table* (GDT). Furthermore, if interrupts are to be handled by the program in Protected mode, an *Interrupt Descriptor Table* is required. There's more — if task switching is to be supported, a *Task State Segment* (TSS) is required for each task.

Fortunately, INT-15h, AH = 87h, keeps it simple. All that is required is a GDT to get the service to work, and it is up to the application program to set this up. The service requires:

CX   =   Number of words to transfer.

ES:SI   =   Physical address of GDT
               (in conventional memory).

AH   =   87h

Switching
the CPU into
Protected
mode

The DOS service takes care of differences between switching the 286 and 386: in the 286 it involves setting the Protect-Enable bit in the *Machine Status Word* register, setting up descriptor tables, and loading the address of the GDT (Global Descriptor Table) into the GDT pointer register.

Machine
Status Word

In both the 286 and 386, getting into Protected mode involves setting certain bits in the Machine Status Word, and in the 386 it is a simple matter of setting a bit in an appropriate way to come back to Real mode. However, the 286 has no mechanism for returning to Real mode, and it has to be done by the incredibly slow method of resetting the CPU, which takes several milliseconds. This is one of the reasons that the 286 has become history.

**INT-15h,
AH = 89h**

INT-15h, AH = 89h, is the service that actually transfers control from the Real mode code to Protected mode code. Since hardware interrupts must occur if the PC is to continue to operate, and the application may need to generate software interrupts (see Chapter 2 for the distinction), an IDT must be in existence; therefore, this service needs more housekeeping.

Just to give an idea of what goes into a GDT, Figure 1.9 shows a basic GDT as required for INT-15h/AH = 87h. The format shown needs a little modification for 386 systems.

**Creation of
a GDT**

A point about Figure 1.9 is that our application needs to create a GDT, and maybe an IDT, and put their addresses into the GDT prior to invoking INT-15h. But when the service performs the transfer to Protected mode, it loads the other descriptors (for DS, ES, SS, and CS) into the GDT.

The DOS service will put *selectors* into DS, ES, SS, and CS inside the CPU. These are just indexes into the GDT, which has the actual addresses.

**Initialisation
of segment
registers**

You may know that when DOS in Real mode loads a program from disk, DOS puts it where there is free memory (see page 14) and automatically sets DS, ES, SS, and CS appropriately (see Figure 1.8). In Protected mode, however, the actual addresses of the segments are put in the Descriptor Table, while the segment registers only have pointers into the table.

**Why have
selectors?**

A small, but vital, question ... **why?** Why put the actual addresses out of the CPU in tables? The answer is simple — segment registers are 16 bits, thus limiting the address range to 1M, while the segment-address in the table is at least 24 bits, thus giving at least 16M address range.

Coding Restraints

Yes, you can write DOS applications that will run in Protected mode. You can see from the above notes that DOS can load code or data above 1M and can also switch into Protected mode and execute the program above 1M.

The requirement is that the program must not expect actual physical addresses to be in the segment registers.

**A problem to
call Real
mode code
from
Protected
mode**

Another implication is that the program cannot call the BIOS and DOS I/O services (normally called by the INT instruction), since these are designed to run in Real mode. Ditto there is a problem with device drivers and TSRs.

It is possible to switch back to Real mode just to run an I/O service or device driver, or Protected mode versions of the BIOS and DOS

I/O services can be provided. This is discussed further later in the book.

Figure 1.9: Basic setup of a GDT.

OFFSET	CONTENT
00-07h	Reserved (should be 0)
08-0Fh	Descriptor for this GDT
10-17h	Descriptor for the IDT
18-1Fh	Descriptor for DS
20-27h	Descriptor for ES
28-2Fh	Descriptor for SS
30-37h	Descriptor for CS
38-3Fh	Descr. temp. BIOS CS

Inside the 286/386/486/etc.

A Pentium is just a fast 386

Mostly I have concentrated on the 386, since the 286 is history. You can consider later processors to be functionally equivalent, just faster. Do not get the idea that later processors, such as the Pentium, are fundamentally different from the 386. Just about all 32-bit code written today will run on a 386. Most architectural differences are to do with speed enhancements.

There are some architectural differences between the 386, 486, 586, 686, and Pentium, but I have focused here on the basic architecture: the 386. This is the common factor underlying them all.

32-bit instruction pointer (EIP)

Everything, almost, has become 32 bits, including 32-bit address and data buses. The Instruction Pointer has grown to 32 bits (see Figure 1.10), which means that the original rationalization for introducing segment registers has been nullified. However, the segment registers are still there, and still 8 bits — the curse of compatibility is still with us!

***Memory and
I/O address
spaces***

Thirty-two bits gives us an enormous addressing capability: 4.3 thousand million bytes (gigabytes). Note that, for compatibility reasons, each address actually addresses 8 bytes of data in memory, even though the data bus is 32 bits.

Addressing of I/O ports is still the same as for the 88/86/286, using the lower 16 bits of the physical address bus, coordinated with the IOR and IOW control lines. 16 bits allows up to 65,536 I/O ports. It is important to note that the I/O address space is separate from the 4.3G memory address space — this differs markedly from the Motorola 68000 family, in which there is no separate I/O space. I/O ports are accessed by the IN and OUT instructions (see page 244).

Figure 1.10: 386 32-bit address and data.

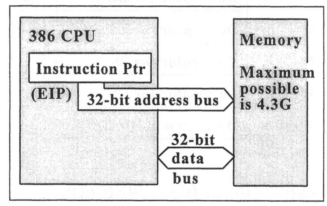

CPU Registers

***Downward
compatibility***

Obviously, if the 386 is to be downwardly compatible it must have the same registers as its older relatives, and yes, they're all there. AX, BX, CX, DX, SI, DI, BP, and SP are the 16-bit registers inherited from the 86 and 286. Incidentally, the 286 has the same register set as the 86 (plus some extra ones for managing Protected mode). It is only with the 386 that significant enhancements of the registers occur: they are all 32 bits, except for the segment registers.

***32-bit
registers***

It is important to understand that the registers you can use in an application can be used as 32-, 16-, and, in some cases, 8-bit registers, for full downward compatibility.

Here are examples:

```
mov   BL,0              ;examples of different size regs.
mov   BH,0              ;8
mov   BX,0              ;16
mov   EBX,0             ;32
```

The "E" prefix denotes a 32-bit register. BX is the bottom half of EBX, and BH and BL are the top and bottom halves of BX.

At this stage, I'll present an overall picture of the registers of the CPU.

The registers shown in parentheses in Figure 1.11 are the portions of the extended registers that are found in the 86 and 286. For example, the 86 and 286 have the Stack Pointer, SP, that is the bottom half of ESP in the 386.

The purpose of each register is somewhat more involved than the tiny descriptions given in the figure, of course. The segment registers are described as being selectors, which is valid for Protected mode. In Real mode they would hold segment (paragraph) addresses. Note that the 86/286 don't have FS and GS.

Only AX, BX, CX, and DX can be operated upon in halves, that is as AH/AL, etc. This is convenient for handling 8-bit data.

Instructions

Learning the basic instruction set

It is somewhat back-to-front, but I have given a thorough coverage of the basics of the instruction set in Chapter 2. Therefore, if this discussion of registers and instructions is "double Dutch" to you, jump to Chapter 2 then come back here. Otherwise, keep reading.

Obviously the 386 has all the instructions of the 86 and 286, but you'll find them enhanced, plus many new ones.

Note that I've put a summary in Appendix A showing which instructions work on the 86, 286, and 386.

Coding specifically for the 386

Once you start to code explicitly for the 386, beware that there is no turning back — your program won't run on the 86 or 286. Most important of all incompatible enhancements is the removal of the 64K segment limitation by means of EIP to access the code segment, ESP to point to the top of stack, and the various other general and data-segment addressing registers (EBX, ESI, EDI, etc.).

```
mov   AL,ES:[BX]        ;16-bit index.
mov   AL,ES:[EBX]       ;32-bit index.
```

Figure 1.11: Registers of the 386.

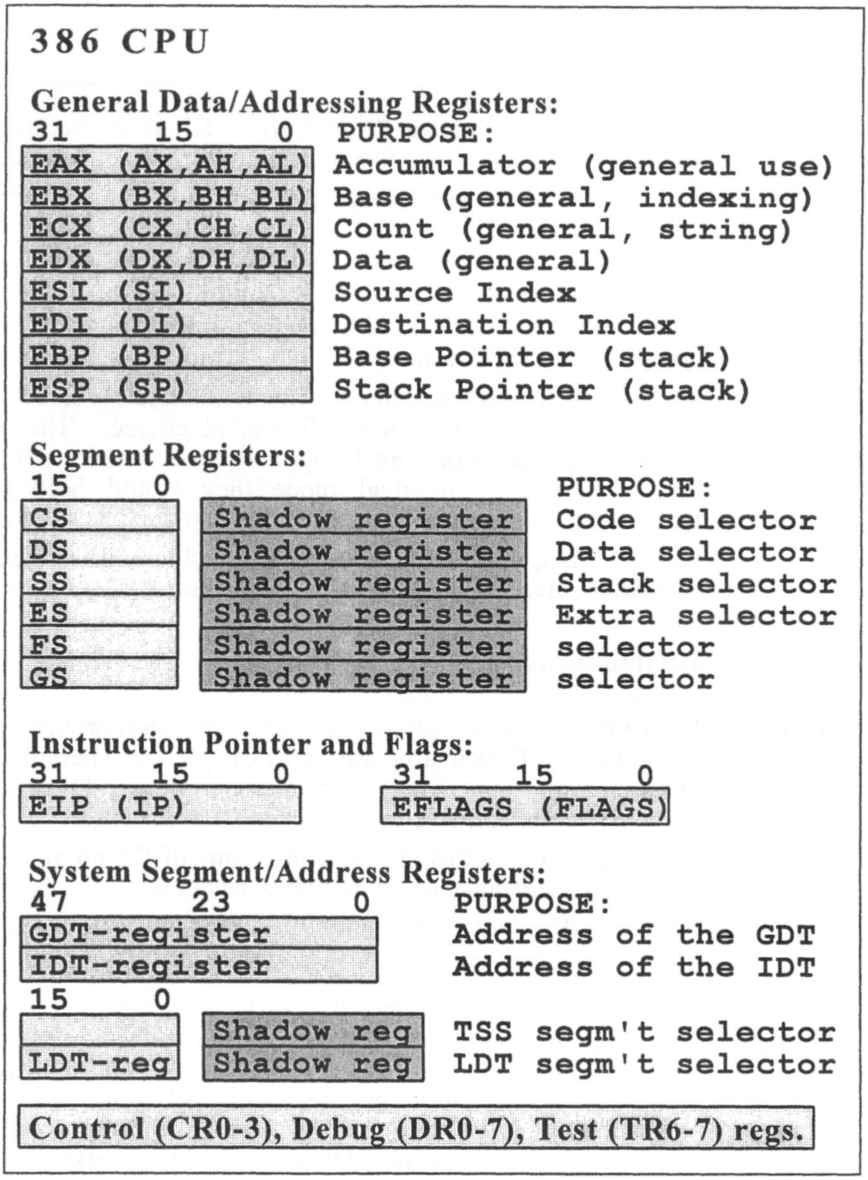

386 CPU

General Data/Addressing Registers:

```
31        15        0    PURPOSE:
EAX  (AX,AH,AL)          Accumulator (general use)
EBX  (BX,BH,BL)          Base (general, indexing)
ECX  (CX,CH,CL)          Count (general, string)
EDX  (DX,DH,DL)          Data (general)
ESI  (SI)                Source Index
EDI  (DI)                Destination Index
EBP  (BP)                Base Pointer (stack)
ESP  (SP)                Stack Pointer (stack)
```

Segment Registers:

```
15        0                          PURPOSE:
CS          Shadow register          Code selector
DS          Shadow register          Data selector
SS          Shadow register          Stack selector
ES          Shadow register          Extra selector
FS          Shadow register          selector
GS          Shadow register          selector
```

Instruction Pointer and Flags:

```
31        15        0      31        15        0
EIP  (IP)                  EFLAGS  (FLAGS)
```

System Segment/Address Registers:

```
47        23        0      PURPOSE:
GDT-register               Address of the GDT
IDT-register               Address of the IDT
15        0
            Shadow reg     TSS segm't selector
LDT-reg     Shadow reg     LDT segm't selector
```

```
Control (CR0-3), Debug (DR0-7), Test (TR6-7) regs.
```

This code shows how to get the single-byte memory contents in the ES segment at offset BX, in the first case, and EBX in the second. Obviously, the first instruction is limited to a 64K segment, due to BX being 16 bits, while the use of EBX extends the limit to 4.3G.

Real and Protected Modes

Memory management

It has already been stated that when in Real mode, the 386 (and 286) operate like the 86, with segment registers having actual segment (paragraph) addresses. The limitation this imposes is that the maximum address range is 1M (plus the extra 64K high memory area — see page 17). Another limitation is that there is no built-in support for memory management.

Multitasking problems

Windows allows more than one program (task) to run at once, and this introduces some incredible constraints. Also, simplicity goes out the "window". Obviously the CPU must be able to divide its time between running the various programs; each must sit in separate areas of memory and none must write to memory where another program is sitting. They must be able to share keyboard input and not scribble all over the screen — each task must only output to its own window. Other resources and I/O must be shared without a fight.

This is asking a lot, but the Protected mode inherited from the 286 will do it, while the Enhanced protected modes of the 386 will do even more.

Memory Management

Distinction between 286 and 386

The 286 has just one Protected mode, also inherited by the 386, and we will look at that first. It employs mapping of the segments to memory via Local Descriptor Tables (LDTs) and a Global Descriptor Table (GDT).

Note that the 386 can work exactly like the 286 but also has other modes: an extension to the descriptor tables, with page tables, and a system with page tables only, known as *virtual-86* mode.

Segmentation Only

Purpose of the LDT

There is only one GDT, but the operating system maintains an LDT for each program currently running (Windows 3.x and 95 are special cases: see footnote on page 32). Think about the LDT — it contains the actual segment addresses, while the segment registers inside the CPU (we will now call them *selectors*) are just indexes into the LDT. When a task switch occurs, the CPU has a simple mechanism for changing to the next LDT, but the selectors don't necessarily have to change, since they only index into the table.

Fundamental reasons for having the LDT are the increased addressing, plus protection. Ok, here is a picture: take a look at Figure 1.12.

The figure gives a fairly good idea of the relative roles of GDT and LDT. When the operating system first creates the GDT, it uses special instructions to put the base (starting) address of the GDT into the 32-bit GDT-register. Thus the CPU will always know where the GDT is.

Purpose of the GDT

So, what purpose does the GDT perform? One major use is to hold the base addresses of all the LDTs. Whenever the operating system creates a new task, it also creates an LDT for that task and makes a new entry in the GDT. This entry has the address of the LDT.

How many LDTs?

Bear in mind that I'm generalising here — Windows 3.x and 95 use one LDT for all Windows applications and separate LDTs for each DOS application, while NT is different again. Seem complicated? — It is, which is why I'm generalising for now!

Descriptors

The GDT has the base addresses of the LDTs, but which one is currently executing? For this, the CPU has the LDT-register, which is just an index into the GDT, pointing to the current LDT descriptor.

Let me use the term *descriptor* from now on. Each entry in a GDT or LDT is called a descriptor.

So, let's suppose the CPU wants to fetch the next instruction of whatever task is currently executing. The CPU will already know where the current LDT is, because it already would have read the GDT entry as indexed by the LDT-register.

Shadow Registers

Incidentally, a most important element is shadow registers. Look back to the picture of the CPU registers in Figure 1.11 (page 24), and you will see that some of them have shadow registers. So does the LDT-register. These shadow registers hold the actual addresses or, more correctly, the *descriptors* read from the table.

When the CPU reads the GDT and gets the descriptor for the current LDT, it puts this into the corresponding shadow register (alongside the LDT-register), so from then on, until a task-switch or until the LDT changes position, the CPU will know where the LDT is, without having to reread the GDT (Figure 1.12).

Figure 1.12: Memory management.

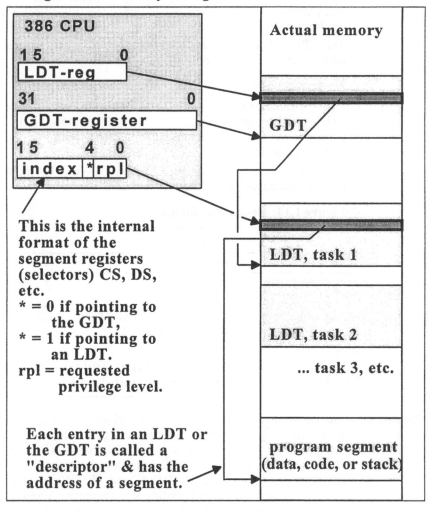

Association between descriptor and shadow register

The next step in this saga is that the CPU can use the selector in the CS register to index into the current LDT and get the actual address, or more correctly the *descriptor*, of the code segment. The IP register (or EIP) will have the offset into that segment from which the CPU will fetch the instruction.

Having read the descriptor from the LDT, the CPU then has the base address of the code segment. To avoid having to look in the LDT every time it wants to fetch the next instruction, the CPU makes use of shadow registers again. Every segment register has an associated shadow register.

The CPU will only have to look in the shadow register to find out the starting address of the segment (plus some other information) and can then go ahead and put together the full 32-bit address for fetching the instruction.

The CPU will add the base address to the offset IP and get a 32-bit address that can be put onto the address bus.

Descriptors

I have introduced the descriptor as being an entry in the GDT or LDT. There are various types of descriptors, but the most common is the normal addressing type that we have been discussing so far.

Each descriptor is 8 bytes in size, and Figure 1.13 shows what a normal descriptor looks like.

Figure 1.13: Descriptor format.

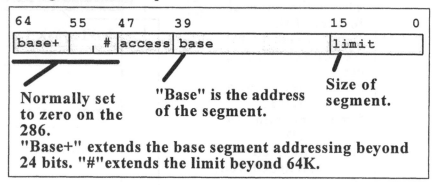

Access field   The *access byte* in Figure 1.13 has various flags and codes. It has a two-bit DPL field (Descriptor Privilege Level) that determines the privilege level of the segment. It has P (Present) and A (Accessed) bits that are used for moving the segments in and out of memory. There are R (Read) and W (Write) bits that set constraints on reading and writing the segment. There is also the C (Conforming) bit and ED. The latter is set if the segment is a stack.

I go into the description of the descriptor in far greater detail in Chapter 12.

386 Paging

There are two paging modes in the 386. One is built on top of the descriptor tables, and the other, called *virtual-86,* does away with the descriptor tables altogether.

I'll look first at the one built on top of the desriptor tables. From our program point of view it looks just like the segmentation mechanism with the GDT and LDTs. The only difference is that the CPU secretly stores the segments in actual memory not in one contiguous chunk, but all over the place as 4K *pages*.

What's wrong with segments?

Why go to this trouble? The operating system has trouble bringing segments in and out of memory because they are all different sizes — if a new segment is to be brought in, space must be found for it, but space released by a segment that has vacated its spot may not be the right size. This is a real problem for the operating system, and it ends up with lots of little unused gaps everywhere. Inefficiency.

By transparently parcelling the segment up into lots of little pages all the same size and storing them wherever there is a space, the mismatch of segment sizes is no longer a problem. We know that a space vacated by a departing page will be exactly the right size to take a new page. No problem.

Page tables and control registers

Well, there is one. To achieve this, more translation tables are required, called *page tables*. The CR registers are used to address these, and the page tables are kept in memory just like the descriptor tables.

The CPU has various extra registers for maintaining the paging mechanisms, most importantly, CR3, which contains the base address of the *Page Table Directory*.

Just for the record ...

Linear address

The address computed from the descriptor table, now renamed the *linear address* (as it is no longer the final physical address), is divided into fields, with bits 22 to 31 being an index into a page-table directory that gives the address of a particular page table. Bits 12 to 21 are the index into this second table, which contains the final address. Bits 0 to 11 are unchanged and become part of the final address.

You will come across the words *linear address* later in the book. Note that sometimes the words *virtual address* are used in various books to mean the same thing, though there is a distinction. The linear address is that 32-bit address that would be the physical address if page tables didn't get in the way.

Virtual-86

This is another paging mechanism that does away with descriptor tables. It was intended to provide the 386 with better Protected mode emulation of the 86 CPU than the 286 can manage, which it does very well.

This mode is fascinating. It also does away with selectors and brings physical segment (paragraph) addresses back into the segment registers! Thus we come full circle, but with a vital difference.

Paragraph addresses are back!

Although the 16-bit segment address is back, and once more programs designed to directly manipulate segment registers can do so. The CPU does compute a 20-bit address consisting of paragraph address plus offset, but this is not put on the external address bus. Instead, it is processed via page tables, that is, translated to some other 32-bit address then put onto the address bus.

Once again, this paging is transparent to the programmer, but it does mean that the program, data, etc. are not where you think them to be judging from the segment registers.

Virtual machine

Virtual-86 mode is useful not just for emulating the old XT computer, but is the very foundation of Windows Enhanced mode. True, each *virtual machine* will have an addressing limit of 1M, but Windows can create many of these (Figure 1.14).

Figure 1.14: Virtual Real mode.

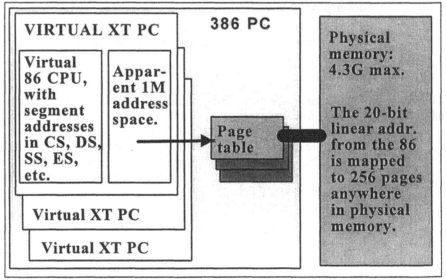

The upper 8 bits of the linear address are remapped

Instead of putting the 20-bit linear address onto the address bus, as for Real mode, virtual-86 mode uses the upper 8 bits of this address as a lookup in the current page table — note that the table entry contains the base address of the page, which is combined with the lower 12 bits of the linear address to form the actual 32-bit address. It is this final 32-bit address that the CPU puts out for a memory access. Refer also to page 274, Figure 11.2.

So what happens if your program writes directly to video RAM at segment B800? This is up to the operating system, which most likely will create virtual screens for each task, setting them up anywhere it wants to in RAM.

Contention Issues

There are various things to think about under this heading, but I have at this stage just addressed the issues of privileges, I/O, and task switching.

The topics are brought up at various points through the book, so look in the Index for other page references.

Privileges

The *dpl* field in the descriptor defines the *privilege level* of that segment. Also you will see back on page 27, Figure 1.12, that the selector has a requested privilege level (rpl).

Four privilege levels

Because it is a 2-bit code, there are four possible levels, zero being the most privileged. The kernel of the operating system will operate up here (zero), while your lowly program will reside at a lower privilege level.[1]

Your program's level is basically reflected in what the rpl is set to, and this must be numerically equal to or less than the segment's dpl to allow access to that segment — otherwise the CPU exits to an error routine and the dreaded UAE (Unrecoverable Application Error) dialog box appears, and that's the end of your program!

I/O Privilege

IOPL field

Privilege levels do have some impact on I/O. If you look at the FLAGS register (see page 244), you'll find 2 bits that hold the *Input/Output Privilege Level* (IOPL). Your application must have a privilege level numerically equal to or less than this to be able to perform I/O. With Windows, the IOPL field is set to zero, most privileged.

IN, OUT, CLI, and STI

However, it is possible for the operating system to give permission for certain I/O to occur, even though the application doesn't have the right privilege. I/O access involves use of the IN and OUT instructions and control of the interrupt flag by CLI and STI

---

[1] Windows 3.0 runs WinApps at level 1, DOSApps at level 3, and DLLs at level 1. Windows 3.1 and later run all three at level 3.

instructions. The interrupt flag is in the FLAGS register and when cleared, prevents hardware interrupts from occurring.

PUSHF,
POPF

If the application has sufficient privilege to perform direct I/O, it can also set and clear the interrupt flag. Although a Windows program does not have the privilege of direct I/O, Windows does allow it, to an extent. If I/O is attempted, the CPU goes to a Windows error (exception) routine, which **does** have the privilege to do what it wants — the routine allows CLI and STI (clear or set interrupt flag instructions) but does not let PUSHF or POPF instructions affect the interrupt flag. This is something to be aware of and a possible source of incompatibility with old DOS code. It also means that an IRET from an interrupt routine may not set the flag as it was prior to the interrupt.

For more information on I/O, refer to page 244.

Task Switching

Considering the complications of multitasking, I sometimes wonder if it is all worth it. Perhaps a more effective solution would have been multiple CPU-boards, each single-tasking. Anyway, we are stuck with the current situation.

Changing
LDTs

Changing from one task (program) to another is a matter of changing to a new LDT,[1] which involves the CPU looking into the GDT and getting the new LDT's address.

However, the "state" of the task about to be suspended must be saved, and the "state" of the incoming task must be restored. This state consists of the CPU and coprocessor registers plus various memory pointers and values, and an incredible time overhead is involved to save and restore this lot.

Task State
Segment

The CPU has to maintain a special segment for each task, called the *Task State Segment* (TSS), into which all of this goes. Then, of course, the CPU must keep track of where these TSSs are, so it maintains descriptors for the TSSs in the GDT. Thus the GDT contains more than just descriptors for the LDTs.

---

[1] Windows 3.x and 95 have only one LDT for all applications, whether in Standard or Enhanced modes, which is a compromise in its design that can potentially cause trouble. This limitation tallies with DPMI version 0.9, which in Windows maintains one LDT per virtual machine, not per task. Windows is seen as a single *client* to DPMI. Windows 95 32-bit applications have individual LDTs.

Interrupts

Real mode interrupts
Like everything else, Protected mode interrupts are a whole new ball game. First, let's review the mechanism in Real mode.

The standard method of doing I/O and file and memory management, plus a heap of other operations, was by the BIOS and DOS interrupt services. These are accessed from an application program by means of the INT instruction, with this syntax:

```
INT   n        ;software interrupt
```

---

where "n" is an integer (whole number) from zero to FF (hex). The usual procedure is that certain registers have to be loaded prior to the INT, depending upon the particular service, and many of the services have subfunctions, usually selected by a value in the AH register.

INT-21h, the main DOS service
The most important of these is INT-21h (h = hexadecimal), which is the main DOS service, with dozens of subfunctions.

A comprehensive list is to be found in my previous book. In this one you'll find extra INT services especially relevant to Windows. It is not that we do away with INT services entirely with Windows, it's just that many of the BIOS and DOS services are designed for DOS and the Real mode and are no longer appropriate.

Windows functions
We access the Windows services by CALL instructions, not INTs, and from the CPUs point of view there **is** a difference. Windows' services, or functions, do all that many programmers would want, though we dig a little deeper in this book and also show how useful the INT services can be.

Real Mode Interrupts

Interrupts, whether from an external source (hardware) or generated internally by the program (software), cause the same reaction in the CPU:

1. The CPU pushes the current Instruction Pointer (IP), Code Segment (CS), and FLAGS register onto the stack.

Interrupt Vector Table (IVT)
2. Then the CPU uses the value "n" as an index into the *Interrupt Vector Table* (IVT), where it finds the FAR address of the service routine.

3. The CPU then loads the FAR address into its CS:IP registers and commences execution of the service routine.

4. Interrupt routines always terminate with an IRET instruction, which has the effect of popping the three values saved on the stack back off, into CS, IP, and FLAGS. Thus the CPU carries on as before, as though nothing had happened.

IRET instruction   Note that when a CALL instruction executes, it works in a similar way, but a FAR CALL only saves CS and IP on the stack, not the FLAGS. Also, if it is a NEAR CALL, only IP is saved on the stack. In addition, the routine called must terminate with RET, not IRET, as the latter pops three values off the stack (expecting FLAGS to be on there as well).

CALL to an ISR   Incidentally, a useful point arises from what I have written above. You can use the CALL instruction to call the BIOS and DOS services, despite the fact that they terminate with an IRET:

```
PUSHF                           ;push flags on stack.
CALL routinename
```

That is, you push the FLAGS on beforehand, using a special instruction, PUSHF (there is also a POPF). You do need to know the address of the routine that you are calling, however, since it doesn't make use of the IVT, as INT does.

Protected Mode Interrupts

Just as segment registers no longer represent real addresses, so too the interrupt mechanism no longer uses the Interrupt Vector Table (IVT). Interestingly, when Windows is running, the IVT is still there, but our applications don't use it. It is still used by Windows, but that's another story.

Structure of the IVT   So, just where is this IVT? Have a look back at page 11. The IVT sits in RAM right down at 0000:0000, occupying the first 1024 bytes. It is set up by the BIOS startup routine and filled in by DOS also.

Interrupt Descriptor Table (IDT)   The fundamental problem is that it contains real segment addresses, which is a no-no in Protected mode (though is ok in virtual-86 mode). Therefore a special table has to be created by the Windows operating system, called the *Interrupt Descriptor Table* (IDT), which contains the linear addresses of the services. Linear addresses are real, but they are actual 24- or 32-bit addresses, without the segment:offset structure.

Using INT within WinApps

There is a fascinating outcome of this. From within a Windows application, you can have an INT instruction — let's say that you want to call the BIOS INT-10h service, which controls the video adaptor. INT-10h is not a service that Microsoft would want you to call from your application, since all control of the video should be done by the Windows functions — but you can do it.

A warning here: some services will crash if called while in Protected mode, and others will behave strangely.

Microsoft has in some cases provided alternative BIOS and DOS services, written especially to run in Protected mode, and when your program executes, say, INT-21h/AH = 35h, the CPU will look up that entry in the IDT (not the IVT) and get the address. Thus it is very easy for Microsoft to substitute its own services into the IDT.

Redirection of IDT to IVT (Protected mode to Real mode)

In many cases (probably most) Microsoft services have not been substituted, and execution goes to the original BIOS or DOS service. Although the Real mode services may in some cases manipulate addresses in the form segment:offset, which will cause the code to crash if the CPU is running in Protected mode, Windows gets around the problem by switching the CPU into Real mode, or into virtual-86 mode, then calling the service.

For such cases, the entry in the IDT points to a special handler, which, apart from changing the CPU to Real mode, must also convert any pointers from selector to segment value. Then the handler will have to look in the IVT to get the address of the Real mode service.

Thus, even the services in the BIOS-ROM will work. At least they will return without crashing the system (in most cases), though whether they do what you want is another matter.

Note however, that there is a difference in accessing interrupts from a 32-bit compared with a 16-bit Windows application. This is a complicated issue and is developed in Chapter 16.

Virtual IVTs

Another fascinating thought occurs about virtual-86 mode, which uses the IVT, but in plural. Although there is an IVT at actual physical address 0000:0000, each virtual-86 task will have its own copy of the IVT, which appears to be at 0000:0000 but is paged anywhere. You need to be aware of this proliferation of IVTs if you want to hook a vector.

Refer to Chapters 10, 11, and 12 for more information, particularly page 282 and thereabout.

Postamble

This chapter mapped out the overall architecture of the x86 processor, and you may have found some of it heavy going. Subsequent chapters are a step back, and topics are revisited in depth.   Chapter 2 is an in-depth treatment of the basics of assembly language.

2

Basic Assembly Language

Preamble

Content of this chapter

This chapter contains an introduction to assembly language for the x86 family of processors. The focus is on 16-bit programming. Later chapters will expand this to 32-bit programming.

Real mode 16-bit programming can be considered an essential step up the ladder of understanding, climbing through 16-bit Protected mode, toward 32-bit Protected mode programming.

Chapter 4 puts this knowledge to use in a first 16-bit Windows application.

Discussion relates to the Microsoft and Borland assemblers, though of course there are other compatibles.

Stack Instructions

Initialisation of the stack

The computer maintains a *stack* somewhere in memory. DOS will set the *Stack Segment* register SS when your program is loaded, and the *Stack Pointer* SP will be initialised to FFFEh, or some value that means the stack is empty. The stack is used by the computer and by your program. For example, whenever an interrupt occurs the CPU pushes the IP, CS, and FLAGS onto the stack, so that when the interrupt routine is finished (terminated by an IRET instruction) the CPU will pop these values back into the respective registers and continue from where it left off.

Purpose of the stack

Thus the stack is used to hold register values to enable the CPU to return from an interrupt and also from a procedure *CALL*.

However you can make use of the stack in your program, by means of the *PUSH* instruction, which pushes a 16-bit value onto the stack, and *POP*, which pops the top value off the stack into a register or memory location. Also *PUSHF* and *POPF* can be used to push the FLAGS onto the stack and pop them off.

Whoa! This is a lot to think about! I've just stated above that there is a memory area called a stack, that it is used by the CPU to store register values for interrupt and CALL-instruction execution, and it is used by the PUSH and POP instructions. You may find it extremely helpful at this point to visualise what is happening. Look at Figure 2.1 and examine the effect of the PUSH and POP instructions.

... temporary storage

In Figure 2.1 you see two instructions, PUSH and POP, that you can use in your program. You can push values onto the stack, and take them off again — why? — one reason is that it serves as a convenient temporary storage.

... CALL/ RET

I also mentioned that the stack is used by the CALL instruction — this is one of the "transfer of control" instructions and is described in the next section.

... interrupt mechanism

I mentioned that interrupts also use the stack — again, explanation is deferred.

Do not worry about these deferred explanations — one thing at a time. Examination of Figure 2.1 will give you an idea about what the stack is, which is satisfactory for now.

Figure 2.1: Concept of the stack.

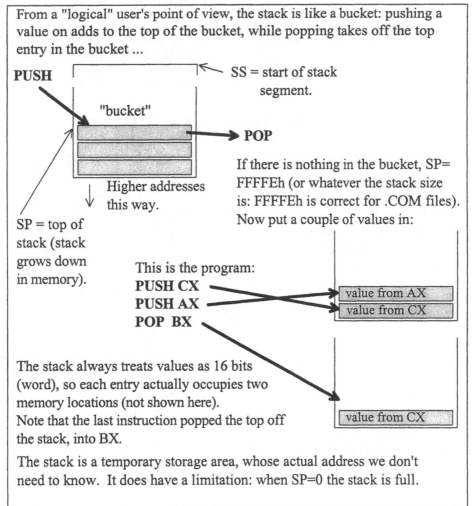

From a "logical" user's point of view, the stack is like a bucket: pushing a value on adds to the top of the bucket, while popping takes off the top entry in the bucket ...

PUSH

SS = start of stack
segment.

"bucket"

POP

If there is nothing in the bucket, SP= FFFFEh (or whatever the stack size is: FFFFEh is correct for .COM files). Now put a couple of values in:

Higher addresses this way.

SP = top of stack (stack grows down in memory).

This is the program:
PUSH CX
PUSH AX
POP BX

value from AX
value from CX

The stack always treats values as 16 bits (word), so each entry actually occupies two memory locations (not shown here).
Note that the last instruction popped the top off the stack, into BX.

value from CX

The stack is a temporary storage area, whose actual address we don't need to know. It does have a limitation: when SP=0 the stack is full.

Transfer of Control

The idea of a computer program is that it is a sequence of instructions: in this book we are looking at machine instructions that the CPU directly understands. Assembly language is just a symbolic (more meaningful) way of writing the machine instructions.

The CPU executes the instructions sequentially — that is, one after the other in order of increasing addresses — but can also jump out of sequence.

LOOP, JMP, CALL, INT, Jx

The topic of this section is those instructions that cause execution to go to some other place in the program. The main ones are: LOOP, JMP, CALL, INT, and Jx. In this section we will examine CALL, JMP, and Jx. LOOP and INT are examined a little bit later:

Figure 2.2: Stack handling for CALL and RET.

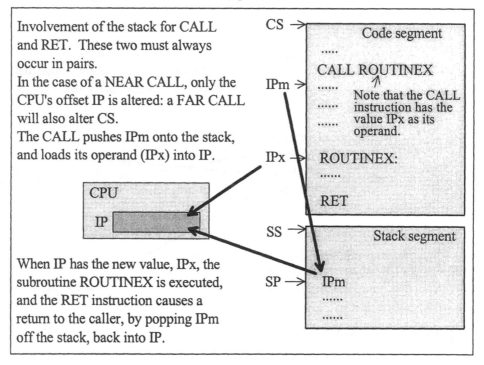

Figure 2.2 illustrates how the CALL and its companion RET use the stack. The basic idea is that the value in the Instruction Pointer, IP, is always the *next* instruction to be executed, so when "CALL ROUTINEX" is executing, IP will have IPm in it. Since the value in IP has to be changed to the subroutine, IPx, the return value has to be saved somewhere: hence the stack is used to save IPm. The RET instruction must always be placed at the end of a procedure, as it pops the top off the stack, back into IP.

If you have programmed in C or Pascal, you know that you don't put a RET, or anything special, at the end of a procedure or function. CALL and RET do go into the code, though, because the compiler translates the high-level source code to machine instructions.

FAR and NEAR

This topic does need some careful thought. Any CALL, RET, or JMP instruction can be a FAR or NEAR jump. What this means is that if the jump is NEAR, the jump is only within the current code segment; that is, only the IP is altered, as per Figure 2.2.

A FAR jump or call, however, can be to anywhere in the entire 1M address range, as both CS and IP are altered. In Figure 2.2, the procedure ROUTINEX is shown as being in the same code segment as the CALL instruction, but it could be somewhere entirely different. Obviously, if ROUTINEX is in a different code segment, then both CS and IP in the CPU would have to be changed to the new values.

Note that it also logically follows that the original values of CS:IP, immediately after the CALL, would both have to be saved on the stack, and RET would have to restore both of them at the end of the procedure.

Note that with what is called 32-bit programming, the distinction between NEAR and FAR just about disappears.

Code labels

One thing that you will notice from Figure 2.2, is that I used a *code label*, ROUTINEX, to name the start of the procedure. This is basically what you expect to be able to do in any high-level language, and you can also do this in assembly language. A code label marks, or identifies, that point in the code, hence a CALL was able to be made to that place.

Code labels with MASM, TASM, DEBUG

With a professional assembler, such as the Borland TASM, or Microsoft MASM, these labels are a normal part of writing a program, but DEBUG is a different story.

DEBUG CANNOT HAVE LABELS!

With DEBUG any instruction that transfers control to another address must contain the actual offset.

What is DEBUG?

What is DEBUG? It is a program that comes with DOS, and from the DOS prompt you will only have to type the name of the program to execute it. DEBUG.EXE is a way of becoming familiar with the instruction set — it allows you to try out the instructions and put together simple programs.

These examples show that DEBUG must have an actual address, not labels:

```
   MOV  CX,9
PLACE1:                    ;this is at 113 (say)
   MOV  AX,0               ;arbitrary instr
   LOOP 113                ;absolute offset (no label)
```

```
...
LOOP PLACE1         ;using a label.
```

However, by writing the code in "proper" assembly language, we do not need to know actual addresses. The second example here shows how a proper assembler can have a symbolic address marker, in this case PLACE1.

JMP instruction

In Figure 2.2, we looked at a CALL instruction, but there is also a JMP (jump) instruction that transfers execution to the address specified in its operand in the same manner as the CALL instruction, but with a major difference: no return address is saved on the stack. This is because JMP is used when you do not want execution to come back.

SHORT, NEAR, and FAR

It was also explained above that the CALL can be NEAR or FAR, but the JMP can be SHORT, NEAR, or FAR.

The example code below shows a JMP to a label. Usually, an assembler defaults to a NEAR jump, as the destination is usually in the same segment.

```
jmp PLACE1
...
PLACE1:                 ;code label.
  mov ax,0              ;arbitrary instruction.
```

At this point, it is instructive to consider how the assembler will assemble this JMP instruction into memory. Obviously, it has to be converted to "machine language", or binary bits. That is what any compiler or assembler does.

Figure 2.3: Generation of machine code, NEAR jump.

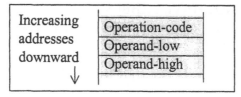

In Figure 2.3 you can see the basic scenario. The first one (or sometimes two) memory location(s) contain the instruction-code, or operation-code, often referred to as the op-code, that identifies this as a JMP instruction (or whatever), while the following zero or more bytes are the operand.

NEAR JMP

In the case of the NEAR jump instruction, the operand contains a 16-bit offset, which is the place to jump to. But, and this is most important, the addressing structure of all the Intel x86 CPUs uses byte addressing, meaning that each address addresses a one-byte (8 bit) memory location.

Therefore, the operand requires two memory locations, as shown in Figure 2.3 as operand-low and operand-high. The Intel x86 convention is that the low-half of the value is stored at the lower address.

FAR JMP

It is also useful to note that if the JMP is a FAR jump, that is, to another code segment, the operand of the instruction will have to contain the destination CS:IP, which is two 16-bit values. Hence it would be 32 bits.

The FAR jump would assemble as the one-byte (or two) op-code, followed by a one-word IP then one-word CS value. Note that the FAR jump can also jump within the current code segment but is slightly inefficient because it is a longer instruction, taking a little longer to execute and using more memory.

SHORT JMP

The JMP instruction has one interesting difference from the CALL: it is able to perform a SHORT jump. This is shown in Figure 2.4:

Figure 2.4: SHORT jump machine code.

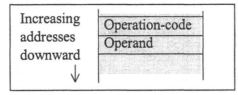

Range of a SHORT jump

This reduces the instruction down to the one-byte (8-bit) op-code followed by a one-byte 2's-complement displacement. This displacement allows jumps to be only +127 to -128 about the current IP position.

In some circumstances, the assembler will automatically make the jump SHORT, but it can also be forced to, by means of the SHORT directive.

Conditional Jump

The conditional-jump instructions test various flags before deciding whether to jump or not. These instructions are always of the SHORT type. This is very important — they can only jump 128 locations away from the current code location. The conditional jump instructions are sometimes confusing for the student, however the concept becomes quite clear with a little practise. Most CPU instructions affect the flags after they have executed, and the conditional jump instructions can be used to test the flags and jump accordingly.

Below is a summary of the conditional jump instructions:

```
JZ          ;jump if previous result was 0
JNZ         ;jump if previous result not 0
JGreater    ;this means "if the SIGNED difference is positive"
JAbove      ;this means "if the UNSIGNED difference is positive"
JLess       ;this means "if the SIGNED difference is negative"
JBelow      ;this means "if the UNSIGNED difference is negative"
JCarry      ;assembles the same as JB.
```

When using these instructions, you do *not* enter the part in italics.

Signed and unsigned compare   Note that when comparing two values, we need to distinguish between whether the values are unsigned or 2's complement.

Here are simple examples:

```
ADD AX,VAL1
  JZ  ZERORESULT ;jumps if previous result=0(zero-flag
  ...                                   ; set)
  CMP AX,56               ;compare instr.
  JA  ABOVE56             ;jumps if AX>56

;Variations ...
  JNC place1              ;jump if Carry flag=0
  JE  place1              ;same as JZ ("Equal")
  JAE place1              ;unsigned jump, if above or equal.
  JBE place1              ;unsigned jump, if below or equal.
```

The ADD instruction, given as an example above, is explained a little further on. Ditto for the CMP instruction.

Note that "ZERORESULT", "ABOVE56", and "place1" are code labels, chosen to have meaningful names.

Addressing Modes

Obviously, the instructions of your program will be accessing registers and memory, and the mechanisms by which this is done are called the addressing modes.

The best way to show this is by example:

```
VAL1 DW 0
  ...
  MOV AX,BX              ;register addressing mode.
  MOV AX,567             ;immediate addressing mode.
  MOV AX,[567]           ;direct addressing mode
```

```
MOV AX,VAL1          ;direct addressing mode.
```

MOV instruction

The humble MOVe instruction is the equivalent of the LoaD-Acc and STore-Acc instructions of the 6800 CPU, for those who have had exposure to that beastie. It simply moves a value from one place to another, in this case copying the value of BX to AX.

Register & immediate addressing

Because only registers are involved in the first instruction of the above example, this is called *register* addressing.

The same MOV instruction appears again on the second line, but note that a value is specified this time. This value is NOT an address; it is an immediate value that is loaded into AX. This is called *immediate* mode addressing

Direct addressing

Now this is different. The square brackets of the third instruction signify "the contents of" and it is the contents of address 567 that is loaded into AX (there is a qualification to the above comment, as the example loads the AX register, which is 16 bits, from a memory location, which is 8 bits).

Note too that with an assembler (not primitive DEBUG though) any address can be replaced by a *label*, so if you had defined address 567 as being represented by label VAL1 (for example), then this would do the same thing:

Both of these are called *direct* addressing.

[ ] syntax

Do note one point about syntax. The last instruction *could* have square brackets around VAL1, and it would be interpreted exactly the same by the assembler (TASM or MASM).

Indirect and indexed addressing

Indirect addressing is somewhat more abstract. It means that the contents of the operand are used as the address. So, the content of BX is the address from which the value is fetched into AX:

```
mov   ax,[bx]            ;indexed addressing mode.
mov   ax,[bx+5]          ;/
mov   ax,[bx+si+5]       ;/
```

That just about covers it, except that indirect addressing does have some options, as shown in the last two instructions above.

The first one adds the contents of BX to 5, and the result is the address, while the second example adds the contents of BX, SI, and 5 to form the address. This modified form of indirect addressing is called *indirect plus displacement* if a constant is specified, or *indexed indirect* if two registers are specifed.

Restrictions on indexed addressing

Note that we often just label these various indirect modes under the title of *indexed addressing*.

Note also, that there are restrictions on the combinations of registers allowed within the brackets: you can have SI or DI, but not both, and you can have BX or BP, but not both. No other registers are allowed.

Segment Registers

Another thought: how do you access data in DS, the data segment? This is the place to keep data, so obviously your program must be able to get to it. Simple: most instructions automatically reference the DS.

For example, the listing below shows how VAL1 is defined and referenced:

```
.DATA
VAL1   DB  0              ;in data segment.
  . . .
.CODE
  mov   ax,VAL1           ;in code segment.
```

Later, you will see more details on how to use the *assembler*, so don't worry about that side of things. Suffice to say that you can define a label in the data segment and reference it from the code segment.

When the program is assembled, the address of VAL1 will be put into the operand of the MOV instruction: note however that this is an offset relative to the DS.

Most importantly, when your program is executed, it must have DS set to the beginning of the data area, as the MOV instruction will automatically use DS to compute the physical address.

Sometimes, especially with pop-up and interrupt routines, the program may be entered with DS not set correctly, so you have to take care of that at the beginning of the program.

Segment override

Although the MOV instruction in the above example automatically referenced the DS register, it is possible to override this. For example you could have data in the code segment, so your program would have this:

```
.DATA
  . . .
.CODE
  jmp place1
VAL1 DB 0                 ;data defined in code segment.
place1:
  mov ax,cs:VAL1
```

Some notes on this:

.COM format
- In the case of .COM programs CS = DS = SS, so the question of override doesn't arise normally. With a .EXE program, data could be kept in the code segment, as long as execution jumps around it: but note also that OS/2 and other operating systems that operate the 286 and 386 CPUs in Protected mode, may be very unhappy with data kept in the code segment/s.

ES register
- Sometimes data is kept in a segment pointed to by ES (or FS and GS in the 386), so ES override might be useful in this situation. The BP register, although a general-purpose register, is treated by the assembler as an offset into the stack segment, SS, by default. Thus, if you want to use BP to access data in segments pointed to by DS or ES, an override is required.

String Instructions

This group of instructions are designed for moving blocks of data from one place in memory to another, and some of them are for searching through and comparing blocks of data. The word "string" does not necessarily imply text, but any block of data.

Mostly you will use the string instructions responsible for moving data around, such as MOVS, LODS, and STOS. Basically, you have the source block in one part of memory and the destination somewhere else, and you have to set certain registers to point to these source and destination areas before using the string instruction.

Concept of the string instructions
The string instructions have an "implied" addressing mode, in that they use certain predefined registers, as shown in Figure 2.5. Figure 2.5 is a picture of memory. DS:SI is where the data is, and ES:DI is where it's sent.

MOVSB, for example, would read a single byte from DS:SI, copy it to ES:DI, and automatically increment both SI and DI, so that the next time the instruction is executed the next byte will be copied.

W/B postfix
All the string instructions can be postfixed with a "B" or a "W". MOVSW would move two bytes of data (one word) and SI and DI would automatically increment by two.

Figure 2.5: Concept of the string instructions.

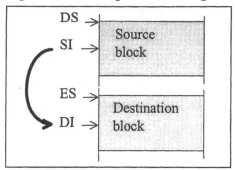

Auto-
increment

String operations make use of SI and DI to point to the source and destination strings respectively, and they are automatically updated each time the string instruction is executed.

Direction
flag, DF

There is a direction flag, DF, that is cleared by instruction CLD, and set by instruction STD. If DF is clear, the string instruction will automatically increment SI and/or DI to point to the next byte or word, and if DF is set they will be decremented. It is normal to operate on a string starting from the lowest address in memory, so use CLD before a string operation (this is the default for the 80x86 family anyway).

DF is one bit of the FLAGS register, shown on page 244.

CLD and STD are described in the Appendices.

REP prefix

REP is a prefix, placed on the same line and before a string instruction. It means "check if CX = 0, if not perform the string instruction, decrement CX, then start again". Example:

```
mov   cx,str_length
rep   movsb                    ;repeat with cx = count.
```

A variation on this is REPNE, which is basically the same but will also terminate if the zero-flag is set.

REP variations are summarised in the Appendices.

LOOP
instruction

Note that the LOOP instruction can do much the same as REP. Again, CX is decremented before CX is compared with zero, so MOVSB will be executed exactly the number of times originally loaded into CX. The loop will terminate with CX = 0. There are some variations on the basic LOOP instruction: have a look in Appendix A.

```
mov   cx,str_length
```

```
again:                  ;code loop does same as above.
  movsb
  loop again            ;loop is an actual instruction.
```

One warning with LOOP is don't initialise CX to zero before entering the loop, as it will then loop around 65,000 times!

When to use LOOP rather than REP? LOOP is not restricted to the string instructions because it is an instruction in its own right, whereas REP is only an instruction prefix designed to work with the string instructions. LOOP can be used wherever a program loop is required, and more than one instruction can go inside the loop: though note that LOOP can only do a SHORT jump.

MOVSB, MOVSW   Transfer contents (byte or word) of source-pointer DS:SI to location specified by destination-pointer ES:DI (hence the name Source-Index and Destination-Index).

CMPSB, CMPSW   These instructions compare bytes or words pointed to by ES:DI and DS:SI and set flags for use by J-condition instructions. For example, to use CMPSB with REP:

```
mov   cx,str_length
rep   cmpsb
jnz   difference_fnd
```

This example will compare the two strings until the end of the string (set by value in CX) OR until a non-equal comparison is reached (in which case CX will point to the position in the string at which the difference was found, and the zero-flag will be clear).

SCASB, SCASW   Use these instructions to compare AL or AX with the value pointed to by ES:DI. Note: they are most often used with REPNE. A typical use is:

```
;setup DS to beginning of PSP (will be for COM files & at
;start of EXE prog). else use ES override....

mov   al,"/"
mov   di,080h         ;length of tail in PSP
mov   cx,[di]         ;(could use override)
mov   di,081h         ;command-tail in PSP.

;we will assume that ES is set to the start of the PSP--
;should be for EXE & COM files.
REPNE   SCASB
jcxz no_slash         ;yes, slash was found...
mov  al,[di]          ;could use override.
```

Command-line tail   The code searches the DOS command-tail in the PSP (see Figure 1.8) to see if there is a "switch" ("/" followed by a letter).

If the loop terminates without finding a slash, CX will equal zero, so the special conditional jump instruction, JCXZ, which tests if $CX = 0$, can be used to detect that no slash was in the string.

Because the string-instruction automatically increments DI each time, at termination DI will point to the next character past the last one tested. If the slash was found, this next character will be the switch.

Note that Windows 3.x and 95 applications still have a PSP.

LODSB, LODSW   The value in the location pointed to by DS:SI is loaded into AL or AX. SI is automatically incremented (+/-1 if LODSB, or +/-2 if LODSW).

STOSB, STOSW   The value in AL or AX is stored at the location pointed to by ES:DI. DI is automatically incremented (+/-1 if STOSB, or +/-2 if STOSW).

STOS and LODS are most useful for video access, as the format of video-RAM in text-mode requires every odd byte to be an attribute character:

```
;...setup ES:DI....
;...setup DS:SI....
 mov cx,string_length
 mov ah,attribute
 ...
next_char:
 lodsb                          ;char-->AL
 stosw                          ;AX-->destination.
 loop next_char
;...this code will send characters to the screen
```

Arithmetic Instructions

PREREQUISITES

These include addition, subtraction, multiplication, and division. I expect you to have a working knowledge of the principles of binary arithmetic: unsigned binary numbers, 2's complement binary numbers, radix conversion among hex/binary/decimal.

For example, suppose I ask you to express -2 as a 32-bit binary number, and also as a 32-bit hexadecimal number. Can you do it? If the answer is yes, then you do have a few clues, so read on. Otherwise look back at Chapter 1, and consolidate with further study if required.

***CMP
instruction***

The CMP instruction has already been introduced but involves arithmetic comparisons, so it will be considered again here.

The example below subtracts 127 from AL, and the result sets the appropriate flags. Decimal is the default with an assembler, unless an "h" is appended to designate hex. DEBUG can only have hex. We will treat 127 as being decimal in this case.

```
cmp   al,127          ;hypothetical subtract.
```

The CMP instruction can be followed by a conditional jump that jumps or doesn't jump depending upon the flags.

Although CMP subtracts the two values, it is only done hypothetically, and the two operands are left unchanged. CMP doesn't care whether the number is unsigned or 2's complement — it just subtracts them. It is the same for all the addition/subtraction arithmetic instructions — it is up to the programmer to decide how to treat the operands and the result.

***2's
complement
versus
unsigned***

This point can be clarified. Since the above example is dealing with 8-bit operands, the range of values depends upon whether we are treating them as 2's complement or unsigned number:

```
Unsigned:    0   <–>   255   or  00 <–> FF in Hex.
2's compl: -128  <–>  +127   or  80 <–> 7F in Hex.
```

So if AL = 128, the example CMP instruction will give a hypothetical result of:

128 - 127 = 1, i.e., the result is +1, or in binary 00000001.

Obviously AL is greater than 127, but that is only if you treat the numbers as unsigned. As a 2's complement number, 128 is actually -128!

```
Unsigned:  0<–>127, 128<–>255   or   00–7F,80–FF in Hex.
2's compl: 0<–>127,-128<–>-1    or   00–7F,80–FF in Hex.
```

So from a 2's-complement point of view, AL is less than the operand 127. That is why there are different conditional jump instructions for signed and unsigned numbers.

Following the "CMP AL,127", we could have any one of the following, depending upon how we want to treat the number:

```
JA  label  ;jump if AL above 127, unsigned.
JB  label  ;jump if AL below 127, unsigned.
JG  label  ;jump if AL greater than 127, signed.
```

```
JL   label   ;jump if AL less than 127, signed.
```

This can be a point of confusion for novice programmers, so be careful. It is a good policy to stick with unsigned compares, unless you have particular reason to do otherwise.

NEG instruction

This is strictly for 2's complement numbers — it changes the sign of an operand. For this example, the result will be -127 in AL:

```
mov   al,127
neg   al
mov   al,-127
```

A useful point to note about the assembler is that you don't ever have to calculate the binary or hex negative 2's complement number; just put a minus sign in front and the assembler will do the conversion. The last line shows this.

INC, DEC instructions

(INCrement, DECrement).   These two do what their names suggest; add 1 to an operand or subtract 1 from it.

Since we have specified an 8-bit operand in the examples below, if INC goes beyond 255 (FF hex), then it will simply roll around and start from zero. Ditto, but the opposite, for DEC.

```
inc   al
dec   al
```

ADD, SUB instructions

Recall from the above notes that ADD/SUB arithmetic instructions don't know whether your operands are 2's complement or unsigned numbers — that interpretation is up to you.   The size of the operands are important in these calculations, and the instruction determines that from the operands themselves.

SUB works just like CMP, setting the same flags (and so can be followed by a conditional jump), but the subtraction is not hypothetical — the result of the subtraction is left in AX.

```
add   al,127
sub   al,127
```

These instructions can handle numbers bigger than 16 bits. Of course so can the 386, since it has 32-bit registers, but for now I'll assume I only have 16-bit registers and I want to add numbers that could possibly have a 32-bit result.

```
add   ax,cx          ;add cx to ax, result in ax.
adc   bx,dx          ;add dx to bx, with carry.
```

For this example we have two 32-bit values in BX:AX and DX:CX. The two lower halves are added, leaving the result in AX. The ADD instruction will set the carry flag if the unsigned result is greater than the limit (FFFF hex).

ADC, SBB instructions ADC means ADd-with-Carry, and adds the carry flag bit plus DX, to BX, with the result in BX. Thus the total result is in BX:AX.

For subtraction of 32-bit numbers, the principle is the same, and there is an appropriate instruction: SBB (SuBtract with Borrow).

DAA, DAS For addition and subtraction of BCD numbers, you need to use DAA and DAS.

The operation of DAA (Decimal Adjust for Addition) is shown pictorially in Figure 2.6. It corrects the result of adding two BCD (packed decimal) values. Operates on the AL register. If the rightmost four bits of AL have a value greater than 9 or the half (auxiliary) carry flag is 1, DAA adds 6 to AL and sets the half-carry flag. If AL contains a value greater than 9Fh or the carry flag is 1, DAA adds 60h to AL and sets the carry flag.

Figure 2.6: Decimal arithmetic.

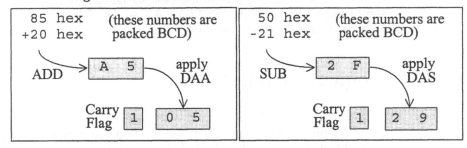

DAS (Decimal Adjust for Subtraction) is the opposite of DAA.

After subtracting two numbers, perform DAS operation on AL. If the rightmost 4 bits have a value greater than 9 or the half-carry flag is set, DAS subtracts 6 from AL and sets the Carry Flag.

MUL, DIV, IMUL, IDIV There are two groups of multiply and divide; MUL and DIV for unsigned numbers and IMUL and IDIV for signed numbers.

One problem we have with multiply is that two 16-bit operands can produce a result up to 32 bits long. Thus in the case of CPUs with only 16-bit registers, the result may have to reside in two registers. The MUL instruction uses AL and AX, or AX and DX, by default.

```
mul   bl          ;al*bl --> ax
mul   bx          ;ax*bx --> dx:ax
```

The first example makes the assumption that the other operand is in AL, so the result will appear in AX. The second example makes the assumption that the other operand is in AX, and the result will be in DX:AX.

Division has problems of its own. The dividend (the operand to be divided) is in either AX or DX:AX, and the divisor is in any other register or variable (8 or 16 bits).

```
div  bl              ;ax/bl --> ah and al.
div  bx              ;dx:ax/bx --> dx and ax.
```

The first example assumes the dividend to be in AX and puts the result in AX in this format: AH = remainder (left over), AL = quotient (result).

The second example specifies a 16-bit divisor, which assumes that the dividend is in DX:AX and the result in DX:AX as follows: DX = remainder, AX = quotient.

A feature built into the CPU is that if there is an error in the calculation, a certain interrupt is generated, and DOS displays an appropriate error message. In the case of DIV, it is possible for the quotient to be too big for AL or AX — DOS will abort your program with a "division overflow" message.

Logical Instructions

Logical instructions basically work on individual bits rather than complete numbers. They relate back to boolean algebra, and as with the arithmetic instructions, I assume a certain background knowledge. You should have a basic understanding of the boolean AND, OR, EXCLUSIVE-OR, and NOT functions.

AND, TEST   AND performs a logical AND on corresponding bits in two operands, leaving the results in one operand.

```
mov  al,01001000b
and  al,00001000b     ;answer al = 00001000b
```

TEST is just like AND but only does the operation hypothetically and doesn't change the operands (this is very similar in concept to the relationship between SUB and CMP).

OR   OR performs a logical OR operation on two operands.

```
mov  al,01001000b
```

```
    or   al,00001000b        ;result al = 01001000b
```

XOR   XOR performs a logical EXCLUSIVE-OR on two operands.

```
    mov  al,01001000b
    xor  al,00001000b        ;result al = 01000000b
```

NOT   NOT complements all bits in an operand (this is not a 2's complement conversion — see NEG).

```
    mov  al,01001000b
    not  al                  ;result al = 10110111b
```

SHL, SHR   SHL (SHift Left) and SHR (SHift Right) do what they suggest, but it is clearer if their operation is viewed diagrammatically (Figure 2.7):

Figure 2.7: Shift instructions.

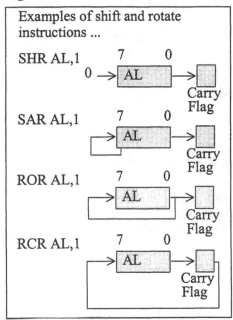

The example of SHR moves all bits in AL one place to the right, and a 0 into the most significant bit (MSB). Note that the least significant bit (LSB) goes into the carry flag, CF.

This instruction is sometimes used to test individual bits, since it can be followed by JC (Jump on Carry set) or JNC (Jump on Carry not set).

A limit with the 8088/8086 is that the "count" operand can only be a value of 1 if in immediate mode, as shown in Figure 2.7. If the shift is to be more than 1 bit, a count value must first be moved into CL:

```
mov   cl,3
shr   al,cl            ;shift 3 bits right.
```

Note that the shift operations can also be on 16-bit (and 32-bit) registers.

SHL does exactly the opposite of SHR, moving zeros into the LSB and the MSB out to the carry flag.

SAR SAR (Shift Arithmetic Right) works like SHR, except it maintains the sign. This is most useful for signed numbers. Refer to Figure 2.7.

ROL, ROR ROL (ROtate Left) and ROR (ROtate Right) work similarly to the shift instructions, except what falls out is rotated around back in the other end. Refer to Figure 2.7.

Thus the contents are never lost, but circulate around the register. ROL is the mirror-image of ROR, sending the MSB to the carry flag and back around to the LSB.

RCR, RCL RCR (Rotate through Carry Right) and RCL work as per ROR and ROL, except the path of the bits goes through the carry flag. See Figure 2.7.

Code and Data Labels

Labels are potentially an area of enormous confusion, so I review them here very carefully. Labels can be used to mark a "place" in the code or to name some data. They are introduced back on page 41.

Code Labels

In the case of a code label, the syntax is that it should start in column 1 and be suffixed with a colon ":", as in this example:

```
    ....
    jmp  place1            ;jumping to somewhere in the program.
    ....
place1:                    ;a code label.
    ......
```

When the assembler assembles the source code, it replaces "jmp place1" with the operation code for a JMP (jump) instruction, followed by the address place1 as the operand to the instruction.

Code labels equate to their address

Thus the assembler *equates place1 to the offset* it is marking. This is a vital point: the assembler simply replaces all occurrences of place1 in the code with the offset address it equates to.

Normally we would be jumping within the current code segment, so place1 equates to an "offset" from the start of the segment; that is, the IP value of that point in the code. A jump within the segment is called a NEAR jump.

NEAR and FAR

Note that it is also possible to jump between segments, which would be a FAR jump, and I have elaborated on this later in the book.

Another very important point is that any transfer-of-control instruction, such as a JMP or CALL, can have various *addressing modes*. These modes are encoded by the assembler as part of the instruction operation code. The above JMP example would be what we call *immediate addressing*, as the operand itself is used as the target address to jump to. Addressing modes have been introduced on page 44.

Procedures

Another kind of label is the *procedure* name, as shown here:

```
    . . . .
    call routine1               ;calling a procedure.
    . . . .
routine1        PROC            ;the procedure.
    . . . . .                   ;body goes in here.
    ret                         ;must have explicit ret.
routine1        ENDP
```

PROC and ENDP

Procedures allow you to organize code into structured modules, that can be called from a main procedure. In some languages they are called *subroutines*. A *function* is a special case of a procedure that returns a value via a register. For example, C functions return a value in the AX register or DX:AX register pair (though when writing C programs you don't know this underlying mechanism of the registers).

The point I want to make here is that procedure names are treated by the assembler just like code labels. In the above example, "routine1 PROC" could have been replaced by "routine1:" (in which case the "routine1 ENDP" would not be needed, since it is a syntactical requirement to match the PROC directive).

Data Labels

Data labels define constant or variable data, including numerical values, strings, arrays, and pointers.

```
str1 DB "message",0        ;defining an ascii string.
var1 DW 56                 ;define word, 16 bits.
ptr1 DW 789
var2 DD 0                  ;define doubleword, 32 bits.
ary1 DB 64 DUP(0)          ;array of 64 bytes.
```

Normally we would think of data as belonging in the data segment, where the code normally expects to access it, but it could just as easily be defined in the code segment, amongst the code, or in the stack. Chapter 4 explores the use of the stack for holding data. Segment override is introduced on page 46.

DB,
DW,
DD,
DUP

DB, Define Byte, DW, Define Word, and DD, Define Doubleword, define 8-, 16- and 32-bit data respectively. For example, var2 is a 32-bit value of 0. `"ary1"` shows the use of the DUPlicate directive, which causes the assembler to assemble 64-byte-size values initialized to 0.

Now for the key points: the assembler equates a data label to its address, just as for code labels. However, depending on the instruction, it assembles a *non-immediate* (i.e., direct, see page 45) addressing mode into the instruction operation code (op-code). This difference is vital.

```
mov  AX,var1      ;referencing a data label.
mov  AX,place1    ;referencing a code label.
```

Major
distinction
between code
and data
labels

The above examples show the difference. At execution time the second MOV instruction will move the actual address of place1 into AX, while the other MOV instruction will use a *non-immediate* mode, moving not the address var1, but its content.

Thus, although `"MOV AX,var1"` assembled with the address of var1 as the operand to the instruction, at execution-time the instruction looks at the content of that address. Make sure you have grasped this distinction before continuing.

Accessing Data

Sometimes, when writing a program, you want to know the address of something, say a point in the program, or the starting address of an ASCII string. I gave an example of how to define a text string (above), and labelled it "str1". The assembler equates str1 to the starting address of the string.

```
mov  AX, str1           ;loads contents.
mov  AX, OFFSET str1    ;loads address.
```

OFFSET override

Unfortunately, because the assembler has assembled the first MOV instruction as non-immediate-addressing, the first MOV here would only load the first two ASCII characters ("me") into AX (two characters are fetched because the destination is AX, which is a 16-bit register).

This is not what we want. We want to load the starting address of the string into AX. What we have to use is an override directive that forces the instruction into an immediate addressing mode. Thus the second example will load the actual operand into AX, which is the required address.

SEG override

Note too that you can get the segment value where that string is stored (which would normally be the data segment), by this override:

```
mov  AX, SEG str1  ;load segment address.
```

OFFSET and SEG only work for *static* data; that is, data that is defined in the data or code segments. It is possible to have *dynamic* or *automatic* data that is created during execution on the stack or heap: getting the addresses of this data involves other techniques, discussed on page 60 (and in Chapters 4 and 5).

Pointers

Data labels can also be pointers. This means that the data content is itself an address. Earlier, I defined "ptr1   DW   789", but the treatment of the content "789" is up to the program. Consider these examples:

```
call  ptr1         ;calls address pointed to.
call  place1       ;calls place1.
```

Immediate versus non-immediate mode CALL

"call ptr1" at execution-time will not jump to the ptr1 data in the data segment — obviously that wouldn't make sense. No, since the CALL instruction has assembled as a non-immediate addressing mode, even though the operand of the instruction is the address ptr1, the instruction looks at the content of ptr1 and uses that. Thus execution will transfer to offset 789 in the code (wherever that is!).

"call place1" is here for comparison. Again the operand will have the address of place1, but the immediate addressing mode will cause execution to go to place1.

Now I'm going to be a little tricky. I will redefine ptr1:

```
.DATA
ptr1 DW place1          ;defining a pointer.
.CODE
  call ptr1
  ...
place1:
  ...
```

Always remember that as the assembler goes through the source code, it simply replaces any data or code labels with the addresses they represent. So where will the CALL instruction transfer execution to?

**NEAR &
FAR
pointers**

The above examples of pointers are jumps within the current code segment, so they are NEAR; however, pointers can also be FAR. This is discussed in Chapter 4; I have also made some references to FAR pointers over the next four pages. Always keep in the back of your mind that for the 386+ the distinction between NEAR and FAR becomes blurred — you will see why.

LES, LDS, and LEA Instructions

As my example code further on in the book makes use of these instructions, some clarification is in order here.

```
.CODE
  mov  DI,OFFSET place2
  mov  ES,SEG place2
  les  DI,place2        ;!!!!!! Example of what NOT to do!
  ....
place2:
  ...
```

Although I have implied that place2 is a code label in the current code segment, let's assume that it is in some other code segment, maybe in a large .EXE program with multiple code (and/or data) segments.

The first two MOV instructions will load the FAR address of place2 into the two registers ES:DI.

**LES with
code-label
operand**

However, the LES instruction will not work. I have put it here to emphasize this point. LES and LDS (also LGS and LFS) are constrained to non-immediate addressing mode only: they are designed to load pointers. What will happen here is a "type mismatch" error, because "place2" is a code label. The operand of these instructions must be a data label, as it is the **content** of the label that is loaded. Read ahead to see code in which it does work.

MOV addressing -mode limitation
Whenever you want to load a segment and/or offset, use the MOV instruction, as shown above, or LEA. However, in some circumstances you *cannot* use the MOV with OFFSET override and must instead use LEA (Load Effective Address). LEA is clarified below, but first, why can't OFFSET always be used?

Restriction of OFFSET directive
The answer is that you would only use OFFSET if place2 is defined in the data (or code) segment, and not if defined as LOCAL (see page 62).

The fundamental reason is a built-in limitation to the addressing modes of the MOV instruction. Automatic data, or any data of a temporary nature (created and destroyed during run-time) as opposed to permanent data assembled into the data (or code) segment, is usually addressed using indexed mode or register-relative mode.

Look at this example:

```
routine2       PROC
   LOCAL       ptr4:DWORD    ;local data created on stack.
   ....
   lea  DI,ptr4
   ....
   ret
routine2       ENDP
```

The assembler will equate ptr4 to [BP-*value*], whereas if ptr4 had been defined in the data segment by something like "ptr4 DW 0", the assembler would equate ptr4 to an offset relative to DS.

LEA compared with OFFSET
BP is something that varies at run-time, so in the first case, ptr4 can only be equated in this way. The problem arises if you compare the above LEA instruction with something like "mov di,OFFSET ptr4" — the latter will not work — it will load the content of ptr4 rather than its offset.

This MOV instruction is translated by the assembler to "mov di, [bp-*value*]", and this *indexed* mode cannot be immediate. It must be non-immediate. So, the golden rule is:

Only use
> **MOV** *reg,* **OFFSET** *label*

if *label* **is defined in the data (or code) segment.**
For temporary data always use
> **LEA** *reg, label*

Some further clarification: the local data label ptr4 only exists within routine2. LEA will load the offset ptr4 into DI.

LES with data-label operand

"LES DI,ptr4" will load the content of ptr4 into ES:DI (non-immediate mode, since ptr4 is a data label — which is the only mode LES can handle).

Note that LDS works like LES, but loads DS instead of ES.

The LEA instruction differs from the other two in that it loads the offset of the label regardless of whether it is a data or code label. "LEA DI,place1", for example, would just load the offset (NEAR address) of place1 into DI, not the segment value.

Local Data

An example is given above, and there is more explanation in Chapter 5.

So far I have been treating labels (code and data) as being equated by the assembler to their addresses. But what of the case of local or automatic data labels that only come into existence when execution enters the procedure in which they are defined?

How the assembler equates automatic data labels

The assembler equates local labels to [BP-*value*], where *value* is known at assembly-time, but the BP register will have a certain value at execution-time. If you want to know more about the special role of the BP register, study Chapter 4. Basically, when execution enters a procedure, BP has an offset pointing to a region in the stack segment (see page 99). Addresses going down from BP can be reserved by the assembler for local data. In the above example, if ptr4 was the only local data, of DoubleWord size (32 bits, or four memory locations), then the assembler would equate ptr4 to [BP-4].

Thus an instruction like "lea DI,ptr4" would actually assemble with the instruction operation code specially encoded to refer to BP for calculation of the address, immediate mode, and with the value of 4 as the operand.

(Again, I remind you that the MOV instruction with BP-relative or index-register-relative addressing cannot be immediate-mode addressing — see the golden rule above).

Type Override

Looking back to that example of a local data label, ptr4 (see page 61), what if I wanted to see what it contains, from within my program?

```
mov  BX, ptr4                 ;wrong!
mov  BX,WORD PTR ptr4
mov  ES,WORD PTR ptr4+2
```

The assembler will be rather rude to you if you give it the first instruction. The reason is that source and destination operands must always have the same *type*.

Type has two aspects to it: size and address.

BYTE,
WORD,
DWORD,
QWORD,
TWORD,
SHORT,
NEAR, FAR

Size can be of type BYTE (8 bits), WORD (16 bits), DWORD (32 bits), QWORD (64 bits), or TWORD (80 bits).

Address can be SHORT (within 128 bytes either way of the current IP; 8-bit signed offset), NEAR (within the current segment; 16-bit offset), or FAR (in another segment; 32-bit segment:offset).

In light of this, take a closer look at that example MOV instruction. BX is a 16-bit register, while the content of ptr4 is DWORD (32-bit). In other words a type mismatch.

Type
mismatch

The assembler will pick this up as a possible error and will tell you so.

Any data values you define must have a size that matches the register. "mov AX,val6" would not work if val6 was defined as "val6 DB 0". Get the idea?

Size
override

The above code shows a solution: *overrides*. We have already looked at the overrides OFFSET and SEG, now you are seeing "WORD PTR". This is a size override. A syntactical note here: in front of "PTR" we can place BYTE, WORD, DWORD, NEAR, or FAR, as appropriate.

Accessing
32-bit data in
halves

The example, using "WORD PTR", tells the instruction to ignore the size-type of the operand and instead treat it as being of size WORD. This override is encoded by the assembler into the instruction op-code, and at execution-time the override only applies to that instruction.

But ... if ptr4 contains a 32-bit value, and by means of the override we are going to stuff it into a 16-bit register, what will actually happen? In the code above I show two MOV instructions with

WORD PTR override. The first will grab the lower 16 bits of ptr4, while the second will grab the higher 16 bits.

Order of storage of data in memory

Make a note of this. All values are stored in memory with the lowest byte at the lowest address and the highest byte at the highest address. That is why I added "+2" to the second MOV instruction.

It may be that in my program I want to see what is contained in ptr4. Any data label defined as having a 32-bit value has a problem with the 8088, 8086, and 80286, because there are no 32-bit registers. So if I wanted to get that value into a register, I would actually have to use two registers. That is why I am forced to use those two MOV instructions with "WORD PTR" overrides, even for the 386 (for compatibility with the other CPUs). In Chapter 4 you will see plenty of examples of this.

If we write code for the 386 and upwards exclusively, then a simple "mov EAX,ptr4" would do the trick.

Storing 32-bit data under two 16-bit labels

There is another way to approach the problem of handling 32-bit data: split it in half.

If you have to store a FAR address, say in a pointer, you can split it into two data labels:

```
.DATA
ptroffset       DW 789h         ;far pointer stored in
ptrsegment      DW 1234h        ;two pieces.
.CODE
   mov  BX,ptroffset
   mov  ES,ptrsegment
```

This may not be practical for data values, but for FAR addresses in the form of 16-bit segment:offset it works fine. It means that source and destination types will match, so no override is required.

... more on order of storage of data in memory

Another little note: just as with the x86 family we always store values with the lowest byte at the lowest address. The same goes for FAR addresses; the offset always comes first, that is at the lowest address.

In the above code I suffixed the values with "h" to indicate that they are hexadecimal values, not decimal. The memory would look like Figure 2.8 after assembly.

Always remember: the lowest byte at the lowest address.

Figure 2.8: Order of storage.

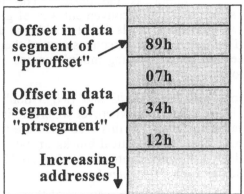

Structures

Whatever language you have experience with, you have probably encountered the concept of *data structures*. These are in fact the foundation of object oriented programming (OOP).

Windows programming makes extensive use of data structures, so it is appropriate to introduce the topic here.

```
.DATA
WINDOW  STRUC    ;Definition of structure...
  field1  DB    "ABCDEFGHIJ"
  field2  DW    0
  field3  DD    0
WINDOW  ENDS
  .....          ;Assembling instances...
win1 WINDOW    <"KLMNO",35,0>
win2 WINDOW    <"PQRSTUVWX",55,234>
  ....
.CODE
  ....  ;Accessing the instances...
  mov  ax,win1.field2
  ....
  mov  si,OFFSET  win1
  mov  ax,[si.field2]
  ....
  mov  ax,[si+10]
```

Object oriented programming
This listing shows how a structure is declared and used. In OOP terminology the definition is the *class*. The instances are *objects* ".[1] A structure is just a convenient way of getting at data. In this case we have data labels field1, field2, and field3. By putting the STRUC and ENDS directives around them, we have a convenient mechanism for creating mutiple copies of those same data declarations.

Instances
The declarations between STRUC and ENDS don't actually get assembled: it is a template, and wherever we create instances, they are what actually assembles. In this case there are two instances: win1 and win2. These are identical blocks of data, able to have their own values, but with identical variable (field) names. In OOP we would call each field a *member*.

The example code shows how we can get at these two instances. The most common method would be the first example. If I had want to access the "field2" field of win2, the instruction would simply be "mov ax,win2.field2".

You can have as many instances as you wish, and as you will see in Chapter 5, structures can be automatic or local to a procedure.

Label Equates

It is extremely useful to understand how the assembler assembles structures. Normally the assembler equates data labels to their offset from the start of the data segment, but fields of a structure are equated to offsets relative to the start of the structure. In the example, field1 equates to 0 and field2 equates to 10. When the instances are created, the names win1 and win2 are treated as normal data labels and thus are equated to offsets from the segment start.

Dot "." operator
In assembly language the "." (period) means exactly the same as "+", so the first code example is really:

```
mov  ax,win1+10    ;same as mov ax,win1.field2
```

The assembler will add the offset of win1 to 10 and assemble the result as the operand, with non-immediate addressing encoded into the operation code. Thus, at execution-time, the content of field2 will get loaded into AX.

Field initialisation
You will see from the listing that the structure declaration initialises the values. These initialisations will be put into the instances, unless overridden.

---

[1] In some languages the structure-definition is called the object, and an instance is called an instance-object.

Overriding by the instances is done by placing values between the "<   >", as shown in the code on page 65. Nothing between "< >" means leave original values as they are. In the examples of win1 and win2, I have overridden the original values, but should I have decided to override some but not all values, I would have put something like this: "<"asdfgh" , , 55>". This will leave field2 alone.

Postamble

There are a host of other considerations for assembly language programming for Windows, but hey, why should I throw it all at you at once? Enough is enough.

3

Opening Windows

Preamble

Triple purpose of this book

You'll find this book a nice way for beginners to learn Windows programming, as well as a look "under the hood" for those with Windows programming experience but with an urge to know more. You can also use it to learn assembly language.

By the very nature of tackling a topic from a fundamental point of view, the "nut and bolts" if you like, the beginner can develop very concrete concepts on which to build. When you have a grasp of what is going on underneath, a lot of what happens "on top" makes more sense. Therefore, a beginner can progress to being "advanced" in the same book, with a solid foundation of understanding.

Content of this chapter, and beyond

This chapter is an introduction to the basic principles of Windows, followed by a complete assembly language program in Chapter 4 — don't worry if the "skeleton" program looks intimidatingly long; this is done to show the nitty-gritty of how an assembly language program works. Chapter 5 shows you how to write an assembly language program that is almost as short as the same thing written in a high-level language such as C.

DOS versus Windows Programming

Other references

So, just how different is Windows from DOS? Below, I have summarized some new concepts you'll need to come to terms with. If you come across a reference to a DOS concept or programming method that you don't understand, refer to a good DOS assembly language book.

There are a dozen or so introductory Windows programming books that could be used to compliment this book, not the least being Microsoft's own *Microsoft Windows Software Development Kit: Reference Vol. 1*, available separately from the SDK.

You do need a book with in-depth coverage of the Windows functions, and again Microsoft's own *Microsoft Windows Software Development Kit: Programmer's Reference Vol. 2* is excellent.

Skeleton program

The next chapter puts together a simple skeleton program, but before we launch into that, let's consider some of the conceptual differences involved. The output on the screen will look different for Windows 3.x and 95. Figure 3.1 is what the skeleton will produce on the screen when running Windows 3.1.

Figure 3.1: Output of skeleton program.

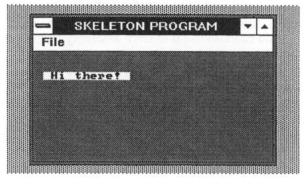

This is a window, amongst other windows, with its own title, system controls, menu-bar, and demo message.

So, a major conceptual difference from DOS is that our program doesn't output to just anywhere on the screen; normally we are constrained to output only within our application's window (or windows).

Interestingly, another major difference is the role of the operating system. Windows does a lot of housekeeping and looks after much of the usual upkeep of the window, such as moving it around, iconizing, and resizing.

In fact, Windows does even more than that, allowing us to program at a more abstract level. Instead of being concerned about the precise hardware details of the I/O device that our program is dealing with, we can use the *Device Independent Graphics* (graphics device interface (GDI)) tools. Translation from our program to the particular device is taken care of by device drivers, and our program can have code that will work on a wide range of different devices, such as various video standards (for example, Hercules, CGA, EGA, and VGA).

Internal Differences

Of course, the results appear on the screen, but the fundamental structure of our Windows program is different from a DOS application. The rest of the chapter is devoted to exploring those differences and the design methodology required to implement them (such as *handles* and *messages*).

Event-driven program structure

A Windows program is what we call *event driven*. The entire structure revolves around this concept. Those of you who have done any programming at all under DOS will know how to read a character from the keyboard. In assembly language, you could use INT 16h, AH = 0. However with Windows we don't do that. In fact INT 16h won't even work — Windows will hang.

The essence of being *event driven* is that for mouse, keyboard, and much other input, we don't write code to explicitly ask for input. Instead we perform a call to Windows, requesting a message, and Windows will send any message that it thinks is relevant to our program.

Thus our program plays a very passive role, taking whatever Windows dishes out.

Application queue

With Windows there is a *system queue* and an *application queue* for each application. Our program calls Windows and asks for the message at the head of our application's queue or waits until a message is put into the queue. Returning from the call, our program then deciphers the message and acts upon it.

There are some little wrinkles in this basic explanation, but that's the gist of it. Technically, Windows 3.x has one application queue for all applications, while Windows 95 32-bit applications have separate queues. This does not affect the programmer. It is an issue for Windows itself, with regard to scheduling of applications.

Multitasking operating system

Another major conceptual change is due to the multitasking nature of Windows. Unlike DOS, where everything usually stays put after it is loaded, code and data can move around. Even video-RAM cannot be treated as being at a particular address — although it actually is, an application may have to output to a "logical" video buffer located somewhere else. Consider another example: the heap. You can request local or global heap space (this is just memory that you can use for storing data), but unlike single-tasking DOS, you cannot just get its address and then write to it. The heap could be moved around by Windows (though you can freeze things also).

These shifting sands impose constraints, such as requiring *handles* to access all screen I/O and of course using selectors instead of segments for data and code manipulation.

Perhaps the newness of this is making you feel uneasy. However, the hands-on examples a bit later should alleviate that.

Building a Windows Application

Library Functions

The Windows routines operate like C functions (though stack handling follows the Pascal convention[1]). The library of functions can be split into three types:

- KERNEL
- GDI
- USER

Whenever you want to do any kind of I/O operation, including everything else involving the operating system, such as various memory management operations, you can call these functions. They are just like the BIOS and DOS INT services, except they are called by the assembly language CALL instruction.

Locations of the DLLs   So, where are these functions actually located? If you look in the C:\WINDOWS\SYSTEM subdirectory (assuming that you installed Windows in the default directory), you will see the three files KRNL386.EXE, GDI.EXE, and USER.EXE. You will also see KRNL286.EXE, which is the version of KERNEL for Windows 3.x running in Standard mode. Windows 95 only has KRNL386.EXE, not KRNL286.EXE. These files provide the API for 16-bit WinApps. For 32-bit WinApps, Windows 95 also has KERNEL32.DLL, GDI32.DLL, and USER32.DLL.

---

[1]   Actually, Windows 3.x follows the Pascal calling convention and Windows 95 and NT follow a mix of Pascal and C convention; that is, parameters are pushed from right to left (C) and stack cleanup is done by the called function (Pascal).

Other functions are available in other .DLL and .DRV files, many of them undocumented, and I'll take you a little bit of the way into this uncharted, but very exciting, territory.

The Mechanics of Assembling and Linking

.ASM, .DEF, .RC, .MAK

It is instructive at this point to consider the path we need to traverse to get from our modest little first program, written using a text editor, to the final .EXE program — that hopefully won't crash.

The steps shown here look pretty awful, but in practise you'll find it's a cinch.

The main problem is that we need to produce many more files than the program source file:

Source files needed

- .ASM   Your source program(s).
- .DEF   Module definition file.
- .RC   Resource script(s).
- .MAK   Make file.

You can produce all of these using a text editor, though there are some special programs that help generate them automatically.

.ICO icon file

In practise, more file types may be required than I list above, but for now we are working toward a simple skeleton program only. An example of another file is the *icon* for your program — the graphic image of this would be in an .ICO file.

Figure 3.2 shows a picture of the steps involved.

I mention the C compiler and .C source file here, but it could be any language, or none if you are writing the entire program in assembly language. In this book we stick entirely with assembly.

.INC, .H Include files

Notice also the .H and/or .INC Include files. Strictly speaking, these are optional, which is why I didn't list them above. The introductory program in this chapter only requires the .ASM, .DEF, .RC, and .MAK files, but in later chapters I have shown the use of WINDOWS.INC.

.H files are used with C programs and .INC with assembly programs. Borland and Microsoft supply utilities to translate .H files to .INC. Functionally, both types are the same; just with different syntax to suit the C compiler or the assembler.

Include files contain equates and definitions that make the program more convenient to write.

Figure 3.2: Steps to generate an executable file.

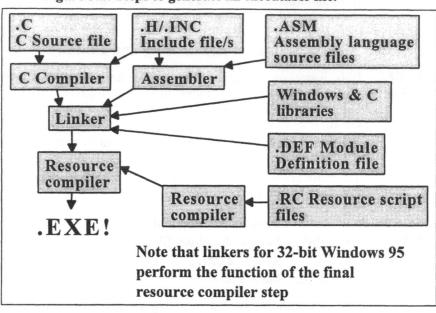

The Link Step

Notice how overloaded the LINK program is! The job of a linker is to combine the various program modules to produce the final .EXE, but in this case there are extra complications.

.DEF Module Definition file

The .DEF Module Definition file defines various program parameters that the linker needs to generate the .EXE file.

What we call *static* library functions can be linked into the .EXE, and become a permanent part of it, which is the way things work in the DOS world.

Dynamic link library (DLL)

However, the Windows library functions get linked in without actually adding to the size of the .EXE. That is, they stay where they are, and are only loaded into RAM memory when the program executes. This keeps .EXEs small. This kind of library is called a *dynamic link library*.

Two Steps for Resources

Resource compiler

The .RC resource file defines parameters connected with the windows, icons, menus, dialog boxes, and segments. The resource compiler is run **twice**, first to compile the .RC file(s) and second to combine the .EXE from the linker with the compiled resources to produce a final .EXE.

After the first compilation, it becomes a .RES file, which has the information in binary form.

With recent LINK programs, there is support to perform the final step by the linker. That is, the .RES file is fed to the linker.

Windows Programming Mechanics

There are some major philosophical differences between Windows programming and conventional DOS-based programming. It is useful to start off with some appreciation of some new terms intrinsic to Windows: *objects*, *handles*, *instances*, *messages*, and *callback functions*. These give us the mechanics of programming in this environment, that is, they are tools that we need to use. Have a look at each one first, then we'll go ahead and put it together into a working program.

Objects

Borland's latest assembler is described as *object oriented*, and there are various C++ compilers around. There is also Turbo Pascal with Objects. So, what are they?

You'll find a chapter on object oriented assembly language later (see page 137), but for now consider just a basic idea. Whatever you can lump together as a whole, as a distinct entity, think of as being an object. Your application's window is an entity on its own, separate from other windows — it is an object. In fact, so too are the distinct elements of that window, such as the various *controls*, the menu-bar, and the *client area* (where you output text and/or graphics to).

You can consider these latter objects as being children of, or related to, the parent window and subject to its dictates, though there are limits as Windows is not a true object oriented environment.

Furthermore, you access any object by getting its *handle*. As you'll see in the skeleton, even writing text to the screen requires you to get a handle for the client area.

Handles

A handle is just an ID, a unique number, that our program can use to access the object. Actually, you probably already have some exposure to the concept. Various PC programming books discuss handles in relation to file access under DOS.

All that has been done in Windows is generalise the concept, so that a handle can be obtained for any object. I am generalising the word *object* here, as Windows literature uses other terms that are still objects but used in a particular context. One that comes to mind is the *device context* — this is also a handle to an object.

DOS file "handle"

Just to elaborate: with DOS, you do a call to open a file or device, and DOS returns a handle. This handle is just a 16-bit number that you can use within the program to read or write the file. Since it is possible to simultaneously open many files, it is convenient to have these handles, a unique one for each file, to read/write the one you want. So, a handle is an ID, an identifier, for that file, device, object, or control.

In Windows programming, just about every resource is referenced by a handle. Even your program has a handle, and indeed so too has each *instance* (see below) of your program.

Instances

Multiple program instances

A fascinating aspect of Windows is that there can be multiple copies of an application running, or at least residing in memory, concurrently.

After all, why not, since this is a multitasking environment? You can, for example, have two copies of your word processor executing simultaneously, and you can jump between them. In such a situation, each copy would be an *instance* of the program. The *current instance* refers to the one you are dealing with at this moment.

There are some interesting considerations from this ability to have multiple copies or instances. Windows is not wasteful and only loads **one** copy of the code into RAM. Windows will, upon entry to each instance, give it a unique handle, but the reality is that there is just the one copy of the code. For this to work, each instance needs to have its own copy of the data segment or segments.

The downside is that your program needs to have some extra statements to handle multiple instances. In practise this is fairly standardized, and you can use the supplied skeleton program as the basis for much more complicated projects, without having to worry about multiple instances.

With 32-bit applications running in Windows 95, multiple instances are treated as totally separate programs, so special instance-handling code is not required.

Messages

Event driven I introduced the basic concept of *event driven* back on page 71; intertwined with this is messages. I also said that Windows sends messages to an application, and the latter has to decipher them and act accordingly. Let us consider this in more detail, since it affects the very soul of our program.

Our program has to call Windows and wait for a message — while waiting, it is in an idle state and other tasks can be executing.

Windows does an incredible amount of housekeeping, including receiving all of the incoming messages and parcelling them to individual queues. Any mouse activity on your application's window, for example, that Windows determines will affect your program will result in the generation of an appropriate message. Windows is always working, seeing everything that happens.

Structure of
WinMain() Below is the application's main function, entered from Windows when the program starts executing. It is called WinMain() — and I've used C syntax — straight from the textbooks:

```
int PASCAL FAR WinMain(hInstance,hPrevInstance,lpCmdLine,nCmdShow)
    HANDLE        hInstance;          //current instance
    HANDLE        hPrevInstance;      //previous inst.
    LPSTR         lpCmdLine;          //command line ptr
    int           nCmdShow            //show-type
...
//...initialization...
//...instance handling...
//...create and display a window...
...
while (GetMessage(&msg,NULL,NULL,NULL))
{
   TranslateMessage(&msg);
   DispatchMessage(&msg);
}
```

32-bit
differences The above code is ok for a 32-bit application as well as a 16-bit application. One difference is the size of the parameters passed to WinMain() — see Table 3.1. A 32-bit application does not have to worry about hPrevInstance. Also, a 32-bit application does not have to name its first entry point WinMain(), but we can continue to do this as a convention. Also, as explained below, and on page 314, the Pascal calling convention is only applicable to 16-bit applications.

C Syntax

The code sample should be readable, even if you don't know C. Note that some of the Windows textbooks give the basic program structure in "classical" C, not ANSI C, and I have stuck with that.

You will notice "&msg" specified as a parameter, and this may need some clarification to those unfamiliar with C. It should become clear later on when you see it in assembly language. This function requires that an address, to which the returned message can be placed, be provided as a parameter. The "&" means "address of", in this case the address of a data area labelled as "msg" (not defined in listing).

Pascal calling convention

You will also notice the PASCAL qualifier in the declaration of WinMain(). This is because Windows 3.x uses Pascal calling conventions, not C conventions. So the override is needed. This is explained in more detail later (see page 112, if the fancy takes you), and a note was made earlier, on page 72.

STDCALL calling convention

You might like to glance ahead to Chapter 13 to see a complete 32-bit application written in assembly language. There, you will see the procedures default to the STDCALL convention (as specified in the .MODEL directive: see page 111). This is a mixture of C and Pascal, in which parameters are pushed onto the stack from right to left, and stack cleanup is performed by the called procedure.

Data label prefixes

I suppose this is as good a place as any in which to introduce the Windows labelling conventions. You have had a first exposure to them in the above listing. What I'm talking about are the prefixes to the parameters. These are put there to clarify the type of data the parameter represents. It would be breaking the flow of the explanation to describe this in detail, but the prefixes used above are "h" to signify type of "handle" and "lp" to signify "long pointer". A more complete list of prefixes and data types is given on page 82.

Message Loop

The WinMain() function contains what we refer to as the "message loop".

Get/ Translate/ Dispatch Message()

Looking at the above listing, it commences with declarations of the passed parameters and their data types. A little further down you'll see GetMessage(). This is the one I've been talking about — it goes back to Windows and waits for a message.

Whenever a message is available on the queue, and also whenever Windows decides the time is appropriate, control will return to your program with the message.

TranslateMessage() is specifically for converting keyboard messages into a more usable form.

It is possible for more processing to be done, but usually nothing much more happens, and strange though it may seem, the next function, DispatchMessage(), sends the message straight back to Windows.

Callback function

Windows then calls another part of your program, named WndProc(), that we know as the *callback function* (see below). It is this function that finally does something with the message.

There is a callback function for each window that your program creates.

Callback Functions

WndProc()

I said above that, having got the message via GetMessage(), your program must then give it back by calling DispatchMessage(). Windows then sends the message to another part of your program, known as a *callback function*. In fact, each window (including windows called *dialog boxes*) has its own unique callback functions.

The name I gave above, WndProc(), is only a suggestion. Unlike the main function, which must always be called WinMain() (though this has become more of a convention only), your callbacks can be called whatever you want. There is a simple mechanism for informing Windows of the names of the callbacks, so it can call them.

This is a C skeleton of a callback:

```
long FAR PASCAL WndProc(hWnd,message,wParam,lParam)
    HWND        hWnd;           //window handle
    unsigned    message;        //type of message
    WORD        wParam;         //more information
    LONG        lParam;         //more information
{
    . . .
    //...case-logic to analyse message,..
    . . .
    //...user-written message-handling...
    . . .
    //...default message-handling........
    DefWindowProc(hWnd,message,wParam,lParam)
    . . .
}
```

There is yet another twist. The message, getting a bit ragged around the edges by now with all that travel, goes to the callback function, which can then process it.   **But** the twist is that most messages are of no interest to your program, and your callback just sends them back to Windows **again**, for final default processing.

Default message handling

DefWindowProc() is a kind of rubbish bin for messages that you don't know what to do with.   And believe me, there are a lot of them.

After sending the message to its final resting place, or handling it in some way within the callback, execution returns to the next statement after DefWindowProc(), which is usually a return from the callback function (designated by "}" above, or by a RET instruction in assembly).   However, this will take execution back to Windows again ....

Figure 3.3: Event-driven structure.

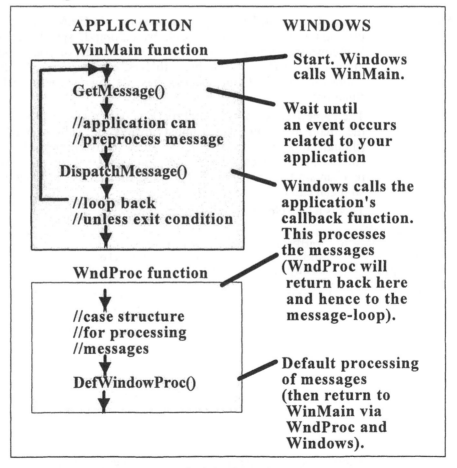

Follow this tangled path right through ....

Windows will return to the statement just after DispatchMessage(), so we are back in the main loop.

The main loop is an endless loop, executing GetMessage(), then TranslateMessage(), then DispatchMessage(), though there is a test for exiting. Figure 3.3 puts the whole lot together pictorially.

A word of advice: don't let this confuse you. See the simplicity behind all of the detail. Windows sends messages to a window, and your program can have as many windows as it wants. The message goes (via WinMain()) to the callback function for that window, where you can respond to it. If you don't know what to do with the message, just call DefWindowProc(). End of story.

Data Types

Hungarian convention

Tabulated in Table 3.1 are prefixes to data and pointer labels. It is known as the *Hungarian convention* and is the voluntary prefixing of data labels with a character or characters to indicate the type of content.

Table 3.1: Data types.

PREFIX	MEANING	SIZE	COMMENTS
b	Boolean value	WORD*	0 = false, non-zero = true
c	Character	BYTE	Extended ANSI character code
dw	Long unsigned integer	DWORD*	Unsigned value
f	Bit flag value	WORD*	16 individual flags
h	Handle	WORD*	Handle of a resource
l	Long integer value	DWORD	Signed value
lp	Long pointer	DWORD*	FAR pointer
n	Short integer value	WORD	Signed value
p	Short pointer	WORD*	NEAR pointer
pt	x,y coordinate point	DWORD*	Unsigned, 2-word value
rgb	RGB color value	DWORD*	Unsigned
w	Short unsigned integer	WORD*	Unsigned value

The asterisk means that these sizes only apply to 16-bit applications. For 32-bit applications, they are all 32 bits.

We should make use of this notation wherever possible while writing programs, as it improves readability.

Reference source

The source of Table 3.1 is Thom Hogan's superb book, *The Programmer's PC Sourcebook*, by Microsoft Press, second edition, 1991. Of course BYTE is 8 bits, WORD is 16 bits, and DWORD is 32 bits. Unfortunately, Thom's book is out of print.

Other prefixes

It is common practise also to use "s" for string, and "sz" for zero-terminated string.

Combinations are allowed, for example "lpsz" means "long pointer to zero-terminated string". For 32-bit applications, the distinction between a long pointer and a pointer is blurred, so the prefixes "p" and "lp" can mean the same.

However, it is a case of "do as I say, not as I do!" For old habits die hard. I do tend to lapse back into non-Hungarian naming of data labels, and where you encounter such lapses, forgive me. Mostly I have not followed the Hungarian notation when writing Windows-aware DOS code, examples of which you'll see in Chapters 10, 11, 12, and 14.

Types of handle

I have described the handle as being used to access almost all resources. However, it is useful to formalize this. Again, the original source of this tabular information is Thom Hogan's book. Refer to Table 3.2 below.

Table 3.2: Types of handle.

NAME	FUNCTION
GLOBALHANDLE	Global memory handle
HANDLE	General handle
HBITMAP	Physical bitmap handle
HBRUSH	Physical brush handle
HCURSOR	Cursor resource handle
HDC	Display context handle
HFONT	Physical font handle
HICON	Icon resource handle
HMENU	Menu resource handle
HPEN	Physical pen handle

HRGN	Physical region handle
HSTR	String resource handle
LOCALHANDLE	Local memory handle
HWND	Handle of a window

One thing that you will notice throughout much of this book is my disregard for upper- or lowercase. For example, I have usually used uppercase for function names. This stems from the dynamic link libraries themselves, in which the functions are recorded (exported) in upper case. Mixed case, in the case of Windows functions, is for readability only. Another factor is that the assembler treats upper- and lowercase alike — well, that can usually be controlled by a switch.

I did have a change of heart in the matter of case sensitivity, and you will find the 32-bit application in Chapter 13 has correct case on everything.

The link step also can be made case sensitive or not, by the use of switches. Note that the command line switches for the linker are themselves case-sensitive (not all linkers, and not earlier Microsoft and Borland linkers), which is not something that you associate with the DOS command line.

4

The Bare Bones

Preamble

Skeleton program

The earlier theory will only really make sense when actual code is shown, so in this chapter I have done a complete application: a skeleton program that just puts "Hi there!" on the screen. Nothing too ambitious, but the skeleton can be built upon for much more ambitious projects.

Assembly for and against

It's quite feasible to write entire applications for Windows just in assembly language, though it is more usual to restrict assembly to critical sections of the program. Although there's no concrete argument against writing the whole thing in assembler, it's a matter of preference and personal requirements. I will show that the argument that assembly programming is more tedious and time-consuming than C is not true.

From the professional's point of view, assembly gives very precise control over what is going on, is more appropriate for low-level and getting-behind-the-scenes development, and is potentially extremely compact and fast.

From the beginner's point of view, looking at how to write the entire program at the assembly level is most useful for learning purposes and gives us useful insights into how Windows works.

The argument in assembly language's favour is developed further in the last chapter (see page 367).

Organisation of this chapter

I have organised this section by example with a simple "Hi there!" introductory program, as shown on page 94. We go through it here step by step and put together the complete application. This program is on the Companion Disk, in directory SKELETN1. Note that I have written the program at the most fundamental level for instructional purposes. However, the next chapter introduces the same skeleton program, but makes use of advanced assembler features, so it is more practical. The program of this chapter has the advantage that it represents the lowest common denominator and should work with just about any assembler.

I recommend that you use this chapter as a theoretical learning tool and focus hands-on experimentation in the next chapter.

32-bit skeleton application

Chapter 13 describes a 32-bit skeleton program; however, I recommend that you follow the steps of the "ladder of learning". The 16-bit applications of this and the following chapter will work fine under Windows 95. By all means refer to Chapter 13 as you study this chapter and the next, as you wish, to see the contrast — you will find the 32-bit code is structurally the same, and very few changes are required to convert a 16-bit application.

Getting Started

Tools Required

Microsoft SDK

So what do you need? Many people will have access to the Microsoft *Software Development Kit* (SDK) and Microsoft assembler (MASM), so this is a good starting point. In my previous book I showed how the SDK and MASM v5.1 could be used to write a complete assembly language program, but I now consider v5.1 to be behind the times. However, I constrained the program in this chapter to work with v5.1, in which case the earliest tools that I can guarantee the program to successfully assemble and link with are in Table 4.1.

Note that SLIBCEW and CWINS are C run-time libraries, and are not required for the skeleton. However, in a situation where you would need them to call C run-time functions, investigate using startup code supplied by the vendors, for correct initialisation (the next chapter shows how to link the Borland startup file, C0WS.OBJ; Microsft's MASM v6.1 supplies APPENTRY.OBJ).

Whenever you see the letter "S" in a library filename, it usually means "Small model", while the letter "W" designates "Windows".

Table 4.1: Earliest versions that will generate an executable.

MASM.EXE	v5.10	**TASM**.EXE	(C++ v2.0)
LINK.EXE	v5.10 (C v6.00)	TLINK.EXE	(C++ v2.0)
NMAKE.EXE	v1.11 (C v6.00)	ditto	
RC.EXE	v3.00 (SDK v3.0)	ditto	(")
RCPP.EXE	(SDK v3.0)	ditto	(")
RCPP.ERR	(SDK v3.0)	ditto	(")
LIBW.LIB	(SDK v3.0)	IMPORT.LIB	(")
SLIBCEW.LIB	(SDK v3.0)	CWINS.LIB	(")
WINSTUB.EXE	(from the SDK)	ditto	(")

Borland &
other tools

The second column of Table 4.1 contains the earliest Borland versions that will work. Other LINK versions should be ok, as long as they are Windows-compatible. MASM prior to 5.10 *should* also be ok.

Installation
of the
development
tools

The normal situation is to have the SDK installed with everything in the appropriate directories. The manuals with the SDK, C v6.00, and MASM explain how the environment variables need to be set so that MASM and LINK can find the appropriate files. Or, you could have one of the other development systems installed, such as Borland C++, that do not need the SDK as a separate entity. Note also the Microsoft C/C++ v7.0 and later is bundled with elements of the SDK.

Actually, the main reason that you require the SDK is for the programs RC.EXE, the import file LIBW.LIB, and Windows-compatible LINK. The SDK does have some other tools, such as a debug version of Windows, but most of these tools are available with recent compilers. There are also a lot of useful manuals with the SDK. Microsoft has gone away from supplying printed manuals, and wherever I refer to a Microsoft manual in this book, it will be on-line; although, in most cases it should also be available for purchase separately. I personally prefer printed manuals.

If such housekeeping (i.e., the correct installation of all the software tools) is too much trouble, get together all the above files, or suitable equivalents, and put them all into the same directory. Problem solved.

Look ahead through this book and you'll see examples of *Make* files for both Microsoft and Borland.

"Legacy chapter"

> I must emphasize again that this chapter is a "legacy chapter". I am using the oldest tools and the most primitive assembly language skeleton. This is *not* what I recommend for actual development, but the very basic skeleton is excellent for learning. I have included all of the meandering through version numbers below, partly to record what I remember, before I completely forget! Should you wish to learn this skeleton and you only have old development tools, or you need to modify or maintain legacy code, you will find this information useful.

Microsoft and Borland version notes

MASM prior to version 6.00 can't handle the high-level language used in subsequent chapters, so I recommend upgrading if you don't have it. The alternative is the long-winded program given in this chapter. In fact, at the time of writing, the latest version is 6.11, and I recommend that you use it in preference to all earlier versions, including version 6.10. Microsoft made some important changes in the upgrade from 6.10 to 6.11!

Other older assemblers *may* be able to handle the code in this chapter.

TASM v2.5

Borland TASM prior to v2.5 should be ok for this chapter, but v2.5 has enhanced features and is the basis, along with TASM v3.00, of the program in the next chapter. At the time of writing, the latest is version 5.0 (see Chapter 13).

Microsoft Quick assembler

Microsoft Quick assembler should be ok for this chapter. I think that Quick assembler version 2.01 can be considered equivalent to MASM version 5.2.

All of this upgrading is difficult to keep up with, but the above notes should prove helpful.

Of course, as mentioned above, with some language products, such as those from Borland, you don't need to have the SDK installed, though I certainly recommend the SDK documentation.

Note that even if you are only interested in writing in-line assembly within your high-level code, consider this chapter to have important buiding-block educational information. Many modern compilers allow in-line assembly, and this is developed further in Chapter 6.

I have gone through the above outline of products and versions and based this chapter on early tools, as not everyone has access to the latest tools. Also, it is actually quite educational to analyse a Windows assembly language program written with an earlier assembler minus the high-level features. Having understood exactly what is happening, high-level features can be introduced later, for much more streamlined programs.

Source Files

The next step is to write the application, for which, of course, you use a text editor. However, it is no longer a case of producing a single .ASM source file — let's call it SKELETON.ASM. The absolute minimum files required are:

- SKELETON.ASM   (program source)
- SKELETON.RC   (resource script)
- SKELETON.MAK   (Make file)
- SKELETON.DEF   (definition file)

Resource and Definition Files

Resource (.RC) and definition (.DEF) files are produced by a text editor, though you can get some help with special paint programs to generate the resource scripts.

.RC file   Resource scripts describe the appearance of what is seen on the screen — dialog boxes, menus, etc. It can also store other information. I wrote SKELETON.RC directly using a text editor, since it is a simple example.

.DEF file   The definition file defines the name, segments, memory requirements, and exported (including callback) functions of the application, and is straightforward enough to write with a text editor. All functions in your program that are to be called by another program must be declared as *exported* — in the case of the callback function, it is called by Windows. The only function that doesn't need to be declared as exported is your WinMain().

Here is the .RC file:

```
# SKELETON.RC
#define        IDM_QUIT          200
#define        IDM_MESSAGE       201
skeleton       MENU
    BEGIN
      POPUP "File"
        BEGIN
        MENUITEM "Quit",   IDM_QUIT
              MENUITEM "Message...",   IDM_MESSAGE
          END
      END
```

Menu-bar

You will be able to figure out what this .RC file does by observing the execution of the program. A menu-bar with only one selection, "File", drops down two menu-items: "Quit" and "Message...". The next chapter has the same "Hi there" program, but written using high-level assembly constructs.

IDM_QUIT and IDM_MESSAGE are arbitrary labels, assigned (almost) arbitrary values. One of these values is passed within a message as an identifier to Windows, if a menu-item is selected.

Message Format

**WM_
COMMAND
message**

Selecting a menu-item generates a WM_COMMAND message, which is one of many possible messages that can be sent to the callback. It is a 16-bit value, and also has other parameters, notably "wParam" and "lParam", that constitute extra data attached to the message.

**message,
wParam,
lParam**

So, this is what constitutes a message:

- message   (16-bit number) (32-bit WinApp: 32 bits)
- wParam   (16-bit number)   (")
- lParam   (32-bit number)   (")

wParam is 16 bits also, hence the "w" (word) prefix. Every message has two parameters attached to it, wParam and lParam, the latter being 32 bits (hence the "l" prefix, meaning "long"). What these parameters contain depends upon the message. The prefixes are just a convenient notation for labels, so that we know what they represent (see page 82). Note that for 32-bit applications, these parameters are all 32 bits (making the "w" and "l" rather confusing, as these prefixes are still used).

Before we delve further in this direction, here is the .DEF file:

```
SKELETON.DEF...
NAME            SKELETON
DESCRIPTION     'Hi there! program'
EXETYPE         WINDOWS
STUB            'WINSTUB.EXE'
CODE            PRELOAD MOVEABLE
DATA            PRELOAD MOVEABLE MULTIPLE
HEAPSIZE        1024
STACKSIZE       8192
EXPORTS         SKELETONPROC
```

Skeletonproc() is the *callback* function, referred to as WndProc() in earlier notes. This is where Windows sends messages to be processed. An application can have a separate callback function for each window, dialog box, or control.

DOS stub   I have explained various aspects of the .DEF file throughout this book, so investigate via the index. Some of the lines are self-explanatory. "WINSTUB.EXE" is a program supplied by the software vendor, that is incorporated into the overall .EXE file, and is executed if you try to run the program from the DOS command line. It just displays a short message and quits.

I have put the DOS stub to very interesting use in Chapter 14.

Make File

Before we go ahead with the application itself, let's consider the Make file. This determines the assemble, compile, and link steps.

With reference to Figure 3.2 on page 74, the first step is to assemble SKELETON.ASM to produce SKELETON.OBJ (any Include files are also assembled). MASM and TASM have various directives to aid with creating Windows applications; however, by writing the program at the most fundamental level I have avoided these, which means that just about any assembler should work. You can see in the listing below how RC.EXE is used to compile SKELETON.RC and how to incorporate SKELETON.RES into SKELETON.EXE. LINK converts the .OBJ to .EXE, and LIBW.LIB provides connection to the Windows functions. LIBW.LIB is *not* itself a library. Note also that LINK refers to the .DEF file.

```
#   SKELETON.MAK...
fn = skeleton
all:$(fn).exe
$(fn).obj : $(fn).asm
   masm $(fn);
$(fn).res : $(fn).rc
   rc -r $(fn).rc
$(fn).exe : $(fn).obj $(fn).def $(fn).res
   link $(fn) /NOD, , , libw , $(fn).def
   rc $(fn).res
```

You create this on a text editor. It requires a certain syntax, and Make programs from different vendors have their own peculiarities. The above will work with Microsoft's NMAKE.EXE and is for MASM versions prior to 6.00. The latest MASM requires modifications to the Make file (refer page 125), though it can be made command line compatible with v5.1.

Borland vs Microsoft Make   Borland's TASM is different again (refer to page 124), because TASM and TLINK have their own command line syntax. Borland's MAKE.EXE also has its own peculiar syntax requirements, but note that the version supplied with C++ version

3.00 (and later) is supposed to be more compatible with NMAKE (this is doubtful — see my comments in Chapter 14).

Why use a Make file? The Make file saves you the trouble of typing in all the assemble, compile, and link steps at the command line. Some integrated environments generate the Make file automatically, so you don't even have to do that much, but there are some sound reasons for learning about and using Make files, not the least of which is flexibility. Some integrated environments generate what is called a *project* file, which is saved with a special extension, and with some products it is possible to convert a project file into a Make file. The fundamental difference in usage is that in the integrated environment you do everything via pull-down menus, while you run the Make file from the command line.

Programmer's Workbench (PWB) Microsoft's *Programmer's Workbench* (PWB) is an example of an integrated environment that works with Make files in its native mode, though the Make files are highly stylised. PWB can, however, read ordinary Make files, and you can open a "project" by opening many of the Make files given in this book.

Explanation of above Make file You can figure out what the above Make file does: it assembles SKELETON.ASM using MASM.EXE, then it compiles SKELETON.RC using RC.EXE, then LINK.EXE links everything together, and finally RC.EXE is executed again to combine SKELETON.EXE and SKELETON.RES (the compiled output from the first RC execution) to produce the final SKELETON.EXE.

Development Cycle

Within Windows You can run the Make file from the DOS prompt, but you can also do it from within Windows. What you should do is open the File Manager and go to the directory containing the application. Then iconize the File Manager and open the Notepad. Use the Notepad to view and edit SKELETON.ASM, and iconize when finished. It is a simple matter to flip between the Notepad and the File Manager.

When in the File Manager, and the directory containing the application is open (and the directory must contain all software tools if the SDK is not installed properly on the PC), select "Run..." from the "File" menu.

In the box, type "NMAKE SKELETON.MAK", just as you would on the DOS command line. After running the Make file, all you need to do to test your program is double-click on SKELETON.EXE in the File Manager.

Other ways   The above is not the only way to do it. There are various reasons why you may want to do everything from DOS and load Windows to test the program, or, have a "DOS-box" open and use <ctrl-esc> to flip between it and Windows. Or, you may be working within an integrated environment, which may have something called a *project file* rather than a Make file. Many integrated environments can generate a Make file from a project file, and can also execute a Make file from within the environment. I have never been entirely satisfied with integrated environments and prefer to be outside one, using the traditional Make file from the command line: but I don't want to prejudice you. If your product has an integrated environment, give it a go. One problem you may have is getting it to handle stand-alone assembly programs.

Programmer's   However, I have rather grudgingly come to like Microsoft's
Workbench   *Programmer's Workbench* (PWB). If you install PWB, you can open almost any of the Make files supplied on the Companion Disk, and thus you will have opened a project. You will however, have to click the "non-PWB Makefile" button. Then you can select "Rebuild All" from the "Run" menu, and see the result in a "Build" window.

PWB can be started from within Windows, and after running the Make file, you can use <alt-tab> to flip over to Windows and try the program.

When can I   This is, of course, just theory if you are reading through the book
"get   linearly — don't worry though, as the hands-on exercises begin
started"?   soon. If you feel the overwhelming desire to try the program, why not? (flick ahead to the next chapter if you want to assemble the simplified skeleton). Copy the appropriate files off the Companion Disk. Then, assuming that you have all the development tools installed, follow the above instructions to assemble, link and test your program. Later on you can learn how it works internally.

Alternatively, you may feel that you don't want to get "bogged down" in a skeleton that is very primitive and would prefer to jump directly into a skeleton that uses the higher level assembler features. In that case, study this chapter theoretically only, and do your hands-on work in Chapter 5. Or, if you really insist on short-circuiting my "ladder of learning", you can get hands-on experience with the 32-bit application in Chapter 13.

Application Structure

It doesn't do much more than put "Hi there" on the screen, but wow, so much red tape! A far cry from the few lines a DOS program would need.

Try to understand as much as possible and identify the major structural elements.

Preliminary Code

```
;SKELETON.ASM Windows assembly language program
.286            ;286 instruction set.
;........
```

WINDOWS.-
INC
The identifiers (equates) shown below would normally be in the WINDOWS.INC Include file (refer page 109). With this skeleton I have minimized the number of files involved.

```
IDI_APPLICATION   .   EQU  32512   ;default icon type.
IDC_ARROW             EQU  32512   ;default cursor type.
OEM_FIXED_FONT        EQU  10      ;font type.
COLOR_BACKGROUND      EQU  1       ;background color
WM_CREATE             EQU  1       ;Windows message
WM_DESTROY            EQU  2       ;   /
WM_PAINT              EQU  15      ;   /
WM_COMMAND            EQU  273     ;   /
WM_LBUTTONDOWN        EQU  513     ;    /
WM_CHAR               EQU  258     ;    /
IDM_QUIT              EQU  100     ;menu-identifiers from
IDM_ABOUT             EQU  101     ;   .RC file.
MB_OK                 EQU  0       ;messagebox identifier.
```

Program listing continues until page 107

Generic program for any assembler

The Windows startup code would normally be in a separate .OBJ module supplied by the compiler vendor; however, in this fundamental skeleton, I have put the startup code into this module. This code is taken from APPENTRY.ASM, which is the source file for APPENTRY.OBJ, supplied by Microsoft. These are a couple of equates used by the startup code:

```
;This is the equates for the startup code...
STACKSLOP  EQU 256 ; amount of stack slop space required
maxRsrvPtrs EQU 5  ; number of Windows reserved pointers
```

Below are the Windows functions that the program calls.   In assembly language we must declare all external functions, which is not an essential requirement in C.

High-level CALL   MASM version 6.00+ is an interesting exception to this, as its INVOKE (high-level CALL) is C-like and doesn't need an explicit EXTRN declaration.   MASM v6.00+ is also C-like in that it accepts the C spelling of EXTERN.   See Chapter 5 for an explanation of INVOKE.   TASM version 5 has PROCDESC, that does the same job as INVOKE (see Chapter 13).

```
EXTRN    INITAPP:FAR
EXTRN    INITTASK:FAR
EXTRN    WAITEVENT:FAR
EXTRN    DOS3CALL:FAR
EXTRN    UPDATEWINDOW:FAR
EXTRN    BEGINPAINT:FAR
EXTRN    ENDPAINT:FAR
EXTRN    DEFWINDOWPROC:FAR
EXTRN    POSTQUITMESSAGE:FAR
EXTRN    REGISTERCLASS:FAR
EXTRN    GETSTOCKOBJECT:FAR
EXTRN    CREATEWINDOW:FAR
EXTRN    SHOWWINDOW:FAR
EXTRN    GETMESSAGE:FAR
EXTRN    LOADCURSOR:FAR
EXTRN    TRANSLATEMESSAGE:FAR
EXTRN    DISPATCHMESSAGE:FAR
EXTRN    LOADICON:FAR
EXTRN    TEXTOUT:FAR
EXTRN    MESSAGEBOX:FAR
EXTRN    SELECTOBJECT:FAR
```

Below is the data segment. Here we define all of the variables, strings, and arrays that the program will use.

```
.DATA
;This must be at beginning of data segment...
    DWORD 0              ; Windows reserved data space.
rsrvptrs WORD maxRsrvPtrs ;16 bytes at top of DATA seg.
    WORD  maxRsrvPtrs DUP (0)      ; Do not alter
hPrev     WORD    0       ; space to save WinMain parameters
hInst     WORD    0       ;/
lpszCmd   DWORD   0       ;/
cmdShow   WORD    0       ;/
;.................................................
.DATA
szwintitle      DB    'SKELETON PROGRAM',0
szskeletonname  DB    'SKELETON',0
hOemFont        DW    0                   ;handle to OEM font.
sout            DB    ' Hi there! '
szabout    DB 'Assembly Language Skeleton',0  ;messagebox
sztitle    DB 'Barry Kauler',0              ;  /
```

Startup Code

The startup code is fascinating, because it is something you normally don't see in a Windows program. It is the code that is first entered when the application is loaded, and it performs various initialisations before calling the entry point of your program, WINMAIN().

DOS3CALL() This code is also the exit point, performing the standard INT-21h/function 4Ch to exit back to the calling program. Look below, but don't be mislead by the DOS3CALL(): this simply does the same as INT-21h, except by a FAR CALL rather than by software interrupt. As far as I'm aware, there is no other difference, except that the CALL is faster.

```
.CODE
;Here is the startup code...
start:
  xor  bp,bp                           ; zero bp
  push bp
  call INITTASK          ; Initialise the stack
  or   ax,ax
  jz   noinit
  add  cx,STACKSLOP       ; Add in stack slop space.
  jc   noinit            ; If overflow, return error.
  mov  hPrev,si
  mov  hInst,di
  mov  word ptr lpszCmd,bx
  mov  word ptr lpszCmd+2,es
  mov  cmdShow,dx
  xor  ax,ax                           ; 0-->ax
  push ax                       ;parameter for WAITEVENT
  call WAITEVENT ;Clear initial event that started this
                          ; task.
  push hInst                    ;parameter for INITAPP
  call INITAPP           ; Initialise the queue.
  or   ax,ax
  jz   noinit

  push hInst                    ;params for WINMAIN
  push hPrev               ;    /
  push WORD PTR lpszCmd+2 ;  / (seg. first)
  push WORD PTR lpszCmd   ;  / (offset second)
  push cmdShow            ;    /
  call WINMAIN
ix:
  mov ah,4Ch
  call DOS3CALL      ; Exit with return code from app.
noinit:
  mov al,0FFh        ; Exit with error code.
  jmp short ix
```

What does the above startup code do? There is an explanation in *Programmer's Reference, Volume 1: Overview,* supplied with the SDK v3.1 (or on the on-line documentation supplied with the SDK). This reference has definitions for each of the above functions, plus explanation of the startup sequence.

It is instructive to consider what the status is when Windows calls "`start:`" — incidentally, scan ahead to the very end of the program, and you'll see that termination is with "`END start`", which is standard practise for DOS programs and defines the starting point of the program.

Register initialisation
"`start:`" is entered with the CPU registers set as per Table 4.2.

However, INITASK() returns its own information in the registers, as per Table 4.3, which are passed as parameters on the stack to WINMAIN().

Table 4.2: Registers at entry to application.

Register	Value
AX	zero
BX	size, in bytes, of stack
CX	size, in bytes, of heap
DI	handle of application instance
SI	handle of previous application instance
BP	zero
ES	segment address of PSP
DS	segment address of automatic data segment
SS	same as DS register
SP	offset to first byte of application stack

INITTASK()
Table 4.3: Register values returned by INITTASK().

Register	Value
AX	1 = ok, zero = error
ES:BX	FAR address of the DOS command line
CX	stack limit, in bytes
DI	instance handle of new task
SI	handle of previous instance of program

DX	nCmdShow parameter
ES	segment address of PSP

INITTASK() also fills the first 16 bytes reserved in the data segment with information about the stack.

WAITEVENT(),   The parameter zero when supplied to WAITEVENT() clears the
INITAPP()   event that started the current task.

INITAPP() initialises the queue and support routines for the application.

WINMAIN()

Below is the rest of the code segment, which has WINMAIN() and the callback function SKELETONPROC(). Functions that are to be called by Windows must be declared as PUBLIC.

```
.CODE
  PUBLIC  WINMAIN
WINMAIN PROC NEAR     ;entry point from Windows.
```

Parameters passed on the stack will be as per the listing on page 77 and will have been pushed on from left to right, with the return address pushed on last (Figure 4.1). You can check this against the startup code above.

```
    push bp              ;save BP so can use to access params.
    mov  bp,sp           ;BP will now point to top-of-stack.
    sub  sp,46           ;mov stack to free region.
```

Figure 4.1: Stack at entry to WinMain().

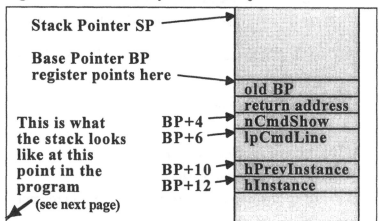

```
        cmp    WORD PTR [bp+10],0       ;hPrevInstance.
                                ;  (=0 if no previous instance).
        jne    createwin
```

Prolog code

One important thing to notice from Figure 4.1 is that after the *prolog* code, BP points to the parameters (so that the program has ready access to them), while SP has been moved away (so that the stack can grow downward in memory without interfering with the parameters or the intermediate area that is to be used for temporary data).

In Figure 4.1, increasing addresses are downward. Note that the return address is not FAR, but NEAR, as WINMAIN() is called by the startup code within the same segment, not directly from Windows.

Note that the old value of BP is saved on the stack. Note that "lpCmdLine" is a 32-bit value and so occupies four memory locations (for explanation of label prefixes, refer to page 82).

First-instance handling

The first instance of the program has to create a window-class data structure and call RegisterClass(). It determined this by testing "hPrevInstance", which is zero if this is the first instance. Note that the handle for this particular instance is "hInstance".

> ALL OF THIS STUFF DOWN TO CREATEWIN IS PRETTY HORRIBLE, SO LET YOUR EYES GLAZE OVER AND READ QUICKLY ONWARD TO **CREATEWIN**:

```
        mov    WORD PTR [bp-46],3               ;wndclass
        mov    WORD PTR [bp-44],OFFSET SKELETONPROC
                                        ;addr of callback
        mov    WORD PTR [bp-42],SEG SKELETONPROC
                                        ;function for window.
        sub    ax,ax
        mov    WORD PTR [bp-40],ax
        mov    WORD PTR [bp-38],ax
        mov    ax,WORD PTR [bp+12]      ;hInstance
        mov    WORD PTR [bp-36],ax
        sub    ax,ax            ;null -- use Windows default icons.
        push   ax               ;       /
        mov    cx,IDI_APPLICATION       ;Default application icon.
        sub    dx,dx                    ;       /
        push   dx                       ;       /
        push   cx                       ;       /
        call   LOADICON
        mov    WORD PTR [bp-34],ax
        sub    ax,ax            ;null -- use Windows default cursor.
        push   ax                       ;       /
        mov    ax,IDC_ARROW             ;Standard arrow cursor.
```

```
cwd                                    ;      /
push dx                                ;     /
push ax                                ;    /
call LOADCURSOR
mov  WORD PTR [bp-32],ax
mov  ax,COLOR_BACKGROUND
mov  WORD PTR [bp-30],ax
mov  ax,OFFSET szskeletonname
mov  WORD PTR [bp-28],ax
mov  WORD PTR [bp-26],ds
mov  WORD PTR [bp-24],ax
mov  WORD PTR [bp-22],ds
lea  ax,WORD PTR [bp-46]    ;wndclass
push ss                     ;this is address of above data
push ax                     ;structure.
```

Register-
Class()

Note that we only have to call RegisterClass() for the first instance of the program. If you double-click on the program icon a second time, the second instance of the program created in memory will not have to register the window with Windows.

```
call REGISTERCLASS;registers this class of window.
or   ax,ax                 ;error test.
je   quitwinmain
```

Displaying
a window

The above block of code registered the "specifications" of our program's window with Windows. Now to display it:

Create-
Window()

Parameters that have to be pushed on the stack prior to calling CreateWindow() are a long-pointer to window class-name, lp to the window title-name, type of window, x and y coordinates, width and height, parent-handle, menu-handle, instance-handle, and an lp to parameters to link with the window.

```
createwin:
  mov  ax,OFFSET szskeletonname
  push ds                  ;long-pointer (far address) of
  push ax                      ;class-name.
  mov  ax,OFFSET szwintitle
  push ds                  ;far address of window-title.
  push ax                  ;      /
  sub  ax,ax               ;type of window (32-bit value).
  mov  dx,207              ;     /
  push dx                  ;    /
  push ax                  ;   /
  mov  ax,150              ;x-coord (16-bit).
  push ax                  ;    /
  sub  ax,ax               ;y-coord (16-bit).
  push ax                  ;   /
  mov  ax,250              ;width (16-bit).
  push ax                  ;   /
  mov  ax,200              ;height (16-bit).
  push ax                  ;    /
```

```
    sub  ax,ax
    push ax                    ;0=no parent for this window.
    push ax                    ;0=use the class menu.
    mov  ax,WORD PTR [bp+12]        ;hInstance
    push ax                    ;
    sub  ax,ax
    push ax                    ;0=no params to pass-on.
    push ax                    ;(32-bit long-pointer).
    call CREATEWINDOW
    mov  WORD PTR [bp-2],ax    ;returns hWnd in AX
                               ;(handle to the window).
                               ;Here we save it temporarily.
    push ax                    ;ShowWindow() requires hWnd
    push WORD PTR [bp+4]       ;and nCmdShow on the stack.
    call SHOWWINDOW   ;Tells Windows to display window.
    push WORD PTR [bp-2]       ;hWnd
    call UPDATEWINDOW ;tells Windows to redraw now.
    jmp  SHORT messageloop ;go to the main message loop.
```

Message loop

Refer back to page 77 for an explanation of the message loop. The event-driven nature of a Windows application means that **GETMESSAGE()** goes to Windows and waits for a message from the queue. After return, key presses are preprocessed by TRANSLATEMESSAGE(), then control is passed to the callback function via DISPATCHMESSAGE() and Windows.

```
mainloop:
    lea  ax,WORD PTR [bp-20]   ;far-addr of message.
    push ss                    ;  /
    push ax                    ;  /
    call TRANSLATEMESSAGE
;.......
    lea  ax,WORD PTR [bp-20]   ;far-addr of message.
    push ss                    ;  /
    push ax                    ;  /
    call DISPATCHMESSAGE
;........
messageloop:
    lea  ax,WORD PTR [bp-20]   ;long-pointer (far addr) of
    push ss                    ;message. (we use the stack
    push ax                    ;region for convenience).
    sub  ax,ax
    push ax                    ;null
    push ax                    ;null
    push ax                    ;null
    call GETMESSAGE
    or   ax,ax                 ;only exit if returns AX=0
    jne  mainloop
;GetMessage() returns FALSE (AX=0) if a "quit" message...
;so here we are quiting....
    mov  ax,WORD PTR [bp-16]   ;return wParam to Windows.
quitwinmain:
    mov  sp,bp
    pop  bp ;restore SP to point to the return address.
    ret  10     ;Causes RET to add 10 to SP after popping
```

```
;ret-address, effectively dumping all params
;(as for PASCAL convention).
WINMAIN ENDP
```

Figure 4.2: Stack at entry to GetMessage().

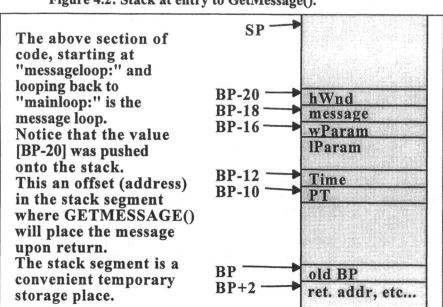

The above section of code, starting at "messageloop:" and looping back to "mainloop:" is the message loop.
Notice that the value [BP-20] was pushed onto the stack.
This an offset (address) in the stack segment where GETMESSAGE() will place the message upon return.
The stack segment is a convenient temporary storage place.

SP

BP-20 → hWnd
BP-18 → message
BP-16 → wParam
         lParam
BP-12 → Time
BP-10 → PT

BP → old BP
BP+2 → ret. addr, etc...

Callback Function

Thus ends WINMAIN(). For the callback function, refer to the listing on page 79. The parameters are passed on to the stack in the order of left to right, with a FAR return address on top.

If this program looks similar to the example in my last book, it's not surprising, since both were originally created from a C skeleton with the compiler set to generate assembly output (see page 151). This listing is, however, substantially different from before.

```
   PUBLIC      SKELETONPROC
SKELETONPROC   PROC FAR
;The function is entered with far-return-addr (4 bytes),
;lParam (4), wParam (2), message-type (2), and
;window-handle (2 bytes) on the stack (ret-addr on top).
;.....
   push ds            ;This is some Standard preliminary
   pop  ax            ;shuffling of the registers.
   nop               ;    /
   inc  bp            ;    / (it is called the prolog code)
   push bp            ;    /
```

```
mov  bp,sp        ;    /
push ds           ;    /
mov  ds,ax        ;    /
sub  sp,146  ;move the stack to a free region
    ;(so as not to mess-up the params).
```

Prolog code

The above *prolog* code may seem strange. It is at the start of all callbacks. However, the above code can be simplified if the application is never to run in Real mode. A Windows application running in Real mode is only possible with Windows v3.0 and earlier and is an unlikely requirement these days.

Alternative simplified prolog code

If the application will always be run in Protected mode, the prolog can be simplified as follows:

```
push bp       ;prolog
mov  bp,sp    ; /   (set up stack frame)
push ds       ; /   (save calling function's ds)
push ss       ; /   (move ss to ds  -- local data segment)
pop  ds       ; /   (    "   )
sub  sp,146   ; /   (reserve local data area)
```

An appropriate modification of the epilog code will also be required. The simplified prolog is more suitable for explanation. You can see that BP and DS are saved. The main task of the prolog is to set DS to the current application's data segment, but this is easy, as SS always points to it, even while execution goes back to Windows. That is, after the application is first entered, SS remains always unchanged and always pointing to the data segment.

After the prolog, the stack looks like Figure 4.3.

Figure 4.3: Stack after executing prolog.

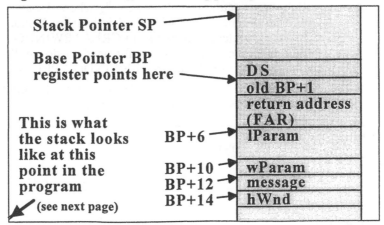

```
    mov   ax,WORD PTR [bp+12]    ;get message-number.
    cmp   ax,WM_CREATE          ;message received after
    je    create                ;CreateWindow() is called.
    cmp   ax,WM_DESTROY         ;message received if a window is
                                ; closed.
    je    quitmessage
    cmp   ax,WM_PAINT           ;message received if Windows has
                ;(already) redrawn any part of the window.
    je    paint
    cmp   ax,WM_COMMAND         ;any selection of the menu will
    jne   notcommand           ;the WM_COMMAND
    jmp   menu                 ;message.
notcommand:
    cmp   ax,WM_LBUTTONDOWN            ;one of many mouse
    jne   notlbutton                  ;messages.
    jmp   break
notlbutton:
    cmp   ax,WM_CHAR                  ;message that a key pressed.
    je    char
;.......
;Default handling of messages....
    push WORD PTR [bp+14]       ;hWnd
    push WORD PTR [bp+12]       ;Message-type
    push WORD PTR [bp+10]       ;wParam
    push WORD PTR [bp+8]        ;hi-half of lParam
    push WORD PTR [bp+6]        ;low-half of lParam
    call DEFWINDOWPROC
    jmp   return                        ;Back to Windows.
```

"Case"
statement

The above code determines the type of message and jumps to an appropriate routine. If the message is not to be handled explicitly by the callback, it falls through to DEFWINDOWPROC() for default handling.

WM_CREATE
message

Follow through the case of **WM_CREATE**. The earlier case logic will bring execution to "create:", where I have obtained the handle to a font. WM_CREATE is sent by Windows when the window is created, in response to CREATEWINDOW(), and for a simple skeleton you do not really need to do anything with this message — just send it on to DEFWINDOWPROC().

Note that even fonts have handles, and to use the OEM font in the program, this is a convenient time to get its handle.

```
create:
  mov   ax,OEM_FIXED_FONT
  push  ax
  call  GETSTOCKOBJECT
  mov   hOemFont,ax                   ;handle to font.
  jmp   SHORT break
;..............
quitmessage:
  sub   ax,ax
```

```
    push ax
    call POSTQUITMESSAGE
    jmp  SHORT break
```

WM_CHAR I implemented the **WM_CHAR** case to show how to respond to a
message keyboard character. See the keyboard tables in Appendix B.
Refer to a Windows programming book on the difference between
ANSI and ASCII.

```
char:
;I haven't bothered to respond to keypresses in any way
;in this simple skeleton ...
    jmp  SHORT break
```

WM_PAINT Even the most basic skeleton will need the following code in
message response to **WM_PAINT**. You will need to put in BeginPaint()
and EndPaint(), even if you don't output anything. WM_PAINT is
sent if anything has happened to the window that will require its
client area to be redrawn. I need a handle (hDC) to the client area
before I can output to it.

```
paint:
    push WORD PTR [bp+14]        ;hWnd is handle of window.
    lea  ax,WORD PTR [bp-42]     ;ps -- far-addr of
                                 ; paint-structure.
    push ss        ;(BeginPaint() will fill the structure).
    push ax                      ;          /
    call BEGINPAINT    ;BeginPaint() returns handle hDC.
    mov  WORD PTR [bp-146],ax  ;hDC -- display-context,
                   ; required before can output to screen.
    push ax                      ;hDC
    push hOemFont
    call SELECTOBJECT        ;attaches hOemFont to hDC.
    push WORD PTR [bp-146]          ;hDC
    mov  ax,8                       ;16-bit x-coord
    push ax                         ;      /
    mov  ax,15                      ;16-bit y-coord
    push ax                         ;       /
    mov  ax,OFFSET sout    ;far-address of string to o/p
    push ds                  ;     /
    push ax                  ;     / (note lo-half pushed 2nd)
    mov  ax,9                ;number of chars in string.
    push ax
    call TEXTOUT
    push WORD PTR [bp+14]        ;hWnd
    lea  ax,WORD PTR [bp-42]     ;far-addr of paint-structure.
    push ss                 ;(was filled by BeginPaint()).
    push ax                      ;       /
    call ENDPAINT
    jmp  SHORT break
;.....................
menu:
```

WM_COMMAND Selection of a menu-item will result in a **WM_COMMAND**
message message, with the identifier in wParam, and zero in the low-half of
lParam.

```
cmp  WORD PTR [bp+6],0        ;low-half of lParam
jne  break                    ;test if a menu-message.
```

Menu-item If our program determines that the message is a
selection? **WM_COMMAND**, we find out more about it by looking at
wParam and lParam.

The low-half of lParam = 0 if the message is a menu-selection, in
which case wParam contains the identifier, and the high-half of
lParam = 1 if an accelerator key has been pressed.

If the low-half of lParam is not zero, then the message is from a
control (such as a scrollbar), and the low-half of lParam = the
handle of the control, and the high-half of lParam = the
notification code.

So wParam can contain the menu-item identifier, the control
identifier, or the accelerator-key identifier.

```
   cmp  WORD PTR [bp+10],IDM_QUIT    ;wParam.
   jne  noquit
   jmp  quitmessage
noquit:
   cmp  WORD PTR [bp+10],IDM_ABOUT
   jne  break                        ;no other menu items.
;............
;displaying a message about this program...
   push WORD PTR [bp+14]   ;hWnd is handle of parent
                           ; window.
   mov  ax,OFFSET szabout  ;far-addr of string to display.
   push ds                 ;    /
   push ax                 ;    /
   mov  ax,OFFSET sztitle  ;far-addr of title of
                           ; dialog-box.
   push ds                 ;    / (see data segment)
   push ax                 ;    /
   mov  ax,MB_OK           ;type of message box.
   push ax          ;  /   (displays single "ok" button)
   call MESSAGEBOX
```

Epilog Finally we have the *epilog* code, which compliments the prolog
code code on page 102. At this stage, BP is pointing to the saved "old
BP+1" which we decrement twice so it points to the saved DS,
which we make the top of stack and then pop to restore the
original DS, followed by the "old BP+1", which we decrement to
restore to its original value.

RET 10 causes RET to add 10 to SP after popping the return
address, effectively dumping all parameters, as required for the
Pascal convention.

```
break:
   sub   ax,ax   ;returns 0 in DX:AX.  (callback functions
   cwd            ;return a 32-bit (long) value).
return:
   dec   bp             ;final Standard manipulation of regs.
   dec   bp             ;   /
   mov   sp,bp          ;  / (it is called the epilog code).
   pop   ds             ;   /
   pop   bp             ;  /
   dec   bp             ;  /
   ret   10             ;removes parameters.
SKELETONPROC    ENDP
;. . . . . . . . . . . . . . . . . . . . . . . . . . . . . . . . . . . . . . . . . . . . . . . . . . . .
   END   start   ;execution entry point of program.
```

Simplified epilog code I showed earlier that there is a simplified alternative for the prolog
code. The matching epilog is similarly simple:

```
   pop   ds             ;epilog
   pop   bp             ;   /
   retf 10
```

So, here again refer to Figure 3.1 to see what it looks like.
Clicking on the "File" menu-item pops down two selections:
"Quit" and "Message...". Selecting the latter results in a message
box looking very much like that shown on page 172.

5

High-Level Assembly

Preamble

*What's in
this chapter* What I have for you in this chapter is the same program from the previous chapter (page 94), but wow is it smaller! One thing you will have noticed from that first program is that it does an incredible amount of stack manipulation: this makes the program both long and very tedious to write.

The Borland and Microsoft (plus other vendors') assemblers have some high-level features that ease the coding burden considerably, even to the point of the program being as short as the equivalent written in C or some other high-level language. That's saying something!

What follows is a breakdown of each section of the previous program, showing how it can be improved ...

Include Files

Equates Refer back to page 94. You will see a whole pile of equates, for example, "WM_PAINT EQU 15". WM_PAINT is simply a meaningful label, a constant, that equates to value 15. This means that wherever the assembler finds the label WM_PAINT, it will be replaced by the value 15.

These semi-English labels are more meaningful to us and therefore make programming easier. Windows has hundreds of these predefined equates, though the example program only uses some of them.

Those people familiar with writing Windows programs in C will recognize this: "#INCLUDE <WINDOWS.H>". It is a statement placed right near the beginning of the program, and has the effect of inserting the file named between the "< >" into the program at that point.

WINDOWS.H contains all of the equates, plus other definitions such as structure definitions. Windows programming also makes extensive use of structures (look back to page 65 for an introduction to structures).

Microsoft versus Borland

.INC files

Instead of explicitly naming all the equates and structures in my program, as I did for the first example program, an assembly language program can also include WINDOWS.H. Or rather, it can't. There is a problem with syntax. WINDOWS.H has a syntax designed to be understood by the C compiler, and this is mostly gibberish to the assembler — however Microsoft introduced with version 6.0 of their Macro assembler (MASM) and Borland with C++ v3.0, a .H-to-.INC translator. Note the convention that all C-syntax Include files have the extension ".H".

Instead of WINDOWS.H, in assembly we use WINDOWS.INC, which is supplied by Borland and Microsoft. Note the convention that Include files for assembly language have the extension .INC, though I can't vouch for this for all software vendors.

Structures defined in .INC file

The listing starts on the next page, and as you look through it, you will see how I have included WINDOWS.INC, and how I have accessed the structures. There are some example extracts from WINDOWS.INC to clarify the explanation.

Assembler version notes

The first listing is designed around Borland TASM version 2.5, so once again I am aiming for the earliest possible version. If you only have MASM version 5.1 or earlier, or TASM prior to version 2.5, which do not have the necessary high-level constructs, you can only assemble the Windows program from the previous chapter. For further discussion of version numbers, see page 88. If you want to make use of the latest features for writing streamlined code, especially if writing for Win32, then the later the version the better.

Complete MASM & TASM skeleton listings for all versions

It is fascinating to watch the game Microsoft and Borland are playing with each other. One tries to leapfrog the other, and Microsoft's version 6.0 was released in response to Borland's version 2.5. MASM version 6.0 has some very nice features, and the releases of 6.10 and 6.11 added enhancements to further streamline coding for Windows. I've put some special notes on compatibility issues for v6.0+ at the end of this chapter (see page 125), and to be completely fair to both vendors and to those readers who have MASM v6.x, I've placed a complete listing of a MASM skeleton program at the end of this chapter.

You will also find the MASM skeleton program on the Companion Disk, in directories \ASMDEMO1 and \ASMDEMO2. The first is a skeleton program that has the startup code inside the program, as is done in the skeleton program of the previous chapter. In the second directory is the same program, but it has the startup code as a separate linkable module. It is the latter case that is listed at the end of this chapter.

You will find the TASM skeleton program on the Companion Disk in directory \SKELETN2. This is the same program listed immediately below. Note that it has a separate linked startup module, C0WS.OBJ. (You may have already noticed that there is nothing apparently logical about the naming of directories or files on the Compnaion Disk. The justification is historical; I have kept the same names as used in the first edition.)

A skeleton written for TASM version 5 is in Chapter 13.

Skeleton Analysis

```
;WINHULLO.ASM-->WINHULLO.EXE Windows demo
.MODEL   SMALL
```

Program listing continues until page 119.
This program works with TASM v2.5+

The ".MODEL" directive is an instruction to the assembler. If you leave it off, the program will still assemble ok. It tells the assembler how many data and code segments this program will need and gives Standard names and qualifiers to the segments.

.MODEL

I have specified "SMALL", which means that the program will have one code segment and one segment with combined data and stack. You have a choice of TINY, SMALL, MEDIUM, COMPACT, and HUGE: your assembler manual will have details on each of these. See page 119 for more information.

If your assembly program has to be linked with a high-level program, you would normally choose the same model that was used for compiling the high-level code. This ensures smooth linking.

```
INCLUDE WINDOWS.INC
IDM_QUIT        EQU    200       ;menu-identifiers:  must be
IDM_ABOUT       EQU    201       ;same as defined .RC file.
```

WINDOWS.-
INC

Here is where **WINDOWS.INC** is inserted. If you look back to page 94 you will see that I have still left in the above two equates. These come from the .RC file (see page 89). If there were enough of these, I would have put them into their own .INC file and included it in both WINASM1.ASM and WINASM1.RC

Unlike C, external functions must be explicitly declared in assembly language. MASM version 6 is a bit different (see page 125), as is TASM version 5 (see Chapter 13).

```
EXTRN   UPDATEWINDOW:FAR
EXTRN   BEGINPAINT:FAR,ENDPAINT:FAR
EXTRN   DEFWINDOWPROC:FAR
EXTRN   POSTQUITMESSAGE:FAR
EXTRN   REGISTERCLASS:FAR
EXTRN   GETSTOCKOBJECT:FAR
EXTRN   CREATEWINDOW:FAR
EXTRN   SHOWWINDOW:FAR
EXTRN   GETMESSAGE:FAR
EXTRN   LOADCURSOR:FAR
EXTRN   TRANSLATEMESSAGE:FAR
EXTRN   DISPATCHMESSAGE:FAR
EXTRN   LOADICON:FAR
EXTRN   TEXTOUT:FAR
EXTRN   MESSAGEBOX:FAR
EXTRN   SELECTOBJECT:FAR
```

Data segment, no major change from before ...

```
.DATA
szwintitle      DB      'HULLO DEMO PROGRAM',0
szwinasm1name   DB      'WINASM1',0
hOemFont        DW      0                ;handle to OEM font.
sout            DB      'Hullo World'
szabout         DB      'Assembly Language Windows Demo',0
sztitle         DB      'Karda Prints',0   ;  /
;....................................................
.CODE
  PUBLIC        WINMAIN
WINMAIN PROC PASCAL NEAR hInstance:WORD,\
       hPrevInstance:WORD,lpCmdLine:DWORD,nCmdShow:WORD
```

High-level PROC

Now for the first major enhancement. If you refer back to page 98 you will see this same section of code and a picture of the stack. The parameters passed on the stack have to be accessed by direct addressing of the stack segment. "cmp WORD PTR [bp+10], 0" for example, to get at "hPrevInstance". However, by declaring all passed parameters as above, they can be accessed within the procedure by name. The example would become "cmp hPrevInstance, 0" — simple hey! The assembler equates hPrevInstance to [bp+10], so it does the dirty work.

PASCAL epilog/ prolog

There's another important aspect to the above high-level PROC — the **PASCAL** qualifier. This eliminates the need to explicitly code the prolog and epilog code. Again, look back at page 98.

The standard prolog code, which is not part of the program listing, is:

```
push   bp              ;save old bp value.
mov    bp,sp           ;set bp pointing to return address.
sub    sp,46           ;operand varies (see notes below).
```

The standard epilog code, which is not part of program listing, is:

```
mov    sp,bp           ;set sp pointing to return address.
pop    bp              ;restore old bp value.
ret    10    ;operand depends on # of parameters to dump.
```

Now back to the program listing:

```
;Define all 'automatic' data...
   LOCAL   hWnd:WORD
;window class structure for REGISTERCLASS()....
   LOCAL   s1:WNDCLASS
;message structure for GETMESSAGE()...
   LOCAL   s2:MSGSTRUCT
```

LOCAL directive

The original prolog code contained "sub sp,46" to move the stack further down in the stack segment, allowing a free area in which to store local data. Once again, we can eliminate the need to explicitly code this. Declare all local data using the **LOCAL** directive, with a syntax as shown above. Incidentally the default type is WORD, so if the data is of type WORD you don't have to declare it.

Note that you cannot initialise this data, since it is only created at execution entry to the procedure, not at assemble time.

For an introduction to local data, refer back to page 62.

Note also a particular problem due to the temporary nature of local data, with regard to getting its address within the program — see page 60.

This local data can be referred to by name, and the assembler will do the job of equating the labels to [bp-*value*]. A most useful side-effect of local labels is that the names are only recognized within the current procedure, not even inside nested procedures. This means that you can use labels elsewhere with the same names (this is a highly qualified statement: refer to page 120).

The syntax is (not part of program listing):

```
LOCAL   label : type [,label : type ] [, ...]
```

**STRUC
directive**

Notice the data types WNDCLASS and MSGSTRUCT above. Structures are introduced back on page 65. Structures used by Windows are defined in WINDOWS.INC, the Include file.

WNDCLASS and **MSGSTRUCT** are the names of structures, and they can also be used in data declarations as the data-type, as has been done with our LOCAL declarations s1 and s2. s1 is merely an instance of structure WNDCLASS, while s2 is an instance of MSGSTRUCT.

For your reference, extracting the definition of WNDCLASS from Borland's WINDOWS.INC (not part of program listing):

```
MSGSTRUCT        STRUC
   msHWND        DW      ?
   msMESSAGE     DW      ?
   msWPARAM      DW      ?
   msLPARAM      DD      ?
   msTIME        DD      ?
   msPT          DD      ?
MSGSTRUCT        ENDS
; And here is the other:
WNDCLASS         STRUC
   clsStyle             DW      ?
   clsLpfnWndProc       DD      ?
   clsCbClsExtra        DW      ?
   clsCbWndExtra        DW      ?
   clsHInstance         DW      ?
   clsHIcon             DW      ?
   clsHCursor           DW      ?
   clsHbrBackground     DW      ?
   clsLpszMenuName      DD      ?
   clsLpszClassName     DD      ?
WNDCLASS         ENDS
```

NOTE:

For 32-bit programming, all of these fields become 32 bits.

The Companion Disk has different Include files. For 16-bit Windows applications there is WINDOWS.INC and WINASM60.INC, and for 32-bit applications there is W32.INC.

There is also an extended window class, with a structure called WNDCLASSEX, that has an extra field. It is used with REGISTERCLASSEX().

Now back to the program listing:

```
    cmp   hPrevInstance,0          ;=0 if no previous instance.
    je    yes1st
    jmp   createwin
;...............................
yes1st:
;Setup the window class structure for REGISTERCLASS()...
    mov   s1.clsStyle,3
    mov   s1.WORD PTR clsLpfnWndProc,OFFSET Winasm1Proc
    mov   s1.WORD PTR clsLpfnWndProc+2,SEG Winasm1Proc
    mov   s1.clsCbClsExtra,0
    mov   s1.clsCbWndExtra,0
    mov   ax,hInstance
    mov   s1.clsHInstance,ax
    call  LOADICON PASCAL,null, 0,IDI_APPLICATION
    mov   s1.clsHIcon,ax
    call  LOADCURSOR PASCAL,null, 0,IDC_ARROW
    mov   s1.clsHCursor,ax
    mov   s1.clsHbrBackground,COLOR_BACKGROUND
    mov   ax,OFFSET szwinasm1name
    mov   s1.WORD PTR clsLpszMenuname,ax
    mov   s1.WORD PTR clsLpszMenuName+2,ds
    mov   s1.WORD PTR clsLpszClassName,ax
    mov   s1.WORD PTR clsLpszClassName+2,ds
```

Registering a window The above block of code is setting up the data structure prior to calling **REGISTERCLASS()**. Compare that with the previous program, page 99. You will see there that we had to explicitly access the stack segment between [bp] and [bp-46], in which the instance of the structure was kept. (Locations greater than [bp] contain the return address and passed parameters, while addresses below [bp-46] is the new working area for the stack.)

WORD PTR override is introduced on page 63.

High-level CALL Now we have another high-level feature, the high-level **CALL**. REGISTERCLASS() only requires one parameter, the FAR address of the s1 data structure.

Refer back to how it was done before: after everything was loaded into the structure in the stack segment, ss:[bp-46] was passed as the FAR address required by REGISTERCLASS(). See page 99 onwards.

Below, we do the same thing but use the name of the structure instead:

```
    lea   ax,s1
    call  REGISTERCLASS PASCAL,ss,ax
    or    ax,ax
    jne   createwin
    jmp   quitwinmain
```

The time has come to create the window on-screen. The high-level **CALL** has various qualifiers and can take multiple parameters.

Note that if the parameters have no defined size, they default to WORD.

Notice the qualifier PASCAL:

```
createwin:
   call CREATEWINDOW PASCAL,ds,OFFSET szwinasm1name,\
        ds,OFFSET szwintitle, 207,0, 150, 0, 400,\
                              300, 0, 0, hInstance, 0,0
   mov  hWnd,ax
   call SHOWWINDOW   PASCAL,ax,nCmdShow
   call UPDATEWINDOW PASCAL,hWnd
   jmp  SHORT messageloop       ;go to main message loop.
```

You may have noticed that I have not used the FAR PTR override for the call instructions: the assembler is smart enough to know from the "EXTRN *functionname* : FAR" declarations that the call should be FAR. The override could be put in, but for the programmer's information only.

PASCAL, C, BASIC, FORTRAN, STDCALL, PROLOG qualifiers

So, what about the **PASCAL** qualifier? The choices here are nothing, PASCAL, C, BASIC, FORTRAN STDCALL, or PROLOG. The qualifiers available vary with different assemblers.

Normally, a CALL instruction just pushes the return address on to the stack, and the RET at the end of the called procedure pops it off.

The PASCAL qualifer will cause the parameters to push on in the correct order and will also remove them, assembling a "RET *number* " at the end of the procedure, as discussed above and on page 107. We require the PASCAL qualifier to call Windows functions.

We would use the C qualifier to call C functions, perhaps some third-party C library we want to use. The effect is the same, but the parameters are pushed on in the reverse order and not removed by the called routine: they are removed from the stack after execution returns from the procedure.

Whatever language we are calling, the result is that the high-level CALL instruction assembles with all of the pushes, pops, and other stack manipulations generated automatically — unassemble such code and you will see something like the program of the previous chapter.

```
;This is the main message loop ...
mainloop:
  lea  ax,s2
```

```
    call TRANSLATEMESSAGE PASCAL,ss,ax
    lea  ax,s2
    call DISPATCHMESSAGE PASCAL,ss,ax
messageloop:
    lea  ax,s2
    call GETMESSAGE PASCAL, ss,ax, null, null, null
    or   ax,ax
    jne  mainloop
;GetMessage() returns FALSE (AX=0) if a "quit" message...
;so here we are quiting....
    mov  ax,s2.msWPARAM    ;return wparam to windows OS.
quitwinmain:
    ret
WINMAIN ENDP
```

Figure 5.1: Stack upon entry to callback.

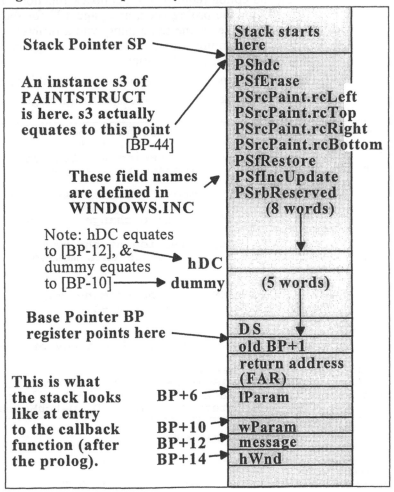

```
;  . . . . . . . . . . . . . . . . . . . . . . . . . . . . . . . . . . . . . . . .
    PUBLIC          WINASM1PROC
WINASM1PROC PROC WINDOWS PASCAL FAR \
            hWnd:WORD,msgtype:WORD,wParam:WORD,lParam:DWORD
    LOCAL           dummy:WORD:5
    LOCAL           hDC:WORD
    LOCAL           s3:PAINTSTRUCT
```

WINDOWS
qualifier

Notice two things here: the **WINDOWS** qualifier, and the "dummy" local variable.   Local declarations can take a repeat-count, which in this case declares five words, the first pointed to by label "dummy".

The WINDOWS qualifier takes care of generation of the special prolog and epilog required for a callback function. Refer back to page 103 for the prolog code, and page 107 for the epilog code. MASM v6 achieves the same thing with a different syntax, while 32-bit programming uses the STDCALL language qualifier and doesn't need further qualification.

Figure 5.1 shows the stack upon entry to the callback function.

Now for the case-logic that processes the messages ....

```
    mov  ax,msgtype             ;get message-type.
    cmp  ax,WM_CREATE           ;msg received after
    je   xcreate                ;CreateWindow() func is called.
    cmp  ax,WM_DESTROY          ;message if a window is closed.
    je   xquitmessage
    cmp  ax,WM_PAINT            ;msg if Windows has (already)
                                ;redrawn any part of the window
                        ;(due to size-change for example).
    je   xpaint
    cmp  ax,WM_COMMAND ;any selection of the menu will
    jne  notwmcommand           ;  produce this message.
    jmp  xmenu
notwmcommand:
    cmp  ax,WM_LBUTTONDOWN       ;one of many mouse
    jne  notwmlbutton           ; messages.
    jmp  xbreak
notwmlbutton:
    cmp  ax,WM_CHAR             ;message that a key pressed.
    je   xchar
;Default handling of messages....
    call DEFWINDOWPROC PASCAL,hWnd,msgtype,wParam, lParam
    jmp  xreturn        ;Back to Windows, which will in turn
                        ; return to after DispatchMessage().
```

Now to process the messages ...

```
xcreate:
  call GETSTOCKOBJECT PASCAL,OEM_FIXED_FONT
```

```
    mov   hOemFont,ax          ;handle to font.
    jmp   xbreak
;....
xquitmessage:
    call  POSTQUITMESSAGE PASCAL,0
    jmp   xbreak
;.....
xchar:
    jmp   xbreak
;........
xpaint:
    lea   ax,s3          ;ps -- far-addr of paint-structure.
    call  BEGINPAINT PASCAL,hWnd,ss,ax
    mov   hDC,ax              ;hDC -- display-contex.
    call  SELECTOBJECT PASCAL,ax,hOemFont
    call  TEXTOUT PASCAL,hDC,10,20, ds,OFFSET sout,11
    lea   ax,s3              ; -- far-addr of paint-structure.
    call  ENDPAINT PASCAL,hWnd, ss,ax
    jmp   SHORT xbreak
;.....................
xmenu:
    cmp   WORD PTR lParam,0      ;low-half of lParam
    jne   xbreak                ;test if a menu-message.
    cmp   wParam,IDM_QUIT        ;wParam.
    jne   notquit
    jmp   xquitmessage
notquit:
    cmp   wParam,IDM_ABOUT
    jne   xbreak                     ;no other menu items.
    call  MESSAGEBOX PASCAL, hWnd, SEG szabout, \
                     OFFSET szabout, SEG sztitle, \
                     OFFSET sztitle, MB_OK
;.....................'....
xbreak:
    sub   ax,ax                      ;returns 0 in DX:AX.
    cwd                      ;return a 32-bit (long) value).
xreturn:
    ret
Winasm1Proc    ENDP
 END
```

.MODEL Directive

I already introduced the .MODEL directive on page 111 and made reference to the TINY, SMALL, MEDIUM, COMPACT, and HUGE memory qualifiers. The .MODEL directive can also take other qualifiers, for example:

```
.MODEL   SMALL,  PASCAL
```

This sets the defaults for the program, and "`PASCAL`" means that all procedures are to be of Pascal-type, which also means that the PASCAL qualifier is not needed in the PROC declarations.

However, high-level CALL instructions still need language qualifiers to pass parameters automatically, so leave the PASCAL qualifier in.

The choices of language qualifier are PASCAL, C, BASIC, FORTRAN, PROLOG, STDCALL, or NOLANGUAGE.

The formal syntax for .MODEL is:

```
.MODEL memorymodel  [ , [language-modifier ] language ]
```

The language modifier is WINDOWS, ODDNEAR, FARNEAR, or NORMAL. The exact syntax may vary with some assemblers. I got this information from the Borland TASM version 2.5 manual. Special notes on Microsoft MASM compatibility are on page 125.

We would not normally put the WINDOWS language modifier in, because WINMAIN() doesn't need it. However, if you were writing callback functions only in assembly language, and perhaps the WINMAIN() in a high-level language, then yes, put it in. This note only applies to 16-bit TASM applications.

Private and Global Data

Traditional assembly language people are accustomed to all labels being global, but with the high-level procedures this is not necessarily the case. Any procedure that uses a language qualifier, such as PASCAL, automatically has private labels — at least that is the case with recent versions of MASM — let us treat MASM as the reference standard. All data and code labels declared inside the procedure are private to that procedure and are unrecognized outside it, which means of course that you can use the same labels elsewhere.

High-level procedures are declared either explicitly, by putting a language qualifier into the PROC declaration, or implicitly in the .MODEL declaration.

Do note that the defaults set by .MODEL can be overriden in individual PROCs and CALLs, as required.

Global labels

So what labels are global? Those declared in the data segment or in WINDOWS.INC. This means that the structures declared in WINDOWS.INC are also global, so instances of them can be made and accessed anywhere. So be careful not to use names that conflict with any of those in WINDOWS.INC.

It is a good move to print out WINDOWS.INC for reference. In some cases you need to know the names of the fields in a structure or an equate, so a printout can be extremely handy.

WNDCLASS If you look back at pages 113 and 114, you'll see how WNDCLASS is used and how it is defined in WINDOWS.INC. Notice the names of the fields: these are different than field names in WINDOWS.H. In the message structure, also shown on the same pages, you can see the fields — msWPARAM, instead of just wParam. Don't worry about upper or lower case, as traditional assemblers don't care. However the "ms" prefix is there to distinguish this global label. I have used WPARAM as a local label within a program, although an assembler would complain bitterly if it found a global with the same name.

Global override Although code labels are local to the procedure, you can declare them as global if necessary:

```
globalplace::          ;a global label (MASM).
```

That's all you need: double colons.

MASM versus TASM Scope

Be careful about differences in the *scope* of labels between MASM and TASM and the various versions of each. It is wise to check your particular manual to clarify this. The above notes are based on reading the MASM manuals, particularly Quick-assembler version 2.01. This version supports high-level PROCs and the LOCAL directive, as discussed in this book. Quick-C with Quick-assembler supports Windows development.

MASM introduced local scoping of labels by default, and looking back through my manuals I see that v5.1 defined all labels as global, so the concept came in after that. Version 5.1 has global code labels only, but local to the module (a module being a source file that will be linked with others). TASM v3.0's VERSION directive claims to be able to emulate MASM versions 4.0, 5.0, 5.1, and 5.2.

Incidentally, MASM version 5.2 **appears** to be equivalent to Quick-assembler version 2.01.

TASM's @@

TASM's native mode is a bit different — if you want a label to have scope only within the current procedure, prefix it with "@@" and put the **LOCALS** directive right at the beginning of the file. This holds true for code labels and all labels defined by high-level PROCs and LOCAL directives.

So, what about TASM's default treatment of labels as global? Quite frankly it's a nuisance. Let me quote the TASM v3.0 manual:

> "All argument names specified in the procedure header, whether ARGs (passed arguments), RETURNs (return arguments) or LOCALs (local variables), are global in scope unless you give them names prepended with the local symbol prefix."

You could have the habit of putting "@@" in front of every label that is to be local to a procedure. This seems ok in principle, except that I encountered assembly errors that do not make sense to me. Apart from my negative personal experience, you can take it as good policy to use "@@" prefixing as much as possible when using TASM. Don't forget to put the LOCALS directive at the start of the file.

Look ahead to Chapter 6 for examples of using "@@". I have had a lot of trouble using "@@" in WINASMOO.INC (Chapter 6) — you can see in the listing on page 168 that I defined "now" as a parameter passed to procedures, which according to Borland's statement above, is global. Yet I have reused it in many procedures, and in each case it assembles correctly. This indicates, though I can't find it mentioned anywhere in the manual, that the local definition of "now" overrides any other local definitions.

So "now" is not really global. It only exists on the stack anyway, so sensibly it is only valid for the life of the procedure in which it is declared. Borland, would you kindly be a little more precise.

The only difficulty with "now" would be if you had a statically declared "now" as well, say in the data segment — then you have a clash. Then it is essential to use "@@" or another name.

Borland has tidied up label scope by using the C-like approach found in MASM version 6 in their TASM version 5.0. See Chapter 13.

Life of Automatic Data

LOCAL directive

I have mentioned TASM's apparently anomalous handling of local symbols. LOCAL data of course exists on the stack and is created on entry to a procedure and destroyed on exit.

However, it will still be in existence at lower level nested procedures. That is, LOCAL data declared at the beginning of procedure A will be "accessable" by procedure B if procedure B is nested within (called from within) procedure A.

What I'm talking about here is the availability of the data, not the scope of the symbols. Scope is discussed above, and varies with MASM and TASM.

There is no problem with the availability of procedure A's data conceptually, because procedure B will be using the stack further down in memory. The stack grows with a procedure call and shrinks upon exit. What is not so apparent is that any LOCAL data declared in WinMain() is also available in a callback. A callback is not what you immediately think of as being nested within WinMain(), but it is, even though the call to it has gone via DispatchMessage() and Windows. See Figure 5.2:

Figure 5.2: Accessing WinMain() local data.

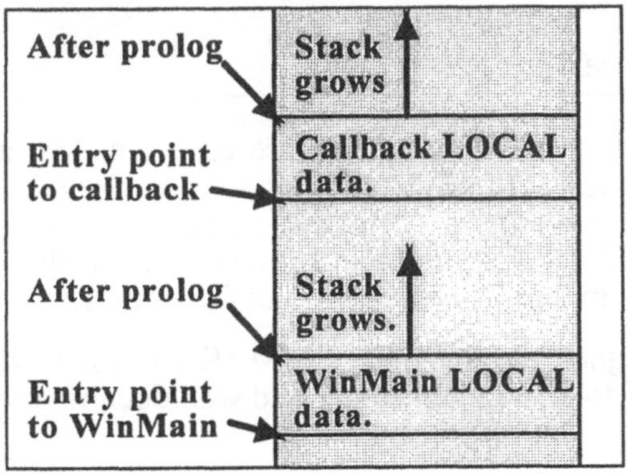

Assembling and Linking

That's it — a complete assembly language program! Of course, you do need those other files to assemble and link. I've listed them here for your convenience. Note that the Make file is designed for Microsoft's NMAKE.EXE, but you should be able to get it going with other Make programs. I prefer NMAKE, even for "making" Borland code. What follows is particular to TASM. Look at the previous chapter for specifics on .RC, .DEF, and .MAK files for Microsoft.

.RC file   This is the WINHULLO.RC file (in \SKELETN2). Nothing new here ...

```
//these (arbitrary) equates could have been in an Include
//file...
#define     IDM_QUIT    200
```

```
#define         IDM_ABOUT  201
winhullo        MENU
  BEGIN
    POPUP       "File"
    BEGIN       MENUITEM "Quit",   IDM_QUIT
                MENUITEM "About...",  IDM_ABOUT
    END
  END
```

.DEF file   Now for the definition file ...

```
NAME            WINHULLO
DESCRIPTION     'Demo ASM hullo program'
EXETYPE         WINDOWS
STUB            'WINSTUB.EXE'
CODE            PRELOAD MOVEABLE
DATA            PRELOAD MOVEABLE MULTIPLE
HEAPSIZE        1024
STACKSIZE       8192
EXPORTS         WINHULLOPROC
```

Note that some LINK programs are case sensitive and some are not. Borland's TLINK.EXE prior to version 2.0 is not, while version 2.0 onwards is. This did create some problems for me, when upgrading. One problem I found was that the line "NAME = WINHULLO" in the .DEF file had to be in capitals.

Borland   I have designed the WINHULLO.MAK file for Borland's TASM
Make file   and TLINK to be comprehensive and well documented.

```
# NOTE this Make file has been modified for Borland C++,
# to be used with TASM and TLINK, however I'm still using
# Microsoft's NMAKE.

# To run this file:    NMAKE WINHULLO.MAK

fn = winhullo
all:$(fn).exe

lpath = \borlandc\lib          #path for libraries
ipath = \borlandc\include      #path for Include files.
epath = \borlandc\bin          #path for EXEs.
sw = /c /n /v /Tw /L$(lpath)    #switches for tlink.
# /n =ignore-default-libs, /Tw =generate Windows exe,
# /L$(lpath) = lib path,   /v =debug-on.
# Note that these paths all assume you are in the same
# drive.

$(fn).obj : $(fn).asm
    tasm /zi $(fn);
```

```
# -r =dont append to exe, -x =dont look in INCLUDE
# environment-variable for incl-files, -i = look in this
# path instead....
$(fn).res : $(fn).rc
      rc -r -x -i$(ipath) $(fn).rc

# c0ws=Windows small start-up-lib, cwins=Windows small
# Standard run-time-library, cs= Standard-run-time lib,
# import=access to Windows built-in library functions.
$(fn).exe : $(fn).obj $(fn).def $(fn).res
tlink $(sw) c0ws $(fn),$(fn),$(fn),import cwins cs,$(fn).def
      rc /30 -x -i$(ipath) $(fn).res

# Note that Borland C++ v2.5 names the Windows library
# CWINS.LIB, while v3.0 names it CWS.LIB.  The BorlandC
# run-time library is CS.LIB, which could be placed
# immediately after CWS, if you need it.  Note that the
# "S" postfix designates the small model.
# Note that if you use the tools from the SDK3.1, such as
# RC.EXE, and you want your program to work with both
# Windows 3.0 and 3.1, put "/30" in second execution of
# RC, as shown.  Also your source program should have
#"WINVER = 0x0300" before the WINDOWS.INC is included, if
# using a WINDOWS.INC derived from WINDOWS.H supplied
# with SDK 3.1.
# (my WINDOWS.INC on the Companion Disk is derived from
# SDK 3.0, which avoids certain problems.
```

When you type this in, there is no need to put in all the comments, but be careful about unnecessary blank lines, and leave a blank line where I have put one. There are certain syntax requirements that can be messed up otherwise. Note that it's on the Companion Disk (\SKELETN2), to save you all that trouble.

The .MAK file shows where it expects all of the files to be located, but you can make changes as necessary. You could even put everything in the one directory, as I suggested, as a quick and dirty option for Microsoft, if the SDK isn't installed (see page 86). Ditto, you could do this with the Borland files, but if you have the complete distribution disks, then why not install properly, in which case the files will load into the above directories by default.

MASM6 versus TASM

MASM version 6.00 is a curious beastie. It was in some respects a disappointment, especially as I acquired it soon after obtaining TASM v3.0 and had been spoilt by the wonderful new features Borland had incorporated into their assembler.

However, while MASM may not be OOP-aware, it does make ground in other ways. Hmmm — MASM v6.0 isn't even

Windows-aware — but its C-like features did (in time) endear themselves to me. Note however, that Microsoft did make MASM much more Windows-aware with the release of version 6.1.

MASM v6.00 is a stand-alone product and as such cannot be used for Windows development (again, corrected by 6.1). The user's manual has barely a line or two on Windows (the documentation for 6.1 is also a disappointment in this regard). The product (6.0) is for DOS and OS/2 development, and those interested in Windows programming are referred to the SDK.

Obviously, if you have v6.0, upgrade it! Note however that you still need the SDK with MASM v6.1. To quote from the *Microsoft Programmer's Guide*, supplied with MASM v6.1 (page 258):

> "MASM 6.1 does not provide all the tools required to create a standalone program for Windows. To create such a program, called an "application," you must use tools in the Windows SDK."

CMACROS Include file

The required tools, such as resource compiler and libraries, aren't there, so you need the SDK or one of Microsoft's recent Windows-aware higher level languages. The 6.x package does have CMACROS.INC, which is required for Windows development, but it is not described in the manual. Once again, the SDK is the place to look.

WINDOWS Qualifier

TASM's WINDOWS qualifier

TASM v3.x (TASM3) has extended the syntax of the language in a very systematic manner, notably with the extended PROC and CALL, and most especially the WINDOWS qualifier. Curiously, Microsoft has only gone partway along that road, with an extended PROC, very much like TASM's, but no WINDOWS qualifier. Microsoft doesn't have an extended CALL either but has opted for something else called *INVOKE*, which is really an extended call.

MASM v6.x's (MASM6's) lack of a WINDOWS qualifier for PROC means that declaring a callback PROC becomes an absolute pain. Rather than resort to CMACROS.INC and PROLOGUE.INC to insert the required prolog and epilog code, how much simpler it would have been if Microsoft had thought ahead just that little bit further.

There is a mechanism, using "OPTION PROLOGUE" and "OPTION EPILOGUE" directives, to overrride the default prolog and epilog, and there is a method for suppressing default prolog and epilog, as well as a method for restoring the default.

The issue of prolog and epilog have become more streamlined with 32-bit applications, requiring only the STDCALL qualifier — see page 78.

MASM6 skeleton program

Anyway, I guess we have to "go with the flow", so the 16-bit skeleton program on the Companion Disk, in directory \ASMDEMO2, and listed at the end of this chapter, uses CMACROS.INC. Note that you can't see it explicitly included in the listing, as that is done indirectly by the WINDOWS.INC file.

The Companion Disk also has a similar program in \ASMDEMO1, which has the startup code in the program, rather than as a separate module.

Prototypes

The program at the end of this chapter can be assembled under MASM6, or more correctly, ML. As the product is not terribly compatible with earlier MASMs, Microsoft has renamed it, though you do have the option of switching on compatibility with version 5.1.

MASM6 has developed features that make it look more like C, most notably the use of *prototypes*. These are skeleton declarations of a procedure, which you place at the beginning of the file, and are used by the assembler for syntax and type checking. These are best illustrated by an example, and an excellent example presents itself in the use of INVOKE.

Borland introduced the equivalent with TASM5, though they have given it a different name: *PROCDESC*. See Chapter 13, page 308.

INVOKE high-level call

MASM6 CALL is definitely low level, so to call Windows functions in the convenient manner that we have become accustomed to in this chapter, we need to use INVOKE instead.

In fact, TASM3's high-level CALL is quite primitive alongside the sophisticated INVOKE, as you'll see.

The first line shows the call to MESSAGEBOX() as we would do it with TASM:

```
;TASM 3.00 high-level call...
   call MESSAGEBOX PASCAL, hwnd, ds,OFFSET szabout, ds, \
                                OFFSET sztitle, MB_OK
;MASM 6.00 high-level call...
   INVOKE MESSAGEBOX, hwnd, ADDR szabout, ADDR sztitle,\
                                MB_OK
```

PROTO declaration

INVOKE does the same job! However if you put it in as shown, it won't work, because something else is required — the prototype.

You can only use INVOKE to call a procedure that has a PROTO declaration, even if the procedure is external, as in the case of Windows functions.

Previously I used EXTRN to declare MESSAGEBOX() as external, and that is still recognized by MASM, but PROTO can be used to replace EXTRN. So, for each and every Windows EXTRN declaration, replace with PROTO, as shown:

```
;TASM (and MASM) external declaration...
  EXTRN MESSAGEBOX:FAR
;MASM 6.00 prototype for INVOKE...
MESSAGEBOX PROTO FAR PASCAL, :HWND, :LPSTR, :LPSTR, :WORD
```

You should find the syntax of PROTO to be self-explanatory. The parameters have to be declared, with their types, and can have arbitrary (or no) names. You can also get away with declaring all types as WORD (16 bits) rather than more specific 16-bit types such as HWND or .BOOLEAN.

Actually, it's not PROTO itself that replaces EXTRN, rather INVOKE defaults to external, in line with C's default behaviour.

Passing 32-bit values

Notice how I passed the FAR address (long pointer) in TASM compared with ML. In the PROTO declaration above, you can see the data type HWND, which is 16 bits, but LPSTR is a 32-bit value (Long Pointer to STRing). With TASM, I passed the segment:offset as two separate items (though it is possible to declare a 32-bit pointer), but this will cause an assembly error with ML, due to a clash with the PROTO declaration.

ADDR, ::

The idea behind this is the extra safety checking that high-level programmers are accustomed to. ML has two very convenient mechanisms for defining a 32-bit parameter. *ADDR* is a directive that will pass the NEAR or FAR address as appropriate. The other mechanism is where we pass a FAR address in two registers. In the skeleton program you see this done often: in ML we combine them with double-colons, for example: "ss::bx". You can see this in action on the Companion Disk and the listing at end of this chapter.

.H to .INC convertor

Microsoft has put a lot of thought into making MASM6 behave like C, despite a very different syntax. There is a utility called H2INC.EXE, that will convert a C Include file (.H) into an assembler Include file (.INC). Most importantly, if used on WINDOWS.H, it will produce the prototypes for the Windows functions, so we don't have to type them in. This WINDOWS.INC is peculiar to MASM6 and don't expect it to be usable by TASM. The reverse is ok however — MASM6 happily reads the

WINDOW.INC that Borland supplies with TASM (and licenced from Microsoft actually).

I used H2INC to generate a WINDOWS.INC for the example program, though note that I had to edit it somewhat (the .INC file) to get it to work with my assembly program.

Callback Design

So, you can very happily go through the earlier TASM program of this chapter replacing CALLs with INVOKEs and EXTRNs with PROTOs (or delete the EXTRNs entirely). However, a major problem still exists: the lack of a WINDOWS qualifer for PROC.

CMACROS.-
INC
This means that you cannot have a high-level PROC declaration for the callback function, and you have to resort to a macro or insert the prolog and epilog code manually. The demo program uses CMACROS.INC to achieve this.

Without the high-level PROC you can't have the LOCAL directive, for convenient creation of data on the stack (CMACROS.INC gets us around this problem).

Because a program isn't going to have too many callbacks, it's not a total disaster, just a nuisance, if you don't want to use CMACROS.INC. The manual approach is to insert the code as follows:

```
push ds                 ;Prolog code for callback function.
pop  ax                 ;  /
nop                     ;  /
inc  bp                 ;  /
push bp                 ;  /
mov  bp,sp              ;  /
push ds                 ;  /
mov  ds,ax              ;  /
...
dec  bp                 ;Epilog code
dec  bp                 ;  /
mov  sp,bp              ;  /
pop  ds                 ;  /
pop  bp                 ;  /
dec  bp                 ;  /
ret  10                 ;  /
```

You can have local data on the stack, but you will have to manipulate the stack directly. To get at all of the data on the stack segment, you could do something like this:

```
hwnd     TEXTEQU <WORD PTR [bp+14]>
msgtype  TEXTEQU <WORD PTR [bp+12]>
wparam   TEXTEQU <WORD PTR [bp+10]>
```

```
lparam     TEXTEQU <DWORD PTR [bp+6]>
dummy      TEXTEQU <WORD PTR [bp-10]>
hdc        TEXTEQU <WORD PTR [bp-12]>
s3         TEXTEQU <[bp- (12 + SIZE PAINTSTRUCT)]>
   LEA  sp,s3                      ;move stack to free region.
```

TEXTEQU,
EQU

Something along these lines will give you access to the labels within the callback. Note that I used TEXTEQU, because EQU cannot be used for text aliasing with ML, a major divergence from earlier MASMs and from TASM. Notice that the text to be aliased must be inside "< >".

Anyway, this is academic.

Other Incompatibilities

PROC syntax
differences

So what else needs changing ...

There is a slight difference in the syntax of the high-level PROC directive. If you look back to the declaration for WINMAIN, you will see that it started like this: "WINMAIN PROC PASCAL NEAR hinstance:WORD .... ". With MASM6 it has to be rearranged like this: "WINMAIN PROC NEAR PASCAL, hinstance:WORD .... ".

With TASM5, Borland has allowed MASM high-level PROC syntax.

STRUC
assembler
differences

I also came across an interesting problem with fields of a structure. Incidentally, MASM6 allows nested structures, which previous versions didn't. Nesting is vital for OOP, though MASM6 is still nothing like OOP-aware.

The problem is that the program of this chapter has a couple of lines like this:

```
;where clslpszmenuname is a 32-bit field of
;structure-instance s1..
   mov  WORD PTR s1.clslpszmenuname,ax
   mov  WORD PTR s1.clslpszmenuname+2,ds
```

I loaded each half of the field with separate instructions. MASM objects to a field being accessed in halves, so I had to resort to:

```
;MASM6's solution ...
   lea  di, s1.clslpszmenuname
   mov  [di],ax
   mov  [di+2],ds
```

Oh, and make sure that your callback procedure name is all capital letters.

Label scope differences

Another major difference is in the *scope* of labels. I have covered that topic beginning on page 120. This is one aspect of MASM6's move toward the code integrity we expect from a high-level compiler. Prototyping is another. I think that many serious programmers will choose MASM on this basis, and it is an area where Borland had to play catch up, with TASM5.

Which one?

All of these comments are, of course, my personal opinion, not the final truth engraved in stone, and I suggest that potential buyers consider most carefully what features are most important. Have a look at reviews in magazines. If OOP is your thing, then look closely at TASM. Do bear in mind that my comments are based upon particular versions, and even "maintenance releases" of the same version number can have significant improvements. Therefore, take all of the above comments with a pinch of salt, and check out the features for yourself before buying.

One interesting point is that MASM6 comes with Programmer's Workbench (PWB), an editor and IDE, as well as CodeView debugger. Borland does not provide an editor or IDE, but the Turbo Debugger is very nice.

I have made some further comments on this on page 309.

MASM Assembling and Linking

Resource compiling and linking are as per Chapter 4, though you can use the Borland TLINK and C0WS.OBJ, CWS.LIB and IMPORT.LIB library files, instead of the Microsoft equivalents (if you wish).

You only need to change one line in the Make file, that of the assembly process:

```
ML   /c   $(fn).ASM
```

where /c suppresses linking (ML normally automatically invokes the linker), and $(fn) is the macro for the filename.

If you want debugging information so the source file can be traced by the Codeview debugger, then you will need the /Zi switch and you will need to use Microsoft's LINK, with the /CO switch— the example Make file on the Companion Disk to show this is named MASM60.MAK.

This is the command line I used to generate WINDOWS.INC from WINDOWS.H:

```
H2INC   /C /Gc   WINDOWS.H . . .
```

MASM6 Program Listing

```
;ASMDEMO2.ASM --> ASMDEMO2.EXE    Windows demo program.
;This skeleton assembly language program has been written
;for Microsoft MASM (ML.EXE) v6.1.   (Do NOT use
;Borland's TASM!).
;It uses PROLOGUE.INC to force the correct Windows
;prolog/epilog on all FAR PROCs.
;This program does NOT have the startup code built-in.
;Note that Borland provide startup object module as
;C0WS.OBJ (small model) and Microsoft provide
;APPENTRY.OBJ with v6.1. One of these must be linked.
;Note that APPENTRY.OBJ should be for the small model, to
;suit this program -- if not assemble APPENTRY.ASM, with
;switches as described in APPENTRY.ASM comments.

.MODEL SMALL

WINVER EQU      0300h
?WINPROLOGUE EQU 1   ;forces win prolog/epil on far procs.
INCLUDE winasm60.INC ;this is not the same WINDOWS.INC
                     ;used by the TASM programs. It is
                     ;generated by H2INC.EXE, and contains
                     ;prototypes. Generated by...
                     ;  H2INC /C /Gc WINDOWS.H . . .
IDM_QUIT    EQU 100      ;menu-identifiers -- must be
IDM_ABOUT   EQU 101      ;same as defined in .RC file.

   EXTRN __astart:FAR      ;startup code, in APPENTRY.OBJ
                           ; (referenced at END)
;..................................................
.DATA
szwintitle      DB      'HULLO DEMO PROGRAM',0
szASMDEMOname   DB      'ASMDEMO2',0
hOemFont        DW      0               ;handle to OEM font.
soutstring      DB      'Hullo World'
szaboutstr      DB      'Assembly Language Windows Demo',0
                                        ;messagebox
sztitlestr      DB      'Karda Prints',0        ; /

;..................................................
.CODE
   PUBLIC      WINMAIN
WINMAIN PROC NEAR PASCAL, hInstance:WORD, \
    hPrevInstance:WORD, lpCmdLine:LPSTR, nCmdShow:SWORD
   LOCAL       @hWnd:HWND
   LOCAL       s1:WNDCLASS
   LOCAL       s2:MSG

   cmp hPrevInstance,0          ;=0 if no previous instance.
   je  yes1st
   jmp createwin
yes1st:
;Setup the window class structure for REGISTERCLASS()...
```

```
      mov   s1.Style,3
      lea   di,s1.lpfnwndproc
      mov   [di],OFFSET ASMDEMOPROC
      mov   [di+2],SEG ASMDEMOPROC
      mov   s1.CbClsExtra,0
      mov   s1.CbWndExtra,0
      mov   ax,hInstance
      mov   s1.HInstance,ax

      invoke LOADICON, null, IDI_APPLICATION
      mov   s1.@HIcon,ax

      invoke LOADCURSOR, null, IDC_ARROW
      mov   s1.@HCursor,ax

      mov   s1.hbrBackground,COLOR_BACKGROUND
      mov   ax,OFFSET szASMDEMOname
      lea   di,s1.lpszmenuname
      mov   [di],ax
      mov   [di+2],ds
      lea   di,s1.lpszclassname
      mov   [di],ax
      mov   [di+2],ds

      lea ax,s1
      invoke REGISTERCLASS, ss::ax
      or    ax,ax
      jne   createwin
      jmp   quitwinmain

createwin:
  invoke CREATEWINDOW, ADDR szASMDEMOname, \
                    ADDR szwintitle,  00CF0000h, 150, 0,\
                    400, 300, 0, 0, hInstance, 0
      mov   @hWnd,ax
      invoke SHOWWINDOW, ax,nCmdShow
      invoke UPDATEWINDOW, @hWnd
      jmp   SHORT messageloop ;go to the main message loop.

;This is the main message loop, in which Windows waits
;for messages
mainloop:
      lea   ax,s2
      invoke TRANSLATEMESSAGE, ss::ax
      lea   ax,s2
      invoke DISPATCHMESSAGE, ss::ax
messageloop:
      lea   ax,s2
      invoke GETMESSAGE, ss::ax, null, null, null
      or    ax,ax
      jne   mainloop

;GetMessage() returns FALSE (AX=0) if a "quit" message...
;so here we are quiting....
      mov   ax,s2.WPARAM       ;return wparam to windows OS.
quitwinmain:
      ret
```

```
WINMAIN ENDP
;................................................
ASMDEMOPROC PROTO FAR PASCAL, :HWND, :WORD, :SWORD, \
                                              :SDWORD
ASMDEMOPROC      PROC FAR PASCAL, ihWnd:HWND, \
         iMessage:WORD, iwParam:SWORD, ilParam:SDWORD
   LOCAL         dummy [5]:WORD
   LOCAL         @hDC:HDC
   LOCAL         s3:PAINTSTRUCT

   mov ax,imessage         ;get message-type.
   cmp ax,WM_CREATE ;message received after CreateWindow()
   je   xcreate            ;function is called.
   cmp ax,WM_DESTROY ;message received if a window is
                          ; closed.
   je xquitmessage
   cmp  ax,WM_PAINT  ;message received if Windows has
         ;(already) redrawn any part of the window (due to
                           ;a size-change for example).
   je   xpaint
   cmp  ax,WM_COMMAND ;any selection of the menu will
   jne  notwmcommand       ;produce
   jmp  xmenu              ;this message.
notwmcommand:
   cmp  ax,WM_LBUTTONDOWN ;one of many mouse messages.
   jne  notwmlbutton
   jmp  xbreak
notwmlbutton:
   cmp  ax,WM_CHAR         ;message that a key pressed.
   je xchar

defhandler:
;Default handling of messages....
   invoke DEFWINDOWPROC,ihWnd,imessage,iwParam, ilParam
   jmp  xreturn

;...............................
xcreate:
   invoke GETSTOCKOBJECT,OEM_FIXED_FONT
   mov  hOemFont,ax        ;handle to font.
   jmp  xbreak

xquitmessage:
   invoke POSTQUITMESSAGE,0
   jmp  xbreak

xchar:
   jmp  xbreak

xpaint:
   lea  ax,s3   ;ss:ax -- far-addr of paint-structure.
   invoke BEGINPAINT,ihWnd,ss::ax
   mov  @hDC,ax       ;hDC -- display-context, required
                       ;before can output to screen.

;For this simple demo, any redraw of the Window will
;cause output of our "hullo world" string....
```

```
    invoke SELECTOBJECT,ax,hOemFont
    invoke TEXTOUT,@hDC,10,20, ADDR soutstring,11
    lea   ax,s3        ; -- far-addr of paint-structure.
    invoke ENDPAINT,ihWnd,ss::ax
    jmp   SHORT xbreak
;. . . . . . . . . . . . . . . . . . . .
xmenu:
    cmp   WORD PTR ilParam,0      ;low-half of lParam
    jne   xbreak                  ;test if a menu-message.
    cmp   iwParam,IDM_QUIT        ;wParam.
    jne   notquit
    jmp   xquitmessage
notquit:
    cmp   iwParam,IDM_ABOUT
    jne   xbreak                  ;no other menu items.
;let's put up a message about this program...
    invoke MESSAGEBOX, ihWnd, ADDR szaboutstr, \
                       ADDR sztitlestr, MB_OK

;. . . . . . . . . . . . . . . . . . . . . .
xbreak:
    sub   ax,ax   ;returns 0 in DX:AX.  (callback functions
    cwd                           ;return a 32-bit (long) value).
xreturn:
    ret
ASMDEMOPROC    ENDP
;. . . . . . . . . . . . . . . . . . . . . . . . . . . . . . . . . .

    END __astart         ;name of startup code.
```

Run-time ***.IF/*** ***.ELSEIF/*** ***.ELSE***	Here is an exercise. Locate the above program in \ASMDEMO2 on the Companion Disk, and assemble and link it. When you have succeeded, have a go at modifying the code with something wonderful available in MASM6. Borland did not catch up with this capability until TASM5.

```
    .IF ax==WM_CREATE        ;*Runtime* IF/ELSEIF/ELSE
      ...                    ;(note that nesting is allowed).
    .ELSEIF ax==WM_PAINT
      ...
    .ELSEIF ax==WM_DESTROY
      ...
      ...
    .ELSE
      ...
    .ENDIF
```

If you can't quite see how to use this, look at the skeleton in Chapter 13.

Run-time high-level IF/ELSEIF/ELSE constructs tidy up your assembly code enormously, and I'm hooked on it. Note that it assumes nothing and does not change any register values. This means that you can jump out from anywhere and jump around inside, like this:

```
 .IF ax==0
    jmp place1         ;goto anywhere, quite legal.
 .ELSEIF ax==1
place1:
  .ELSEIF ax==2
    jmp place2
 .ENDIF
place2:
```

Your mission, should you decide to accept it, is to introduce the high-level decision constructs to the above example program.

You will also find other high-level constructs in the MASM6 and TASM5 manuals, such as DO/WHILE.

6

Program Design

Preamble

This chapter is about interfacing assembly language with C and C++ and about one aspect of program design that is an outcome of the interface with C++ — objects. I have not gone into any general methodology of software design.

History of OOP and assembly

Programmers are migrating from C to C++. Ditto with other languages, and of course the new kid on the block is Java. You have got to think in terms of objects. Early in 1991 I put a lot of thought into object oriented assembly language, including the presentation of a paper.

I developed techniques for OOP, but found the assemblers of that time to be somewhat inadequate. So about mid-1991, I wrote to Borland in the USA explaining in detail what was wrong with their assembler and what it needed to be able to handle objects. Then, in February 1992, I was fascinated to learn that Borland had released a new assembler that they advertized as "object oriented". I like to think that I was one of their inspirations.

A rationale for OOP

Why should you even bother with objects when programming at the assembly level? The answer is very simple:

1. To interface with OO languages such as C++.

2. To "improve" the development and maintenance of the assembly language code.

The much-touted advantages of OOP also apply to assembly language. Do you want reusable and maintainable code? Do you want to program faster and debug faster? Then go for objects.

What is OOP?

In a nutshell, OOP is just the use of structures. In C++ the STRUCT declaration is almost exactly the same as CLASS. The reason is very simple: a class is only a structure (with some bells and whistles!). Look back to page 65 to clarify what structures are and how they are used, and you've already grasped the principle of objects. Objects are just instances of a structure, or the actual copies of the structure that are created. In Chapter 5, I used structures in a skeleton program.

OOP *terminology* is what confuses everything.

In the second half of this chapter I have shown the impressive power of OOP when applied to assembly language, but for now you need to know a few basics ...

Object Addressing

Windows assembly, generic Windows, 386/486 architecture, OOP, C++!

C++ has a lot of terminology that can be very intimidating. Yet the underlying concepts are quite simple.

It is also quite true that you can read an explanation from a C++ manual or textbook a dozen times, and not fully grasp it. But if you were to see how that concept is implemented at the assembly level, it would become clear.

This is one reason why I am in favour of this book being used as a prerequisite, not just to Windows, but also to C++ programming.

Class

The way we write a program using OO techniques is by grouping data and code that naturally belongs together into a *class* (structure definition). A structure need not contain just data; it can also have pointers as fields in the structure (or a pointer to another structure of pointers), and this is one of the key features of the OO technique.

Calling a Function

Object = instance

With C++ there are objects, and a procedure or function (now called a *function-member* or *method*) is part of a class. The objects are instances of a class. Data is also part of the class. An instance is a complete copy of the class, with possible unique initialisations, created in memory.

For now we will focus on just one implication of this: how functions are called.

CALLing a function member (method)

After all, that's something we want to do all the time while writing a program. A simple CALL instruction is what we are familiar with, and of course, as you saw in Chapter 5, there are high-level qualifiers for calling Pascal or C procedures/functions and for passing parameters. This simplifies the stack manipulation, but now, with procedures that are part of a class, we have something more to consider.

Say that you have a procedure in a program, and for argument's sake give it a name: TEXTOUT(). Also say that it uses the Pascal stack-handling convention, for no other reason than consistency, since the external windows functions do.

Our problem is that we want to call this function from somewhere else in the program. No problem, you think: just do this:

```
call    TEXTOUT    PASCAL,param1,param2
```

(Assume also that it requires two parameters.)

Yes, this will work, or at least will get execution to the TEXTOUT routine, but there are other factors to consider ...

- THIS
- Polymorphism

Object pointer

"THIS" is a keyword in Borland assembly language and C++. It is just an equate:

THIS = address of current object.

Borland C++ often uses the SI register to hold THIS. Generally, an "object pointer" points to the current object or whatever object we wish to deal with.

32-bit coding

A little note on the side that will help as you study this chapter. The use of SI to hold THIS applies to 16-bit code. For 32-bit code, it becomes ESI. Quite simple. In general, convert any examples in this chapter to 32-bit code by prefixing the registers with "E". Any reference to FAR pointers may not be relevant because the 32-bit addressing can address the whole 4.3G with just the offset.

Also, when writing 32-bit Windows applications, use the STDCALL language qualifier (see .MODEL on page 111), not PASCAL. STDCALL convention is that parameters are pushed right to left, with stack cleanup in the called function.

"Polymorphism" means that TEXTOUT() can in fact be many different routines, all with the same name.

At this point some code will help:

```
.DATA
WINDOW STRUC
   active   DB  0                  ;example data-member.
   TEXTOUT  DW  textoutmain        ;example function-member.
   ...
WINDOW ENDS
...
WINCLASSA STRUC         ;sub-class of WINDOW.
   WINDOW   < >         ;Inherits everything
WINCLASSA ENDS          ;from WINDOW.
...
WINCLASSB   STRUC       ;Ditto, but a function override.
   WINDOW   < , textoutdlg >
WINCLASSB   ENDS
...
;creating instances ...
window1 WINCLASSA    < >
window2 WINCLASSB    < >
window3 WINCLASSA    < >  ; etc ....
...
.CODE
   lea  si,window1
   call textoutmain PASCAL,par1,par2,si
   call [si].TEXTOUT PASCAL,par1,par2,si
   ...
   lea  si,window2
   call textoutdlg PASCAL,par1,par2,si
   call [si].TEXTOUT PASCAL,par1,par2,si
   ...
textoutmain  PROC PASCAL p1,p2,THIS
   ...   ;this is the textoutmain procedure ... etc...
```

An object combines code and data

Further down in the code you would have to have the two procedures: textoutmain() and textoutdlg().

Look very carefully at the above listing. First I defined a class (structure) called "WINDOW", with a data-member "active" and a function-member "TEXTOUT". The latter is a pointer to a procedure[1] called "textoutmain".[2]

---

[1] The purists are probably very unhappy with my interspersion of the words "procedure" and "function" as though they mean the same thing. For our purposes they do. So there!

[2] Most assemblers do not let you put a forward-reference into a structure field. It must be done when the instances are created. In this example, "textoutmain" would have to be placed in the "< >" portion of each instance-declaration. This is messy. TASM v3.0 is the first truly object oriented assembler, and has a mechanism for allowing forward-references, as shown in the second half of this chapter.

TEXTOUT could be a routine that sends text to a window, but there could be many such routines designed for different output mediums.   In this case I have arbitrarily created a class, WINCLASSB, that overrides the pointer with textoutdlg(), while WINCLASSA does not.

The polymorphic principle

The key point here is that I can call TEXTOUT, but because it is a pointer, the actual routine that gets called depends upon what is stored in that field.   In the case of instance window1 it is textoutmain(), and in the case of window2 it is textoutdlg().

You could imagine two windows of different types on the screen, requiring different textout routines.   C++ uses THIS to specify which instance (object) is currently being referenced.

Each sub-class (and indeed each instance of a class) can have its own TEXTOUT function, so our code must be able to distinguish. Look again at the above listing to see how I have done it.

I have disassembled a lot of C++ code to find out what makes it tick.   Borland usually put the value of THIS into SI, which may be worth noting if you have to interface with C++ code.   When coding at the assembly level, we need to think carefully where we want to store THIS, if anywhere at all.

Notice that I also used SI[1] to hold THIS (see the code examples in previous listing).

Object pointer passed on the stack

Whenever Borland C++ calls a function-member, it always passes THIS on the stack (last parameter), so that the called function knows which object it is dealing with.

Notice that in the PROC declaration, I gave the passed THIS parameter the same name — in practise you would have to use a different name, because the assembler will object to one of its keywords being used as a label.

Early Binding

The first call in the above listing is an example of *early binding*. Why?   Because I have hard-coded the address of the function I want to call into the CALL instruction, in this case textoutmain().

---

[1]   A warning here, though, is that if your instances are LOCAL and if you use a memory model in which data and stack segments are different, then there are potential problems with using SI.   A memory access to the stack segment requires BP-relative addressing or an SS: override if using SI.

It is possible for the object to be located in some other segment entirely, and in that case THIS would have to equate to a FAR address, such as ES:[SI].   This comment does *not* apply to 32-bit programming, which uses a FLAT memory model in which there is only one segment.

This will be an immediate-mode instruction and is fast, but it is a **deviation** from "pure" OO principles.

C++ will normally compile a C++ program into calls having early binding, except for the case where the call is to take polymorphism into account.

Look at the rest of the line. I passed two parameters, arbitrarily named "par1" and "par2". At the end I passed the address of the object that is to be acted upon. Further down in the actual code for textoutmain(), see how I used a variable THIS to receive that address. This is important: we must always pass the address of the object to the function.

Late Binding

The second call in the above listing (page 140) is an example of late binding. The meaning of this is "call the TEXTOUT function in the instance window1". Another way of writing it is:

```
call [OFFSET_window1 + TEXTOUT]
```

This is non-immediate and will call the function pointed to at offset_window1 + TEXTOUT, which in this case is textoutmain().

The end result is the same as for early binding, except that this one call instruction will call whatever TEXTOUT function we want, simply by setting SI appropriately beforehand.

```
lea  si, window2
call window2.TEXTOUT   PASCAL,par1,par2,si
call [si].TEXTOUT   PASCAL,par1,par2,si
```

This code calls textoutdlg(). The last two lines are actually the same, due to the way in which the assembler treats the window2 label in this context, but I recommend that you stick with the latter to avoid confusion. THIS passed on the stack must always be the register, not the label, so be consistent and use SI in both places.

This implements polymorphism.

C++ Binding

Examine this C++ code ...

```
class WINDOW    //Everything here does exactly the same
    {           //as the assembly language on page 140.
      public:
        int active;
        virtual void TEXTOUT ( int,int );
```

```
                // Define any other members here ...
      }
void WINDOW :: TEXTOUT( int param1, int param2 )
      { // actual code for function here.  This function is
        // the equivalent of textoutmain() in the assembly
        // listing.
      }
      ...
class  WINCLASSA : WINDOW     // subclass of WINDOW.
      {                  //(inherits active and TEXTOUT members).
      }
class  WINCLASSB : WINDOW      // override TEXTOUT()
      { void TEXTOUT (int param1, int param2 )
      }
void WINCLASSB :: TEXTOUT(int param1, int param2)
      { //actual code for function here. This function is
        // the equivalent of textoutdlg() in the assembly
        // listing.
      }
      ...
//create instances ...
WINCLASSA       window1, window3;
WINCLASSB       window2;
   ...
main ()
{
//code example of early binding ...
   ...
   window1.TEXTOUT( value1 , value2 );
   ...
//code example of late binding ...
   WINDOW  *ptr;
   ptr = &window1;
   ptr -> TEXTOUT ( value1 , value2 );
   ...
}
```

" :: "
operator

The program starts by declaring a class called WINDOW and the data and function members it has. I only put in two members: active and TEXTOUT. After that I put in the actual code for TEXTOUT. Notice the syntax for doing this — the "::" means that this function belongs to the class named to its left, which is WINDOW.

Subclassing
with override

Because I wanted this code to do exactly what the assembly listing does (page 140), I created two subclasses — WINCLASSA and WINCLASSB.   WINCLASSA is identical in every way to WINDOW, but in WINCLASSB I have overridden TEXTOUT. Notice that I didn't have to give the new procedure a different name.

Then I declared three static instances (permanently in the data segment). I could have made them automatic simply by moving them down into main().

" -> "
object
pointer

The code within main() shows how easy it is to call the function associated with a particular object. A call to this function means that TEXTOUT() will execute but will automatically work on the data and functions that are part of the referenced object. This is because the THIS pointer is passed on the stack (see page 140).

The example of late binding may look rather complicated. "ptr" is a label that is a pointer to data of type WINDOW. The "*" simply declares that it is a pointer. The data type tells C++ that ptr can only be used to address objects (instances) of WINDOW.

The next line sets ptr to point to window1.

The following line uses ptr to call window1.TEXTOUT(). This line corresponds exactly with the assembly language code:

```
call [si].TEXTOUT   PASCAL,value1,value2,si
```

(I have used the PASCAL qualifier here, rather than C, for consistency with later examples. It does cause some differences, such as reversed order of stack pushing and stack clean-up. For more specific details see ahead to the section "Interfacing With C++" on page 147.)

Compiler
optimisation

It is interesting to analyse how the compiler decides whether to compile early or late binding. When the compiler sees that the call is fixed (that is, to a particular routine) and will not change at run-time, it optimises and compiles early binding. Note that any function that is to be called by late binding must be declared as "virtual" in the C++ source code, but such a declaration does not mean that the compiler will do so.

The compiler will compile a call using late binding if the function is virtual, and if the call involves THIS as a pointer. The call immediately above is an example in which THIS is contained in SI, so its value is not actually known at assembly-time. Therefore late binding is required.

In my assembly language example I gave window1 and window2 different routines for TEXTOUT ...

Assembly Language Binding

Manual optimisation

Binding has been discussed over the previous few pages; however, further clarification is in order.

In assembly language, we have full control over whether to use early or late binding, since we don't have a compiler to make such a decision for us. Look back once more to the listing on page 140.

The example of a call to textoutmain() by early binding (the call immediately after the LEA instruction) is ok, because SI will always be the same when execution reaches the CALL instruction.

However, what if the code has multiple entry points to the CALL?

```
    . . .
    lea    si,window1
    jmp    redraw
    . . .
    lea    si,window2
    jmp    redraw
    . . .
redraw:
    call   [si].TEXTOUT PASCAL,x,y,si
    . . .
```

In this case you **must** do a late-binding call, because the SI value can have different values at execution-time. The CALL will automatically call the correct routine.

Use of THIS

THIS is a pointer to the current object and is already introduced and discussed at length earlier in this chapter. However, this section will consider the rules of usage of THIS.

I have explained how SI is passed on the stack to the function. Why pass it on the stack, since SI will be the same value upon entry to the function anyway and can thus be accessed from the register?

C++ does it that way, but your assembly program doesn't necessarily have to. However, it may be wise to stick with C++ conventions to enable smooth linking with C++ code.

Accessing the right data

You can see back in the class definition for WINDOW (page 140) that I put an example data field labelled "active". Perhaps this is a flag indicating whether this is the active window or not — whatever.

Obviously the instances "window1" and "window2" will have their own copies of "active", so the TEXTOUT function must access the "active" field in its own instance.

Thus if you have:

```
lea  si,window1
call [si].TEXTOUT PASCAL,x,y,si
```

SI would be passed to the function to let it know which object to communicate with. For there is a general rule with OOP:

**A Function should never access a
data-member of another object.**

***Encapsulation
of data within
an object***
A function should only write to (and even only read from) data-members of the current instance, as pointed to by THIS.

In OOP terminology, this is the principle of *encapsulation*. The data belonging to a particular object should only be accessed by functions belonging to that object, and only if THIS is set to that object. C++ does allow you to get around this, but think of it as the ideal to be aimed for.

In assembly language you can break all the rules, but you should try not to write OO code that accesses data belonging to other objects. If your function wants to access some data elsewhere, the proper way to do it is to change THIS to that object and then call a function that is part of that object. If no such function exists, then you will have to write one.

***Structure of a
function
-member***
Referring back to our earlier call to TEXTOUT, with THIS set to window1, the actual procedure called will be textoutmain(), which could have the following structure:

```
textoutmain    PROC PASCAL,x,y,now
  mov  si,now
  ...
  mov  al,[si].active
  ...
  lea  bx,window3
  call [bx].TEXTOUT PASCAL,x,y,bx
  ...
  ret
textoutmain    ENDP
```

This skeleton shows how data-member "active" is accessed. Since the data of any other object should not normally be accessed directly, I have put in some code to show how to change to another object and then call a function belonging to that object. Upon return, SI is still set to this function's current object.

Examine this code and you may be surprised. What function is actually being called by the CALL, and why is it ok? That's for you to think about.

Interfacing with C++

Pascal versus C++ stack handling

Although I have standardized on the Pascal calling convention for most of this book, for compatibility with Windows functions, standard C handles the stack somewhat differently. It is not something that will cause much trouble, since you can take care of everything by use of the high-level CALL instruction and PROC directive.

However, in the case of passing THIS to the called routine, you will need to know whether it is pushed on first or last.

With Pascal, the high-level CALL pushes the parameters on in the order in which they are listed; that is, the leftmost one first. The high-level procedure that is being called will automatically remove the parameters from the stack before returning to the calling level.

With C, the high-level CALL pushes the parameters onto the stack in the reverse order, so the leftmost one gets pushed on last. The called procedure does not clean up the stack before returning, and the parameters must be removed from the stack after return to the calling level. Code for the latter operation is generated automatically by the high-level CALL by the "C" qualifier.

C++ also pushes THIS onto the stack last in the case of calls to function members. Thus your PROC declaration will need to show THIS as the leftmost parameter if it is called as a function-member from C++ code.

Fortunately, there is an easy way to figure out the interfacing requirements between C++ (or C, or any other language) and assembly language, and that is to utilize the compiler's ability to generate assembly language output.

Compiling to ASM O/P

Most high-level compilers will do this by means of a switch on the command line. The compiler will produce an assembly language listing of the C program, showing the exact correspondence of lines of C to the equivalent assembly code. This is highly educational, but it is particularly useful for linking between C and assembly.

**Compile a
.ASM stub**

The trick is to write the assembly language module into the C program in the form of a stub or skeleton. That is, it won't do anything except have the data transfer C instructions. Compile it, and look at the assembly listing for that routine. Extract that routine into a separate assembly language file, and delete the original stub.

**Name
-mangling**

This works fine and is surprisingly easy to do. The method overcomes some serious hurdles, especially that of *name-mangling*. It is a C++ feature that the source code can have the same name for different functions, and other labels can also have identical names. The compiler gets around this problem by "mangling" the labels — applying an algorithm so that even labels of the same name will have new unique names. The problem is that, if you are writing an assembly language module that must access labels in C++ modules, you can't reference them by name — you can only reference them by their mangled names.

The only way to know the mangled names is by the stub method described above, because the assembly language output will show all labels in their mangled form.

In-Line Assembly

A completely different approach is not to write the assembly language module as a separate file, but to write it in-line with the C code. You have to have a compiler that supports this, and of those that do, the in-line assembler is not quite so fully-featured as the stand-alone assembler. You lose in one way, but gain in another. What you gain is seamless integration with the C program. You can write the assembly code with full access to the C labels, and the registers that you use are automatically saved and restored by the compiler upon entry to and exit from your assembly module.

Here is a simple example:

```
class  WINDOW
{
  public:
  int  active;
  virtual  void  TEXTOUT (int);
};
void  WINDOW :: TEXTOUT (int x)
{
  asm  mov    si,this
  asm  mov    dx,[si].active //addr relative to DS.
  asm  mov    ah,2
  asm  int    21h
};
```

```
// WINDOW  window1; //static object, in data segment.
main ()
{
  WINDOW window1;    //automatic object, in stack seg.
  window1.active = 07;
  window1.TEXTOUT (0);
}
```

*"_asm"
keyword*

I have shown here how the function-member TEXTOUT(), belonging to class WINDOW, can be written in in-line assembly code with data members fully accessible. THIS is also available to the assembly code, and I have put it into SI for convenient usage.

Note that I preceded each line with the "asm" keyword; however, it is also allowed to have a single "asm" keyword followed by an opening "{" brace and then multiple lines of assembly code not requiring the asm keyword, terminating with a closing "}".

*Static versus
automatic
instances*

I have shown two ways of creating the instance window1. The commented-out example is static, because it is outside main(), while the other is automatic, because it is created on the stack, for the duration of execution within the function.

See how I have addressed the data-member "active" from assembly code. Actually, this is dependent upon memory model and whether the object is static or automatic. For the SMALL (and FLAT) model the SS and DS registers are the same, so there is no problem. For those models in which SS and DS may be different, the code given here would be ok for a static object, but SS override will be required for automatic data. This can be taken care of by using BP instead of SI, since BP by default references the stack segment.

*What the
above
program
"does"*

By the way, the above program passes the value 07 to TEXTOUT(), which sends it to the screen. 07 is the "bell" character, so you get a beep to indicate success.

Although a parameter is passed to TEXTOUT(), I haven't used it within the assembly routine. I put it in to show that it is an option. Note that the compiler will give a warning (at least Borland's BCC compiler does) that the passed parameter is unused.

In-Line Dos and Don'ts

While we are on the topic of in-line assembly, I might as well cover the major do's and don'ts.

I've grouped these below for easy reference:

- The "asm" keyword differs for different compilers. Borland C++ will accept "asm" and the latest version accepts "_asm" for compatibility with Microsoft's C/C++.
- Notice in the example that I chose to use the "asm" keyword at the start of every line, rather than use the "{ }" opening and closing braces. I prefer doing it this way because the in-line assembler cannot define code labels (at least Borland's C/C++ can't). By using the keyword on every line, at the termination of each line the compiler regains control and a label is allowed. For example:

```
    asm   je    place1
    asm nop
place1:
    asm mov   ax,val1
```

- You have complete access to all data and code labels in the C program, barring the usual C constraints.
- Note that the compiler saves and restores some CPU registers upon entry and exit from an in-line assembly section. Compilers differ in what they save and restore.
- You cannot use the ";" (semicolon) to start a comment. Instead you have to use the standard C delimiters. For example:

```
    mov   ax,val1          //moves val1 into AX.
```

- But also note that you do not use the ";" to separate in-line statements, not even the last one.

The ASM Stub

Object pointer

If you refer back to the program listing in the section "C++ Binding", on page 142, you will see the creation of an instance "window1" and the use of a pointer "ptr" to implement late binding.

Recapitulating:

```
WINDOW   *ptr;
ptr = &window1;
ptr -> TEXTOUT (val1, val2);
```

So that you are absolutely clear on what this compiles down to, here is the actual assembly language generated:

```
mov  si,OFFSET window1
mov  ax,val2        ;notice the order of pushing.
push ax
mov  ax,val1
push ax
push si             ;notice that THIS pushed last.
mov  bx,[si+4]      ;4 is the offset of the pointer to
call [bx]           ;TEXTOUT(), in object window1.
```

Looking back again at the code from "C++ Binding" on page 142, you will see the definition of TEXTOUT(). But if TEXTOUT() is to be the assembly language module, you would leave it in the C program for now, as a stub. You would put in the skeleton code, as follows:

```
void WINDOW :: TEXTOUT (int val1, int val2)
{
    int x;
    box1 . draw ( 1,2 );        //member of another object.
    this -> dosomething () ;    //hypothetical function.
    x = active;                 //data of current object.
}
```

Calling a member, current object

This code shows various ways of getting at data. Dosomething() is an example of calling a function-member belonging to the current object, though I haven't actually defined such a function.

"active" is a data-member of "window1" and I have accessed it in the stack. Notice also how I can access functions of other objects.

... and a different object

"box1" is some other object belonging to a different class, say "BOX". The choice here is arbitrary. It has an arbitrary function called draw().

Compile and Assemble Steps

If we use Borland's BCC compiler, the command line to compile to assembly is as follows:

```
BCC  -c -S  filename.CPP
```

Where "-c" suppresses linking and "-S" generates .ASM output. Note that case is important with the switches.

Mangled names

The *filename*.ASM file that you get will not have any high-level assembly language features in it, so you have to look through it and extract the useful information. Then you can put together your own assembly module. It will look something like this:

```
.MODEL SMALL                      ;must match C++ module.
   PUBLIC         @WINDOW@TEXTOUT$qii
   EXTRN          @WINDOW@dosomething$qv:NEAR
   EXTRN          @BOX@draw$qii
   EXTRN          _box1
.DATA
   x  DW 0                 ;local data.
.CODE
@WINDOW@TEXTOUT$qii    PROC   C  now,val1,val2
;
;how to get at the passed parameters ...
   mov  si,now          ;actually at [bp+4]. Addr of window1.
   mov  ax,val1         ;actually at [bp+6]
   mov  bx,val2         ;actually at [bp+8]
;
;to access another function, another object ...
; _box1.draw (1,2) ...
   lea  ax,_box1
   call  @BOX@draw$qii  C  ax,1,2  ;early binding.
;
;to access a function, current object ...
; this -> dosomething () ...
   call  [si].@WINDOW@dosomething$qv  C  si
                  ;late binding.  no other params to pass.
;
;getting at data-member of current object ...
; x = active ...
   mov  ax, [si].0    ;offset is 0, since field is first in
   mov  x,ax          ; object.
;
   ret
@WINDOW@TEXTOUT$qii    ENDP
   END
```

The skeleton program gives you the mangled names and how to access the data and function members. Then you can go ahead and flesh out the assembly module.

Your next step would be to remove the stub from the C++ module and compile as follows:

```
   BCC   filename1.CPP  filename2.ASM
```

Or, if the fancy takes you, it can be done in steps:

```
   BCC  -c  filename1.CPP
   TASM  /ml  filename2     (.ASM file)
   TLINK  filename1  filename2
```

Note that Borland C++ does have a mechanism to suppress name-mangling for linking with Standard C modules, but I found it too limited for assembly work. It doesn't work for data and function members.

Note also that C++ does have an EXTERN declaration, so that any function that is referenced in the C++ module but is defined in the assembly module can be declared as EXTERN. However, this also has limitations and is optional anyway.

The Amazing 9-Line Program

So you think assembly language programming for Windows is difficult — think again!

Coding development over previous chapters

The "high-level" assembly language program of Chapter 5 is not much longer than one written in C or any other conventional high-level language. In the first half of this chapter, I introduced objects and some details about the inner working of C++ and how to interface to it — now, applying OO techniques brings an assembly language skeleton program down to just nine lines!

A 9-line skeleton

OOP and assembly language go together in a most natural way, with the result that coding becomes a breeze. Here is an OO skeleton program:

```
;WINASMOO.ASM --> WINASMOO.EXE
INCLUDE WINDOWS.INC
INCLUDE WINASMOO.INC
.DATA
window1 WINDOW {   }
.CODE
kickstart:
   lea  si,window1            ;addr of window object.
   call [si].make PASCAL,si   ;make the window.
   ret
 END
```

There are eleven lines there, but take off the comment line and put the code-label on the same line as the following instruction, and it becomes nine lines.

This program is the most basic skeleton, putting only a window on the screen and nothing else. In a moment I'll show you how simple it is to add the menu-bar and message box, as per skeletons from previous chapters. But first have a look at the above.

Kickstart:

In the data segment I created an instance of a WINDOW structure called "window1". In the code routine called "kickstart:" I set THIS to window1 and then called make( ), which, as its name suggests, creates the window and puts it on the screen.

You may have noticed that the syntax for creating the instance of WINDOW doesn't look much like that for structures (see page 65), but don't worry about that for now.

Hiding the "red tape"

There is a trick here: I have taken all the "red tape", the complexity, of the Windows program and hidden it away in the Include file WINASMOO.INC. This hiding of the unnecessary complexity and exposing only what is needed can only be done by using OO techniques.

My object oriented Include file is a world's first. Nobody has done this before. No Microsoft or Borland documentation will tell you how to do this. The Microsoft documentation is appalling from the assembly language programmer's viewpoint. The Borland manuals keep getting thinner too. Mind you, the simple program you see above didn't just materialize in my mind. I just about tore my hair out at times.

Simple C++ classes for Windows

I came across a very interesting article by John Dimm titled "A Tiny Windows Class Library" in *Programmer's Journal*, USA, Dec. 1991. I also studied Norton and Yao's *Borland C++ Programming for Windows*, Borland/Bantam, USA, 1992. A few ideas come from these and other sources, but I ended up doing my own thing, and what is presented in this chapter is quite simple and elegant.

It is written in Borland TASM version 3.0, for the simple reason that this assembler is specifically designed for OOP. However, I must emphasize that the code is very general and with some modification will work with earlier versions of TASM and with MASM. I have pointed out the divergence from non-OOP assemblers within this chapter. The disadvantage of the non-OOP version is that it is awkward, cumbersome, and verbose. The OOP version is easier to use, conceptually simpler, and requires fewer lines of code.

Look on the Companion Disk for various example OO programs.

A Skeleton Program

You might like to recall how complicated and enormous was the skeleton from Chapter 5. Now, here is the same thing ...

```
;WINASMOO.ASM --> WINASMOO.EXE

INCLUDE WINDOWS.INC
INCLUDE WINASMOO.INC
;.......
```

```
.DATA
window1 WINDOW  { szclassname= "WINASMOO", sztitlename= \
        "Main Window", paint= w1paint, create= w1create,\
        command= w1command }
;........
.CODE
kickstart:
   lea  si,window1                ;addr of window object.
   call [si].make PASCAL,si    ;make the window.
   ret
;....
w1paint PROC PASCAL
   LOCAL   hdc:WORD
   LOCAL   paintstructa:PAINTSTRUCT
   lea  di,paintstructa
   call BEGINPAINT PASCAL,[si].hwnd, ss,di
   mov  hdc,ax
   call SELECTOBJECT PASCAL,ax, [si].hfont
   call TEXTOUT PASCAL,hdc,10,20, cs,OFFSET sout,16
   call ENDPAINT PASCAL,[si].hwnd, ss,di
   ret
sout DB "Demo OO Program!"
w1paint ENDP
;....
w1create:
   call GETSTOCKOBJECT PASCAL,OEM_FIXED_FONT
   mov  [si].hfont,ax
   ret
;....
w1command:
   cmp  WORD PTR [si].lparam,0 ;lo half
   jne  notmenu
   cmp  [si].wparam,200    ;IDM_QUIT. Is "Quit" selected?
   jne  notquit
   call [si].destroy
   ret
notquit:
   cmp  [si].wparam,201    ;IDM_ABOUT. selected?
   jne  notabout
   call MESSAGEBOX PASCAL, [si].hwnd, cs,OFFSET szmsg, \
                        cs,OFFSET szhdg, MB_OK
notabout: ret
notmenu:  ret
szmsg   DB      "Created by Barry Kauler, 1992",0
szhdg   DB      "Message Box",0
   ret
 END
```

What you will recognize from this is all of the **essential** functionality from the skeleton of Chapter 5 without the red tape.

Overriding class defaults

In the data segment, I have created an instance window1 of the structure WINDOW. Now, if I had just ended that line with "{ }", the window would have the defaults as defined in WINASMOO.INC. However, any of the defaults can be overridden to create any kind of window. You need to know precisely how to do this, of course, but for now just look at the overrides in the above example.

I have given the window a class-name of "WINASMOO" and I have given it a title to appear in the title-bar at the top of the window. If you remember back to Chapter 3, and in particular page 77, you'll know that whenever anything happens over your application's window while the window is active, such as a menu-item being selected or key being pressed, then Windows will send a message via the message loop in WinMain() to the window's callback function. It is then up to the callback function to process the message.

Overrides

Overriding
PAINT
message

WINASMOO.INC handles all the messages in a default manner, but should you want to process any message, just put in an override when creating the instance of the window. All Windows messages are prefixed with "WM_", such as "WM_PAINT", or "WM_COMMAND". In my skeleton program I wanted to override default handling of WM_PAINT, so I put "paint = w1paint", where "w1paint" is my routine (see above). You will find the code for WM_PAINT handling is just about identical to that of Chapter 5.

COMMAND
message
override

Ditto for WM_COMMAND. I put in my own routine called "w1command", because I wanted to respond to menu-bar selections. I also overrode WM_CREATE. It's that simple.

PROC ...
ENDP
syntax
notes

One thing you will notice with my routines "w1command" and "w1create" is that I didn't put in PROC — ENDP directives. These are not essential, and the routines work perfectly well without them. Putting them in would make no difference. In fact, putting the PASCAL qualifier on would also make no difference, since no parameters are being passed.

However, notice that I did put PROC PASCAL — ENDP around the "w1paint" routine. The reason for this is that I wanted to have LOCAL data, and only TASM's "high-level" PROC automatically takes care of LOCAL declarations. The simple act of putting the PASCAL qualifier onto the PROC directive transforms it into a "high-level" PROC.

Leave off PROC [PASCAL] — ENDP if you wish, but put it on if your routine has LOCAL data. The only effect of the high-level PROC will be to correctly handle LOCAL data within the procedure.

This is a syntactical deviation from the main discussion, so I will weave my way back to the next step.

Kickstart

"kickstart:" is where the ball starts rolling. Of course the entry point to your program is at WinMain(), but this function is inside WINASMOO.INC. WinMain() takes care of all the red tape and ends up calling "kickstart:". "kickstart:" must always be in your object oriented program. Again, I've left off the PROC [PASCAL] — ENDP, for the sake of brevity and simplicity.

A static instance of the WINDOW structure already exists in the data segment, so the first thing that kickstart() does is get the object's address. The next thing it does is actually create the window and display it on the screen. You will remember from previous chapters that this was a particularly long-winded process.

Look back to page 116 and you will see that the application calls Windows CREATEWINDOW() function to create the window, then SHOWWINDOW() and UPDATEWINDOW() to actually show it on the screen. All of this is red tape and is hidden away.

Message Handling

Processing the CREATE message

After creating the window, Windows sends a WM_CREATE message to the window's callback function. I used this message to get the handle to a particular font that I used in the program (yes, even fonts have handles!). Hence I put in the w1create() routine.

Processing the PAINT message

Whenever Windows redraws any portion of the client area of the window, it lets the callback know by sending WM_PAINT. This is so the callback can redraw the client area or the portion that requires redrawing. The UPDATEWINDOW() function also generates a WM_PAINT message.

I wanted to put out a simple text message, in this case "Demo OO Program!". It also uses the font that I previously got a handle for, rather than the default font. There is a bit of red tape involved to output the message, and some temporary data storage is required. "hDC" is the handle to the window's client area, that is, the area of the window that we can output to, and this handle must be obtained before we can gain access to the window client area. It is normal practise to release this handle immediately after use, which has been done by ENDPAINT().

Processing the COMMAND message

The other thing I did in my skeleton was respond to the "File" menu-item, with its "Quit" and "About ..." sub-items. The normal way to define these is by the resource file .RC, and I have used exactly the same one as before. The WM_COMMAND message needs to have its lparam and wparam analysed to determine what kind of command has been sent to the callback, and this example shows that if lparam = 0 then the command has come from the window's menu-bar. In such a case, wparam is analysed to see which item has been selected from the menu-bar.

Handling
QUIT
menu-item

Notice that selection of "Quit" results in calling destroy(). Notice also that it is prefixed with "[si].", as are all the other parameters of the window. You can understand this from the principle of structures. The SI register contains the address of the object or the instance of the structure. "Destroy" is a field in this structure. Fields can, in OOP terminology, be data-members or function-members (*methods* in Pascal terminology). The field "destroy" does not contain data, but a pointer to a routine.

To effectively use this object oriented approach, you need to know the fields of the WINDOW structure and the purpose of each ...

The WINDOW Object

TABLE
directive

Here is the structure definition of WINDOW. It is actually located inside WINASMOO.INC ....

```
.DATA
WINDOW TABLE    {
   VIRTUAL      definewndclass:WORD =   WINDOWdefinewndclass
   VIRTUAL      create:WORD =           WINDOWcreate
   VIRTUAL      paint:WORD =            WINDOWpaint
   VIRTUAL      command:WORD =          WINDOWcommand
   VIRTUAL      timer:WORD =            WINDOWtimer
   VIRTUAL      resize:WORD =           WINDOWresize
   VIRTUAL      mousemove:WORD =        WINDOWmousemove
   VIRTUAL      lbuttondown:WORD =  WINDOWlbuttondown
   VIRTUAL      lbuttonup:WORD =        WINDOWlbuttonup
   VIRTUAL      char:WORD =             WINDOWchar
   VIRTUAL      defaultproc:WORD =  WINDOWdefaultproc
   VIRTUAL      destroy:WORD =          WINDOWdestroy
   VIRTUAL      make:WORD =             WINDOWmake
   VIRTUAL      wndproc:WORD =          WINDOWwndproc
   VIRTUAL      hwnd:WORD =             0
   VIRTUAL      wmessage:WORD =         0
   VIRTUAL      wparam:WORD =           0
   VIRTUAL      lparam:DWORD =          0
   VIRTUAL  classstyle:WORD = CS_VREDRAW+ CS_HREDRAW
   VIRTUAL      sziconname:BYTE:32 =    0
   VIRTUAL      szcursorname:BYTE:32 =  0
   VIRTUAL      hbrbackground:WORD =    COLOR_BACKGROUND
   VIRTUAL      szclassname:BYTE:32 =   0
   VIRTUAL      sztitlename:BYTE:32 =   0
   VIRTUAL      hmenu:WORD =            0
   VIRTUAL      hwndparent:WORD =       0
   VIRTUAL      wheight:WORD =          200
   VIRTUAL      wwidth:WORD =           250
```

```
VIRTUAL        y_coord:WORD =              0
VIRTUAL        x_coord:WORD =              150
VIRTUAL        createstylelo:WORD         0
VIRTUAL   createstylehi:WORD = WS_OVERLAPPEDWINDOW
VIRTUAL        hfont:WORD =               0
}
```

This doesn't look like any structure definition you've seen before! Instead of using STRUC, I have used TASM's TABLE directive, which has some advantages but a different syntax.

OO limitations of STRUC

The Borland programmers will probably gag when they see how I have used their TABLE directive, but I found it useful to define both data and procedures. I wanted to retain a program that would work with other non-OO assemblers with only minimal change. The above TABLE can be replaced with the conventional STRUC, but the latter has disadvantages, the two most glaring being:

- it cannot initialise fields with forward references; and
- initializing fields of instances is rigid and awkward.

However, it can be done — check out "Object Oriented Programming in Assembly Language" by R. L. Hyde, *Dr. Dobb's Journal*, March 1990, p. 66-73, 110-111.

The TABLE directive only exists with TASM version 3.0, not before. I have only bitten off a little bit of the new TASM's OO capability; however, my end result is quite simple and elegant.

TASM User's manual: limitations & TABLE description

Despite a wonderful new assembler, Borland's manual has only about two and a half pages devoted to Windows programming and only two demo programs on disk. The OOP neophyte will find the TASM manual to be quite daunting, with all of the OO terminology. The manual supplied with TASM version 5 has even *less* documentation. This book addresses all of these problems.

Not only do I demystify OOP, but I show how to write windows programs effortlessly.

The Borland manual describes the use of TABLE to define function-members (methods) for an object, with the data-members defined separately. There are certain reasons for this, but I wanted a system that is conceptually simple.

Notice the VIRTUAL qualifier in front of every field declaration. Don't worry about this — just pretend it isn't there.[1]

---

[1]   Readers with some knowledge of OOP will know that VIRTUAL is a qualifier used with functions, but I've also put it in front of data-members, because I have used TABLE in a way that Borland never intended (or thought of!).

Analysis of the table

Look at the first field. "definewndclass" is a NEAR pointer to a procedure "WINDOWdefinewndclass".   What this actually means is that when an instance of the structure is created, the field will be as per Figure 6.1.

Window1 equates to the offset in the segment at which the structure-instance starts.

Definewndclass equates to 0, being the first field, while create equates to 2. This is exactly as in any normal structure. The contents of the fields are addresses of the procedures; in this case they are default procedures defined within WINASMOO.INC.

Figure 6.1: definewndclass pointer.

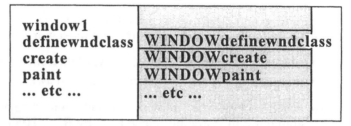

These addresses are forward references, which is why STRUC won't accept them, and why I have used TABLE. With STRUC you have to put them in when creating the instance, which I used to do by means of a macro.

Further down in the WINDOW structure you'll see plain old data, which should be easy enough to understand.

TABLE overrides

Apart from being able to declare forward references, the beauty of the TABLE directive (and TASM's new extended STRUC) is the declaration of overrides when creating instances.

A TABLE or STRUC declaration is only for the assembler's use, and is not actually assembled.   It is the instances that get assembled. A static instance is one that you declare in the data segment. You could declare automatic instances on the stack or dynamic instances on the heap. Dynamic instances involve an extra level of complexity, so let me shelve that one for now.

The skeleton program on page 154 declared a static instance as follows:

```
.DATA
window1 WINDOW  { szclassname= "WINASMOO", sztitlename= \
        "Main Window", paint= w1paint,create= w1create, \
        command= w1command }
```

Overriding the message handlers
The instance-declaration of a conventional structure would require a very rigid syntax of comma delimiters. Here, all you have to do is override only those fields you want, and you can put them in between the "{ }" in any order. No commas are required for unchanged fields.

You need to have an understanding of each field of the WINDOW structure to make use of them in the program.

Create(), paint(), timer(), etc., down to destroy(), are the main WM_ messages that Windows sends to the callback function. I have only implemented the WINDOW structure with these, although there are many more. The others all go directly to Window's default handling.

You are quite welcome to expand the structure with more messages.

Overriding the major "hidden" functions of the Include file
Definewndclass(), defaultproc(), destroy(), make(), and wndproc() are major routines within WINASMOO.INC. By putting them in the structure, you can override them for any sub-class or instance. For example, wndproc() is a pointer to the callback function for that window — it basically performs a case-switch, calling the appropriate message-handler create(), paint(), timer(), etc. But, there is nothing to stop you from overriding it and defining your own special calllback, say for example, to handle a dialog box.

These pointers are probably ones that it would be wise to leave alone until you've become familiar with this software.

Data members of the WINDOW class
The rest of the structure comprises various data associated with the window. Here are explanations:

hwnd	handle of this window
wmessage	the message sent to this window
wparam,lparam	data associated with the message
classstyle	parameter used by REGISTERCLASS()
sziconname	ditto. What will look like when iconized
szcursorname	ditto. What cursor like over window
hbrbackground	ditto. Color of client area
szclassname	ditto. ASCIIZ name this class of window
sztitlename	param. used by CREATEWINDOW()
	ASCIIZ title appear at top of window
hmenu	ditto. Menu or child identifier
hwndparent	ditto. Handle of parent window
wheight	ditto. Height of the window
wwidth	ditto. Width of the window
y_coord	ditto. Top-left y-coordinate
x_coord	ditto. Top-left x-coordinate
createstylelo	appearance features of window
createstylehi	ditto.
hfont	application-specific

So override any of these parameters to make your window look and behave exactly as you want.

Creating multiple windows

You are not limited to just one window. As well as being able to have multiple instances of your program quite happily sharing the same screen, any one instance can have multiple windows. It is simply a matter of declaring another instance and calling make(). Make() doesn't have to be called in the kickstart() routine, though that's where you would create the main window. No, you can make windows from anywhere in the message-handling routines.

Nor are you limited to the one WINDOW structure. You can declare sub-classes, which inherit the fields from WINDOW, but with their own extra fields. These sub-classes can also override any of the parent-classes' fields.

It is time to peek further inside WINASMOO.INC ...

WINMAIN( )

WinMain() is hidden inside the Include file

WINASMOO.INC has the job of hiding the "red tape" of a Windows program. It must handle multiple instances of a program; that is, if you double-click on the program's icon more than once. It must handle multiple windows within the one instance.

Any one window would have its own instance of the window structure or object, as I did by creating "window1". For a second window, I could create an instance of WINDOW called "window2".

Program listing continues until page 171

WinMain() looks similar to code that you would find in a conventional WinMain() function, with some curious differences. Well, look at the whole lot ...

```
;WINASMOO.INC + WINASMOO.ASM -->                    ;
WINASMOO.EXE  Windows OO program.
;remember that Windows funcs only preserve SI,DI,BP & DS.
.MODEL SMALL
LOCALS                     ;turns on "@@" prefix for auto local
                           ;name-mangling (Borland only).
;......
;These are Windows functions ...
EXTRN    UPDATEWINDOW:FAR, BEGINPAINT:FAR
EXTRN    ENDPAINT:FAR, DEFWINDOWPROC:FAR
EXTRN    POSTQUITMESSAGE:FAR,
EXTRN    REGISTERCLASS:FAR, GETSTOCKOBJECT:FAR
EXTRN    CREATEWINDOW:FAR, SHOWWINDOW:FAR
EXTRN    GETMESSAGE:FAR
EXTRN    LOADCURSOR:FAR, TRANSLATEMESSAGE:FAR
```

```
EXTRN    DISPATCHMESSAGE:FAR, LOADICON:FAR
EXTRN    TEXTOUT:FAR
EXTRN    MESSAGEBOX:FAR, GETDC:FAR
EXTRN    RELEASEDC:FAR
EXTRN    SELECTOBJECT:FAR, GETWINDOWWORD:FAR
EXTRN    SETWINDOWWORD:FAR, SENDMESSAGE:FAR
EXTRN    DESTROYWINDOW:FAR
;..........................................................
.DATA
MAIN             TABLE {
  VIRTUAL        initinstance:WORD =    kickstart
  VIRTUAL        hinstance:WORD =       0
  VIRTUAL        hprevinstance:WORD =   0
  VIRTUAL        ncmdshow:WORD =        0
  }
;...
WINDOW TABLE  {
  VIRTUAL definewndclass:WORD =    WINDOWdefinewndclass
  VIRTUAL        create:WORD =          WINDOWcreate
  VIRTUAL        paint:WORD =           WINDOWpaint
  VIRTUAL        command:WORD =         WINDOWcommand
  VIRTUAL        timer:WORD =           WINDOWtimer
  VIRTUAL        resize:WORD =          WINDOWresize
  VIRTUAL        mousemove:WORD =       WINDOWmousemove
  VIRTUAL        lbuttondown:WORD =     WINDOWlbuttondown
  VIRTUAL        lbuttonup:WORD =       WINDOWlbuttonup
  VIRTUAL        char:WORD =            WINDOWchar
  VIRTUAL        defaultproc:WORD =     WINDOWdefaultproc
  VIRTUAL        destroy:WORD =         WINDOWdestroy
  VIRTUAL        make:WORD =            WINDOWmake
  VIRTUAL        wndproc:WORD =         WINDOWwndproc
  VIRTUAL        hwnd:WORD =            0
  VIRTUAL        wmessage:WORD =        0
  VIRTUAL        wparam:WORD =          0
  VIRTUAL        lparam:DWORD =         0
  VIRTUAL classstyle:WORD = CS_VREDRAW + CS_HREDRAW
  VIRTUAL        sziconname:BYTE:32 =   0
  VIRTUAL        szcursorname:BYTE:32 = 0
  VIRTUAL        hbrbackground:WORD =       COLOR_BACKGROUND
  VIRTUAL        szclassname:BYTE:32 =  0
  VIRTUAL        sztitlename:BYTE:32 =  0
  VIRTUAL        hmenu:WORD =           0
  VIRTUAL        hwndparent:WORD =      0
  VIRTUAL        wheight:WORD =         200
  VIRTUAL        wwidth:WORD =          250
  VIRTUAL        y_coord:WORD =         0
  VIRTUAL        x_coord:WORD =         150
  VIRTUAL        createstylelo:WORD =   0
  VIRTUAL createstylehi:WORD = WS_OVERLAPPEDWINDOW
  VIRTUAL        hfont:WORD =           0
  }

main1            MAIN  { }  ;create static instance.
pwindowDW        0          ;ptr to current window object.
pwindowflag      DB   0     ;=0 pwindow not valid.
;..........................................................
.CODE
```

```
; . . . . . . . . . . . . . . . . . . . . . . . . . . . . . . . . . . * . . . . . . . . . .
    PUBLIC          WINMAIN
WINMAIN PROC PASCAL NEAR @@hInstance:WORD, \
   @@hPrevInstance:WORD,@@lpCmdLine:DWORD,@@nCmdShow:WORD
   LOCAL          msg:MSGSTRUCT        ;see WINDOWS.INC
```

Notice the use of the "@@" prefix. This keeps these labels unique
to this procedure. Refer back to page 121.

```
    lea   si,main1
    mov   ax,@@hinstance        ;save params in main1 object.
    mov   [si].hinstance,ax ;     /
```

It should make sense so far. In the data segment I defined two
structures, MAIN and WINDOW. There will only be one instance
of MAIN in the application, called "main1" (see above). The
application is entered from Windows at WinMain(), and I have
used main1 to save the parameters.

This is what is happening now ...

```
    mov   ax,@@hprevinstance          ;     /
    mov   [si].hprevinstance,ax       ;     /
    mov   ax,@@ncmdshow        ;     /
    mov   [si].ncmdshow,ax     ;     /
; ...
    call  [si].initinstance ;call kickstart() **no pascal**
    or    ax,ax
    jne    messageloop
    ret
```

You should be able to recognize the message loop below. Usually
WinMain() will have instance initialisation and window creation
code in here, but I have shifted it out to make(), via kickstart().
This enables me to make as many windows as I want and also
enables me to bring out only the essential part of the program to
the "front end". This diversion is implemented via the above
CALL.

```
loopback:
    call TRANSLATEMESSAGE PASCAL, ss,di
    call DISPATCHMESSAGE PASCAL, ss,di
messageloop:
    lea  di,msg
    call GETMESSAGE PASCAL, ss,di, null, null, null
    or   ax,ax
    jne  loopback
    mov  ax,[di].msWPARAM  ;return wparam to windows.
    ret
WINMAIN ENDP
```

One callback for all windows

There is nothing new about the message loop. Remember how Windows calls GETMESSAGE() to get a message from the application's queue, then calls DISPATCHMESSAGE() to send it on to the callback function. Because each window has its own callback function, we have to design the program so that the message will end up at the correct callback — except that in this program there is a trick. There is only one callback function, called exportwndproc().

It is a common practise with Windows programming to reuse one set of code with different data for each window.

Most Windows programs can have multiple instances, that is, multiple copies running simultaneously without conflict, even though they use the same code. Each time you double-click on the application's icon, a new data/stack/heap segment is loaded, but the original code segment is used. This practical functionality is enabled in the .DEF file by specifying the data as MULTIPLE (see page 177).

The same principle can be applied to multiple windows within the one instance.

Callback

One callback, but each window is a separate object

Now this is interesting. Despite the fact that a program can create as many simultaneous windows as it wants, there is only one callback function, exportwndproc(). Exportwndproc() determines which window has sent the message, which is easy enough, because its handle, hwnd, is passed to the callback, then it gets the address of the corresponding window object, which it loads into SI.

This is conceptually quite simple. Any activity related to the active window on the screen will result in Windows sending a message. The callback can use the same code for all windows, except for overrides — all it needs to know is the address of the object (the data and pointers) for that window ...

```
    PUBLIC        exportwndproc
exportwndproc  PROC WINDOWS PASCAL FAR \
               @@hwnd:WORD,@@message:WORD,\
               @@wparam:WORD,@@lparam:DWORD
    LOCAL dummy:WORD:5
; . . . .
    cmp   pwindowflag,0       ;Make() controls this flag.
    jne   normalwndproc
    call  DEFWINDOWPROC PASCAL, @@hwnd,@@message, \
                @@wparam, @@lparam
```

```
     ret
;....
normalwndproc:
   push si       ;callback must preserve si.
   push di       ;and di
   call   GETWINDOWWORD PASCAL,@@hwnd,0
                        ;0=offset in Windows internal data.
   mov   pwindow,ax    ;get addr of current window object.
   mov   si,ax         ;don't use LEA
```

**Saving &
restoring a
pointer to a
window
object**

Don't worry about pwindowflag for now.
GETWINDOWWORD() is a Windows function that returns
information about the window that Windows has stored internally.
The intention here is that I have the handle to the window, hwnd,
and I want to know the address of the object for that window.

In the case of my simple skeleton program, there was only one
window anyway, and I created the window1 object for it (refer to
page 153).   There is a bit of a trick here, because when I used
make() to create the window, I also gave the address of the object
to Windows for Windows to store as part of its own record about
that window.   GETWINDOWWORD() enables me to retrieve any
information that Windows has about that window, plus the extra
information I gave it.

This is a mechanism for associating a particular set of data, in this
case object window1, with a particular window.

I stored my special data at an offset of 0 in Windows internal data
structure, so here I get it back, returned in AX.   I then put the
address into the global pointer "pwindow", and into SI.

```
   mov   ax,@@message        ;save params in window object.
   mov   [si].wmessage,ax
   mov   ax,@@wparam
   mov   [si].wparam,ax
   mov   ax,WORD PTR @@lparam
   mov   WORD PTR [si].lparam,ax
   mov   ax,WORD PTR @@lparam+2
   mov   WORD PTR [si].lparam+2,ax

   call  [si].wndproc PASCAL,si

   pop   di
   pop   si
   ret
exportwndproc   ENDP
```

Having got the address of the object, I then save the parameters
that Windows passed to the callback into the object.

I then called wndproc(), whose address is actually in the object.
By default it is WINDOWwndproc(), shown below.   You can

override this to provide your own wndproc() for a particular window, such as a dialog box, but in most cases you will leave well enough alone. wndproc() works fine for normal windows, and has a very simple task — it just implements a CASE statement to call the appropriate message handler. These message handlers (paint, create, timer, etc.) are all pointed to via the object, and can be overridden for any particular window. Any "WM_" message not catered to in the CASE statement results in a call to the default routine, and I've even provided for overriding this.

```
WINDOWwndproc PROC PASCAL now
    mov  si,now                      ;current window object.
    mov  dx,0                        ;hi return flag. set default 0.
    mov  ax,[si].wmessage            ;get message
    cmp  ax,WM_CREATE                ;msg rec'd after CreateWindow()
    jne  case2
    call [si].create                 ;**note no pascal**
    jmp  SHORT   endx
case2:
    cmp  ax,WM_DESTROY               ;msg if a window closed.
    jne  case3
    call [si].destroy
    jmp  SHORT   endx
case3:
    cmp  ax,WM_PAINT                 ;msg if Window redrawn.
    jne  case4
    call [si].paint
    jmp  SHORT   endx
case4:
    cmp  ax,WM_COMMAND               ;any selection of the menu.
    jne  case5
    call [si].command
    jmp  SHORT   endx
case5:
    cmp  ax,WM_LBUTTONDOWN           ;a mouse msg.
    jne  case6
    call [si].lbuttondown
    jmp  SHORT   endx
case6:
    cmp  ax,WM_CHAR                  ;msg that a key pressed.
    jne  case7
    call [si].char
    jmp  SHORT   endx
case7:
    call [si].defaultproc
endx:   ret                          ;return dx:ax flag (maybe).
WINDOWwndproc  ENDP
```

I could have been a bit more impressive and emulated the case statement with a dual-column table and a program loop to find a message that matches, which would be better if a lot of messages are to be handled. The above code is ok though.

MAKE( )

Now for the part that actually creates the window; herein are some secrets that make the program work. By referring back to Chapters 4 and 5 you will see the code that remains from before, such as REGISTERCLASS(), CREATEWINDOW(), SHOWWINDOW() and UPDATEWINDOW().

The data structure WNDCLASS is there, or rather an instance of it. It needs to have data put into it, and rather than do it in-line I have called the function definewndclass() to do it. Compare this with the listing starting on page 112 — look back there also to see how WNDCLASS is defined in WINDOWS.INC. The data for this structure is from the window object (pointed to by SI).

```
WINDOWmake  PROC  PASCAL  now
   LOCAL wndclassa:WNDCLASS
   mov  si,now
   xor  ax,ax              ;clear ax (default return value)
;....
;does this window already exist? ... check hwnd ...
   cmp  [si].hwnd,0
   je   nexist
   jmp  endhere
;...
;is it a child? ... this make() can't handle a child ...
; (needs slight mod to handle normal child window)
nexist:
   cmp  [si].hwndparent,0
   je   nochild
   jmp  endhere
nochild:
   lea  di,[si].wndclassa
   call [si].definewndclass PASCAL, di, si
   call REGISTERCLASS PASCAL, ss,di
;.....
   mov  pwindowflag,0      ;disable wndproc() processing.
   lea  bx,[si].szclassname
   lea  ax,[si].sztitlename
   call CREATEWINDOW PASCAL, ds,bx, ds,ax, \
   [si].createstylehi, [si].createstylelo, [si].x_coord,\
   [si].y_coord, [si].wwidth, [si].wheight, \
   [si].hwndparent, [si].hmenu, main1.hinstance, 0,0
```

pwindowflag　STOP! Go no further. Look at what I have done above. Just before CREATEWINDOW(), I cleared "pwindowflag". You must remember that this program is capable of handling multiple windows, but with only one callback function.

Therefore the callback must be able to determine which object is associated with the window, to access all the data and pointers for that window. However, at the moment, the cart is before the horse. CREATEWINDOW() will send some messages to the callback, but I do not put the address of the object into Windows internal record until *after* CREATEWINDOW().

Exportwndproc() used GETWINDOWWORD() to retrieve the object address, but I put it in below by using SETWINDOWWORD(). This latter function can only be called after CREATEWINDOW(), because it requires the handle that CREATEWINDOW() returns.

Since CREATEWINDOW() itself sends messages to exportwndproc(), the latter has to test pwindowflag and disable normal processing until it is set.

```
    mov    [si].hwnd,ax         ;save handle in window object.
    mov    di,ax
    or     ax,ax                ;exit if handle is 0.
    jz     endhere
    call   SETWINDOWWORD PASCAL,di,0,si
                    ;store addr of window object in Windows
                    ;internal data (at offs.0)
    mov    pwindowflag,1 ;enable callback normal processing.
;...
;Callback disabled above, but my callback needs
;WM_CREATE. So send it now...
```

One deviation leads to another — a problem arises because CREATEWINDOW() sends the WM_CREATE message to the callback, which my exportwndproc() has ignored due to pwindowflag being cleared.

However, now that SETWINDOWWORD() has done its job, pwindowflag has been set. I have used SENDMESSAGE() to resend the WM_CREATE message. Now it goes to the callback (via all the usual rigmarole — the application queue and the message loop) and is processed in the normal way, calling the create() routine.

```
    call   SENDMESSAGE PASCAL, di, WM_CREATE, 0, \
                    0,0      ;last 2 are incorrect!
;....
    call   SHOWWINDOW PASCAL, di ,main1.ncmdshow
    call   UPDATEWINDOW PASCAL, di
    mov    ax,1
;....
endhere:           ret
WINDOWmake         ENDP
```

If you have done much Windows programming, you may have noticed something missing — a test for hPrevInstance followed by a conditional jump. Actually it isn't really needed!

**Window class
data structure** Continuing the program listing ...

```
WINDOWdefinewndclass   PROC PASCAL pwndclass,now
    push di
    push si
    mov  si,now
    mov  di,pwndclass ;pointer to wndclassa (see make())
;....
;Setup the window class structure for REGISTERCLASS()...
    mov  ax,[si].classstyle ;get specs from object and load
    mov  [di].clsStyle,ax   ;into wndclassa structure....
;...
    mov  [di].WORD PTR clsLpfnWndProc,OFFSET exportwndproc
    mov  [di].WORD PTR clsLpfnWndProc+2,SEG exportwndproc
;....
    mov  [di].clsCbClsExtra,0
    mov  [di].clsCbWndExtra,2
;....
    mov  ax,main1.hInstance
    mov  [di].clsHInstance,ax
;....
    cmp  [si].sziconname,0
    je   noicon
    lea  ax,[si].sziconname
    call LOADICON PASCAL, main1.hinstance, ds,ax
    jmp  SHORT yesicon
noicon:
    call LOADICON PASCAL,null, 0,IDI_APPLICATION
yesicon:
    mov  [di].clsHIcon,ax
;....
    cmp  [si].szcursorname,0
    je   nocursor
    lea  ax,[si].szcursorname
    call LOADCURSOR PASCAL, main1.hinstance, ds,ax
    jmp  SHORT yescursor
nocursor:
    call LOADCURSOR PASCAL,null, 0,IDC_ARROW
yescursor:
    mov  [di].clsHCursor,ax
;...
    mov  ax,[si].hbrbackground
    mov  [di].clsHbrBackground,ax
;....
    lea  ax,[si].szclassname
    mov  [di].WORD PTR clsLpszMenuname,ax
    mov  [di].WORD PTR clsLpszMenuName+2,ds
    mov  [di].WORD PTR clsLpszClassName,ax
    mov  [di].WORD PTR clsLpszClassName+2,ds
    pop  si
    pop  di
    ret
WINDOWdefinewndclass ENDP
```

The above routine simply copies data from the object into wndclassa.

***Default
message
handling***

What follows are the default routines (function-members, or methods) that the WINDOW structure is initialized to. As you can see, they don't do much, and if not overridden, all you will get on the screen is a blank window. It will have a system menu, so you can quit the program, and it can be minimized, etc. — all of this functionality was set by REGISTERCLASS() and CREATEWINDOW().

```
WINDOWdestroy:
   call POSTQUITMESSAGE PASCAL,0
   ret
;......
WINDOWcreate:
WINDOWpaint:
WINDOWcommand:
WINDOWlbuttondown:
WINDOWlbuttonup:
WINDOWchar:
WINDOWtimer:
WINDOWresize:
WINDOWmousemove:
WINDOWdefaultproc:
   call DEFWINDOWPROC PASCAL, [si].hwnd, [si].wmessage,\
                             [si].wparam, [si].lparam
   ret
```

Inheritance

***Example OO
program with
a control***

The next example shows how to create a control. You will need to refer to a Windows programming book to learn all about controls; however, this example will give you some idea.

A control is a child window, that is, a window that resides within the client area of the parent window and normally sends its messages to the callback function of the parent.

The example creates a simple "button", with the title "OK" inside it. When the mouse is clicked over the button, it disappears. Pressing any key brings it back. Not much, but it does illustrate some useful principles. Figure 6.2 shows what it looks like.

The button that is added by this program is the one on the main window. The message box is also a type of child window, created by MESSAGEBOX(). Controls can be all sorts of things, including edit boxes, check boxes, buttons, and scrollbars.

Figure 6.2: Simple OO demonstration program.

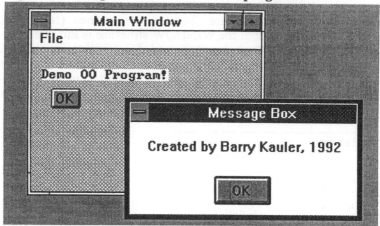

Control class

Since a control is just a window, why not use the WINDOW structure and make()? Well, yes, it can be done, except that controls do have some special requirements.

If you think in terms of conventional programming, you would probably delve into make() and see how to patch in the handling of such a special case. Unfortunately, this is one of the major problems with such programming; the continual patching of code to handle special cases. If your code works, the process of patching is liable to make it less stable and predictable.

Better to leave well enough alone. We have a functional make() for normal windows, so let's think like OO programmers. We could simply create another instance of WINDOW, say "window2", and override the make() with a new routine.

That is ok if all we ever want to do is create one control, but it is nicer if we think in the long term. Why not create another class, call it CONTROL, and let it inherit everything from WINDOW, but with any necessary overrides?

Complete OO program with a control

This is what has been done with my program, and the new make() routine can become part of WINASMOO.INC, along with the new class. First, here is the final program:

```
;WINASMOO.ASM --> WINASMOO.EXE
INCLUDE WINDOWS.INC
INCLUDE WINASMOO.INC
IDM_QUIT        EQU     200
IDM_ABOUT       EQU     201

.DATA
```

```
window1  WINDOW { szclassname="WINASMOO", \
     sztitlename="OO Demo", paint=w1paint, \
     create=w1create, command=w1command,  createstylehi=\
     WS_OVERLAPPEDWINDOW + WS_CLIPCHILDREN, \
     char=w1char, sziconname = "icon_1" }
control1 CONTROL { szclassname= "BUTTON", sztitlename=\
       "OK", x_coord=20,y_coord=40,wwidth=30,wheight=20,\
   \     hmenu=IDOK, createstylehi=WS_CHILD+ \
       WS_VISIBLE, createstylelo=BS_PUSHBUTTON }
.CODE
kickstart:
   lea  si,window1             ;addr of window object.
   call [si].make PASCAL,si    ;make the window.
   lea  si,control1
   call [si].make PASCAL,si    ;make child window
   ret
;
w1paint PROC PASCAL
   LOCAL         hdc:WORD
   LOCAL         paintstructa:PAINTSTRUCT
   lea  di,paintstructa
   call BEGINPAINT  PASCAL,[si].hwnd, ss,di
   mov  hdc,ax
   call SELECTOBJECT  PASCAL,ax, [si].hfont
   call TEXTOUT  PASCAL,hdc,10,20, cs,OFFSET outstring,16
   call ENDPAINT  PASCAL,[si].hwnd, ss,di
   ret
outstring        DB "Demo OO Program!"
w1paint ENDP
;
w1create:
   call GETSTOCKOBJECT  PASCAL,OEM_FIXED_FONT
   mov  [si].hfont,ax
   ret
;
w1command:
   cmp  WORD PTR [si].lparam,0 ;lo half=0 if a menu select.
   jne  notmenu
   cmp  [si].wparam,IDM_QUIT  ;Is "Quit" selected?
   jne  notquit
   call [si].destroy
   ret
notquit:
   cmp  [si].wparam,IDM_ABOUT   ;Is "About.." selected?
   jne  notabout
   call MESSAGEBOX  PASCAL, [si].hwnd, cs,OFFSET szmsg, \
                    cs,OFFSET szhdg, MB_OK
notabout: ret
notmenu:
   cmp  [si].wparam,IDOK  ;button child window selected?
 ;note that lo-word of lparam has handle of control
 ;window, hi-word of lparam has notification code.
   jne  notbutton
   lea  si,control1           ;since si points to window1.
   call DESTROYWINDOW PASCAL,[si].hwnd ;kill button
 mov [si].hwnd,0 ;must clr hwnd, if want to make() later.
notbutton:
```

```
    ret
szmsg   DB          "Created by Barry Kauler, 1992",0
szhdg   DB          "Message Box",0
;
w1char:
;let's bring back the button if any key pressed...
   lea  si,control1              ;since si points to window1.
   call [si].make PASCAL,si
   ret
;...................
   END
```

"IDOK" equates to 1 and is defined in WINDOWS.INC. It is a convenient identifier to pass to the parent callback in the wparam of the WM_COMMAND message.

Pressing the button results in this message.

Make() for CONTROL class

Now for the new make() routine:

```
;Here are extensions for handling controls ...
;......
.DATA
CONTROL TABLE {
   WINDOW,
   VIRTUAL make:WORD = CONTROLmake
   }
;.........
.CODE
CONTROLmake    PROC PASCAL now
   mov  si,now
   xor  ax,ax          ;clear ax (default return value).
;....
;does this window already exist?  check hwnd...
   cmp  [si].hwnd,0
   jnz  ending
;Is it a child? ... all controls are child windows ...
   cmp  [si].hwndparent,0
   jne  nending
;so, we have to give it one ... (this involves an
;assumption)...
;pwindow still points to the parent window object, so...
   mov  bx,pwindow
   mov  ax,[bx].hwnd
   mov  [si].hwndparent,ax
;....
nending:
   lea  bx,[si].szclassname
   lea  ax,[si].sztitlename
   call CREATEWINDOW PASCAL, ds,bx, ds,ax, \
   [si].createstylehi, [si].createstylelo,[si].x_coord,\
   [si].y_coord, [si].wwidth, [si].wheight,\
   [si].hwndparent, [si].hmenu,main1.hinstance,0,0
   mov  [si].hwnd,ax ;save handle in window object.
```

```
;...
ending:ret
CONTROLmake      ENDP
```

Comparison between CONTROL and WINDOW classes
You can treat the control object just as you would a window object, using all the same data and function members. To make this statement almost completely true does actually require a little more refinement — message, wparam and lparam data members of the control object are not actually used, so it would be wise to put in some testing to avoid them being accidentally accessed — though this is unlikely. Ditto for most of the functions.

The problem with inheritance is that I can't throw away the previous structure's fields. All I can do is redefine them. Actually, although there is redundancy here, it is possible for a control to have its own callback, which means that all of the fields would be of use.

One immediate refinement could be to override all of the message handlers for the CONTROL class, so that they just return without doing anything.

Anyway, I've kept this code as simple and as elegant as possible.

Getting it Together

OOP overhead
One thing you may be starting to appreciate is that Windows adds an incredible processing overhead — even a simple key press has to go through so many steps before it reaches the destination. Then we go and make things even worse by using OO techniques, that add yet another layer of processing. If you want code that rockets along, for a video game for example, you will want to know mechanisms for speeding things up. OOP may make the coding easier, but it may be going against a fundamental reason why we are using assembly language. Let me post this as a thought for now.

Make file
Oh yes, the WINASMOO.MAK file has a couple of minor changes from before, so here is the listing:

```
# NOTE this Make file has been modified for Borland C++,
# to be used with TASM and TLINK, however I'm still using
# Microsoft's NMAKE, as Borland's MAKE has some strange
# quirks ... though the version supplied with TASM v3.0
# claims to have improved compatibility with NMAKE ...
# this I haven't yet tried.
# To run this file:    NMAKE WINASMOO.MAK
fn =    winasmoo
all:$(fn).exe
```

```
lpath = \borlandc\lib          #path for libraries
ipath =\borlandc\include       #path for Include files.
epath =\borlandc\bin           #path for EXEs.
sw = /c /n /v /Tw /L$(lpath)       #switches for tlink.
# /n =ignore-default-libs, /Tw =generate Windows exe,
#  /Lc:$(lpath) =lib path,  /v =debug-on.

$(fn).obj : $(fn).asm $(fn).inc
   tasm /zi /p /w+  $(fn);

# -r =dont append to exe, -x =dont look in INCLUDE
# envir-variable for incl-files,  -i =look in this path
# instead....
$(fn).res : $(fn).rc
      rc -r -x -i$(ipath) $(fn).rc

# c0ws=start-up-lib, cws=Windows-runtime-lib,
# cs=Standard-runtime,  import=access-builtin-libs
$(fn).exe : $(fn).obj $(fn).def $(fn).res $(fn).inc
   tlink $(sw) c0ws $(fn),$(fn),$(fn),import cws,$(fn).def
      rc -x -i$(ipath) $(fn).res

# Note that Borland C++ v2.5 names the Windows library
# CWINS.LIB, while v3.0 names it CWS.LIB.  I used the
# latter above. The C runtime library is CS.LIB, which
# could be placed immediately after CWS, if you need it.
#  Note that the "S" postfix designates the small model.
```

**Program
custom icon**

So, that's WINASMOO.MAK — much the same as before. The
.RC and .DEF files can be the same as for previous skeletons,
though of course if you want to try experimenting with OOP you
might like to try adding on to the .RC file.

Most Windows programs will want to have their own icon, rather
than one of the defaults, and I have done this with the extended
program example (the one with the child control button). Icon
images have to be created with a special paint program — I used
Borland's Resource Workshop — a lovely product — to design my
icon, which I then saved as WINASMOO.ICO.

**Resource
script**

Resource Workshop then automatically added an extra line into
my WINASMOO.RC file:

```
//these (arbitrary) equates could have been in an include
//file...
#define IDM_QUIT     200
#define IDM_ABOUT    201

winasmoo     MENU
   BEGIN
     POPUP "File"
       BEGIN
            MENUITEM "Quit",    IDM_QUIT
            MENUITEM "About...", IDM_ABOUT
```

```
        END
     END
ICON_1   ICON   winasmoo.ico
```

The icon resource is arbitrarily named "icon_1", so when I created "window1" in my program, I put in the override sziconname = "icon_1".

Definition file

There is a useful note that I can make about the .DEF file, so here it is:

```
NAME          WINASMOO
DESCRIPTION   'Demo OO asm program'
EXETYPE       WINDOWS
STUB          'WINSTUB.EXE'
CODE          PRELOAD MOVEABLE
DATA          PRELOAD MOVEABLE MULTIPLE
HEAPSIZE      1024
STACKSIZE     8192
EXPORTS       exportwndproc
```

Multiple instances

What I would like to point out in particular here are the specifications for the data segment. PRELOAD means that it loads when the program is first loaded. MOVEABLE means that it can be moved by WINDOWS. MULTIPLE means that every instance will have its own copy of the data segment. The latter point is important if you want the program to support multiple instances. I have designed the code to support multiple instances with the same ease that it supports multiple windows within the same instance, but this only works if each instance has its own complete copy of the data/stack/heap. Note that all instances will use the same code segment, which is no problem at all.

This works because code cannot be changed. Even though you can keep data in the code segment, and I have done so in the skeleton program, you cannot change it. Windows sets the attribute of code segments such that they cannot be written to, and your program will crash if you try. Most interestingly, though, there is a way around this, because Windows has a function that gives you a DS selector for a code segment (see Chapters 10, 11 and 12).

SMALL model

Note that my OOP code is designed for the SMALL model. The major limiting factor is the pervasive use of NEAR pointers. It would probably be easier to design a completely different Include file for other memory models. It should be easy to upgrade to 32-bit code though.

Virtual Method Table

TASM v3.0 encourages the classical implementation of objects, in which the pointers to procedures (*Virtual Method Table*, VMT) are not stored physically with the data of each object instance, but somewhere else (which is why they invented the TABLE that I have misused). There are arguments for and against this. Any one class can have one VMT, and instances could all access a single instance of the VMT. This would be efficient in terms of memory but would not allow individual overrides by each object instance. As mentioned earlier, I decided on an approach that allows easy conversion to non-OOP assemblers, is conceptually simple, and offers some flexibility advantages that the VMT doesn't.

Improving Make()

Make() has been presented in this chapter in a simple, uncluttered form, as has the rest of the code. The .INC file can be massaged in various ways to do more. For example, make() can be made to handle normal child windows with only minor modifications. Thus the same WINDOW class could be used for parent and child windows. The alternative would be to create another class, called, say, CHILD, just like I did for CONTROL. The product is evolving all the time, and you may find some interesting new stuff on the Companion Disk or my Web site.

Postamble

You can have a lot of fun playing with these tools. You may think of improvements — let me know.

7

PC Hardware

Preamble

This *could* be an enormous chapter. I'm an electronic engineer, so the hardware is my forte, and I could keep writing for some time. However, the publisher only agreed to a book of around 400 pages, and I'm already pushing it!

Very few assembly language books delve deeply into the hardware, and certainly no Windows books do. Well, many Windows programming books do cover, more or less, the CPU architecture and memory management, as I have done in Chapter 1. For systems programming, it is very helpful if you understand something about the hardware beyond the CPU, i.e., the other chips on the motherboard and plug-in cards, how they work together, and how to utilise them.

CPU Bus

Look at any block-diagram of a computer system, and you are likely to see more than one distinct bus shown. In a nutshell, the bus carries the address and data, and the bus that is directly connected to the CPU, or processor, is called the CPU, system, or processor bus.

The other possible buses perform the same basic task, i.e., carry address and data, but they are optimised for some specific purpose, such as for connection to I/O (input/output) plug-in adaptor cards. Anybody who has been around PCs for awhile will have heard of the *ISA* bus — this is an example of such a special-purpose bus.

The best starting-point is to consider the structure of the bus that is directly connected to the CPU.

Address, data, and control buses

First, we can analyse the CPU bus by breaking it into three logical groups of lines. Really, the bus is a big bunch of wires, with certain wires carrying the address, some carrying data, and some performing control functions — this is shown in Figure 1.5 on page 13.

In fact, each of these groups is sometimes referred to as a bus in its own right.

Difference between memory and I/O access

Intuitively, you can imagine that if the CPU is to access memory, it would have to send the correct address to memory on the address bus, and the data transfer would take place over the data bus. But what about I/O? If the CPU wants to send data to an output device, for example a printer, there is the same scenario of these three buses.

The CPU has to put the appropriate address of the printer output port onto the address bus, and then the CPU will have to put the data onto the data bus.

The essential point here is that the address and data buses are being used for two different purposes. So how do the various chips that are connected to the bus know whether the current operation is an I/O-port access or a memory access? After all, they are all wired onto the same bus, as Figure 1.3 shows.

Control Bus

To understand the problem introduced above of how the bus performs access to two different kinds of chips — memory and I/O — it is necessary to have a closer look at the control bus. First, look at Figure 7.1. Also look at Figure 7.2.

Machine cycle

For a memory access, say, to read the next instruction, the CPU goes through what is called a *machine cycle*, which simply means it reads or writes memory. There is also such a thing as a "null cycle", in which the CPU is doing something within itself for that clock-period.

When the CPU wants to access the memory, it puts an address onto the address bus at the beginning of the cycle, then it puts ALE low to let the rest of the system know there is a valid address.

Depending upon whether the CPU wants to do a read or write operation, it pulses MEMR* or MEMW* low. In the case of a memory read it would send MEMR* low, which tells the memory chips that they are supposed to send data to the CPU.

The memory responds by putting the data on the data bus, and the CPU reads what is on the data bus near the end of the cycle — the exact moment when the CPU reads the data bus is when MEMR* goes high.

Figure 7.1: CPU bus showing some of the control signals.

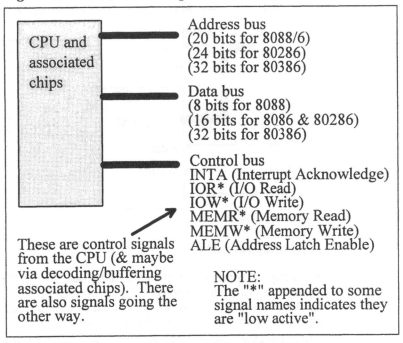

Figure 7.2: Generalised CPU machine cycle.

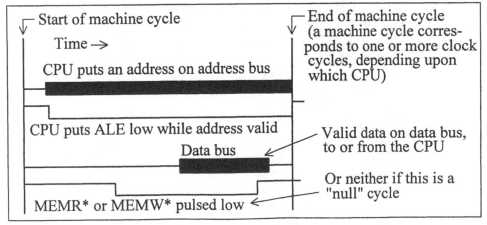

There is still a loose end to the above description. How does memory determine which data to put on the data bus? The CPU is sending out an address asking for the data at a particular memory location. Figure 7.3 shows what the circuitry looks like at the memory end.

Figure 7.3: Interface, CPU to memory.

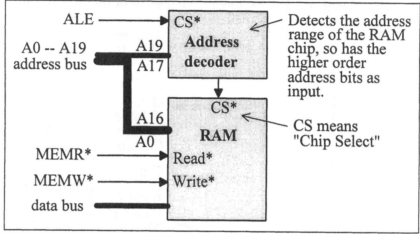

Address Decoder

Basically, a memory chip has a data bus, an address bus, chip select input(s), and read/write control input(s). This example RAM (Random Access Memory) chip has an active-low chip select line coming from an *address decoder*.

This decoder detects the presence on the address bus of the appropriate addresses for this particular memory chip — this chip is being addressed, it "selects" the memory chip.

Note that the address decoder itself has a CS* (chip select) input — ALE is connected to this. It ensures that the address decoder only operates when there is a valid address on the address bus.

Assuming that the RAM is addressed correctly, the CPU tells it via MEMR* and MEMW* which way the data is to go.

Notice that only A17 to A19 go to the address decoder — this is an example circuit only, and specific circuits may differ from this, but generally it is only necessary for some of the address lines to go to the decoder. This is because the memory chip resides at a range of addresses — the lower order address bits go directly to the chip to select a particular memory byte.

Get the idea? The higher address lines select the chip, while the lower lines select a particular location on that chip.

There are three address lines into the decoder in this example, A17 to A19. Say that the decoder is designed to detect an input of 101 binary:

```
BIT: 19 18 17 16 15 14 13 12 11 10  9  8  7 ...  0
      1  0  1  0  0  0  0  0  0  0  0  0  0 ...  0
      1  0  1  1  1  1  1  1  1  1  1  1  1 ...  1
```

This means that the RAM chip occupies address range A0000h to BFFFFh, and the size of the RAM would have to be $2^{17} = 128K$ bytes.

I/O Ports

If you peek back at the diagram of the control bus for the CPU (Figure 7.1), you will see that there are a couple of lines called IOR* and IOW*. These are for reading and writing I/O ports. Unlike some CPUs, such as the 6800 family, that do not distinguish between memory and I/O operations, the Intel 86 family have special instructions and special control lines for I/O.

Figure 7.4 is a typical I/O circuit. Notice its similarity to the memory interface shown in Figure 7.3. A major difference is that IOR* and IOW* go to it, instead of MEMR* and MEMW*.

Figure 7.4: Interface, CPU to I/O port.

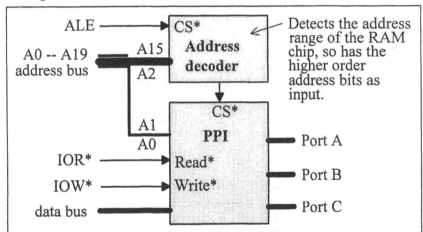

Whenever the CPU executes a read-port instruction (IN), it performs an I/O read machine cycle that looks just like the timing diagram for memory access, except IOR* gets pulsed low. Now we have fully answered the question regarding the dual purposes of the bus.

Programmable More special chips are used for the interface between the buses
Peripheral and the external world. By external I also mean the keyboard, disk
Interface drive, etc. Notice that the I/O chip in Figure 7.4 is labelled "PPI".
This is the name given to a chip used in early-model PCs. PPI
means Programmable Peripheral Interface, and it is a simple
general purpose I/O chip, with three external 8-bit ports, as shown.

The functionality of the original PPI is still in the latest PCs — it
is just contained within a larger chip. We refer to big chips as
VLSI technology (Very Large Scale Integration).

Notice that the PPI in Figure 7.4 has only two address lines going
directly to it. That is because it only has four ports, or registers.
Three of them are ports A, B, and C, and the fourth is a
configuration port.

I/O Instructions

Although the address bus is used to select I/O ports, only A0 to
A15 are used, so the address range is only 64K. With the I/O
instructions, data is always via the AX register. The I/O port
address must be placed in DX before executing the I/O instruction
if the address is over 256.

Examples:

```
IN   AL,2Fh   ;A byte from port-address 2Fh loaded into AL.
IN   AX,2Fh   ;Input a word from 2Fh to AX.
OUT  5,AL     ;Contents of AL written to port 5.
```

Keyboard Interface

This section talks a little bit about interrupts in general, since
interrupts are tied in with how the keyboard interfaces to the
computer. I have introduced interrupts on page 33, and in further
depth on page 250.

Hardware Refer to the circuit of Figure 7.5. The keyboard scancode is routed
description to port A on the PPI chip, when PB7 = 0. The address of port A is
60h, port B is 61h, port C is 62h, etc. The keyboard also generates
an interrupt to the 8259 Interrupt Controller chip, causing INT-9.

With AT-class PCs, including most 386, 486, and Pentium PCs,
we can still visualise the operation as following this pattern. There
are two microcontrollers, an 8031 on the actual keyboard, and an
8042 on the motherboard. The latter implements the functionality
of the original PPI with some changes. For example, port C (62h)

has completely fallen by the wayside.   The 8042 has itself been consumed into larger VLSI chips.

Figure 7.5: Keyboard interface.

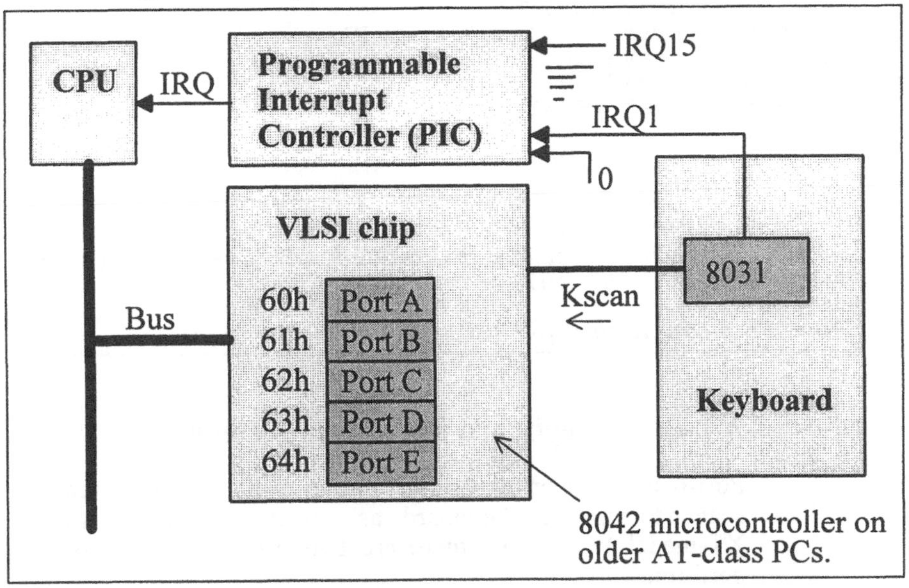

Scancodes   Each key generates a unique *scancode*.   The keyboard outputs a scancode when a key is pressed and again when it is released (and of course generates an interrupt each time).   The difference is determined by PA7 = 0 when pressed, and PA7 = 1 when released.

INT 9   Note that it is the job of the BIOS routine INT-9 to convert the keyboard scancode to ASCII and place it in the input buffer.

A small detail to keep in mind is that the keyboard interrupt goes into the IRQ1 input of the Interrupt Controller chip, hence to the CPUs interrupt input, IRQ.

IRQ to IVT mapping   Question — how does the CPU know that a keyboard interrupt is "INT-9" (i.e., to look at the ninth entry of the interrupt table for the address of the keyboard-handler routine)?

Answer — The CPU and the Interrupt Controller communicate automatically over the data bus, and take care of this detail. INT-8 to INT-F correspond to IRQ0 to IRQ7.

With the AT-class PC though, a view under the hood shows that the 8031 sends a *Kscan* byte for each key press/release, which the 8042 converts to the normal scancode.   Thus, it may be that we never have to encounter Kscan codes, unless our work involves directly programming the 8031.

Keyboard   This is the basic structure of INT-9 in the BIOS, as pointed to by
housekeeping entry 9 in the IVT:

```
...disable keyboard...(AT)
in   al,60h    ;read scancode from PA.
push ax        ;save it.
in   al,61h    ;read PB.
or   al,80h    ;set PB7=1
out  61h,al    ;          /
and  al,7Fh    ;clear PB7.
out  61h,al    ;          /
pop  ax
...INT-15h... (AT only)
...Check for keyboard commands Resend,Ack,Overrun..(AT only)
...Update LEDs... (AT only)
...process key...
...issue End Of Interrupt (EOI)...
```

AT-Class Keyboard Port Enhancements

Port-60h has been expanded beyond that of merely reading the
scancode from the keyboard, as was its sole role in the earlier
XT-model PC. Now, there are two groups of functions it can
perform.

Controlling   Port-60h is now capable of sending commands, mostly directed to
the 8031   the 8031 controller on the actual keyboard.
and 8042
Port-60h can also be used to receive other data, which works in
conjunction with port-64h. Basically, port-64h is for sending
commands to the motherboard 8042 controller, and if any of those
return data, it is read at port-60h. Therefore, you use these two
ports in a particular sequence — an OUT to port-64h, followed by
an IN from port-60h.

Status of   Port-64h can also be read, and it provides status information about
the 8042   the 8042, or whatever chip is being used as the AT-class
motherboard keyboard controller, as shown in Table 7.1.

How to read   A most important point that you should note from Table 7.1 is that
and write   you must test bit-1 before performing any OUT to ports 60h or
port-60h &   64h, and you must test bit-0 before doing an IN from port-60h.
-64h
In fact, a curious piece of information is that on a "Type 1" MCA
PC, you must wait seven microseconds after bit-0 becomes
logic-1, before reading port-60h. MCA is IBM's own proprietary
expansion bus system. Fortunately, it implements ports 60h and
64h much the same as in AT machines. MCA is just about history.

Table 7.1: Port-64h input.

BIT	MEANING
7	=1: Parity error on serial link from keyboard
6	=1: Receive timeout error from keyboard
5	=1: Transmit timeout error to keyboard
4	=0: Inhibit keyboard, from keyboard lock switch
3	=0: Data was just sent to 8042 via port-60h
	=1: Data was last sent to 8042 via port-64h
2	=0: Power-on caused reset
	=1: 8042 self-test completed successfully
1	=0: A write can be made to port-60h or 64h
	=1: No writes allowed to port-60h or -64h
0	=0: A read from port-60h will not be valid
	=1: Data available, use port-60h to read them.

Testing for the XT model

There are a whole lot of commands that you can send to port-64h. Of course, this presumes that you are not using an IBM-XT PC. If your software is to run on AT-class machines only (including MCA, EISA, PCI), then you may have to state that fact with the documentation, and/or your software could perform a simple test. For example, the AAh command to port-64h is a self-test, and if the keyboard controller passes the self-test, it will return the value 55h in port-60h. The XT would not respond to this at all. Of course, what you read from port-60h in an XT could accidentally (though very unlikely) be scancode 55h.

Some of these commands result in data returned via port-60h, but, as noted above, you must read port-64h, in a loop, testing bit-0.

Further references

Further details, such as the commands that port-60h can send to the 8031, are to be found in *The Undocumented PC* by Frank Van Gilluwe, Addison-Wesley, 1994.

For further details on keyboard interrupt handling, refer to Chapter 10.

PC Expansion Buses

If you look under the lid of a PC, the plug-in cards are most obvious. These may include video, printer, serial communication, and disk adaptor.

Some PCs will have some of these on the motherboard rather than as plug-in cards.

The socket into which these boards plug is basically an extention of the CPU bus, with address, data, and control lines, but usually it is in a somewhat modified form.

Some expansion bus standards have become history, such as MCA, VESA local bus, and EISA, so I won't mention them further. The ancient *ISA* (Industry Standard Architecture) standard is remaining popular and is on just about all new PCs. New PCs usually have another bus for high speed known as the *PCI* (Peripheral Connect Interface) local bus.

Industry Standard Architecture (ISA)

8-bit ISA bus

Early PCs use an 8088 CPU, which, despite advertisements, is only an 8-bit CPU, since it is based on the size of the data bus. Hence the ISA bus also has only an 8-bit data bus.

Some early PC compatibles have an 8086 CPU, which internally is identical to an 8088 but has an external 16-bit data bus. As far as I am aware, these machines still have only an 8-bit ISA bus.

16-bit ISA bus

The advent of the AT-model PC, with an 80286 CPU having a 16-bit data bus, saw the introduction of the ISA bus with a 16-bit data bus.

So that 8-bit cards would still work, the older connector was retained, but a second connector, that the 16-bit cards used, was placed end-on to it.

Although 8-bit cards will work ok in a 16-bit ISA system, they will not run quite so fast as 16-bit cards. This is something to be aware of when shopping around — a display adaptor card, for example, could be 8 or 16 bits.

Reference book

There are other books with a stronger hardware focus that will give you further details, such as the functions of the pins on an ISA bus connector and timing diagrams. One such book is *Interfacing to the IBM Personal Computer* by Lewis Eggebrecht, Sams, USA, 1990.

Auto-configuration of plug-in cards

A plug-in card gets an opportunity to execute configuration code stored on ROM on the card during the power-on sequence. One of the typical things that this code does is "hook" interrupt vectors. For example, a video card may hook the BIOS INT-10h interrupt.

In such a case, the address in the IVT will point to the new code that replaces it. This "redirection" is done by DOS itself, by device drivers and TSRs, and by plug-in expansion cards (that may have their own ROM with startup code and new BIOS routines).

Video is a very good example of this. Most PCs have plug-in video adaptor cards that are the interface between motherboard and monitor. This card plugs into an expansion bus connector. The original video services provided by the BIOS-ROM are at entry 10h in the IVT, however, it is normal for the video card to execute some code during start-up, that replaces the address in the the IVT with a new address that points to code in ROM on the video board.

Figure 7.6 shows the effect of an adaptor card. During the power-on sequence, the BIOS startup code sets up the IVT at the beginning of RAM and puts ISR pointers into entries zero to 1Fh. Entry 10h is the video-handling ISR, and this entry points to an ISR in the BIOS-ROM.

Figure 7.6: BIOS extensions during power-up.

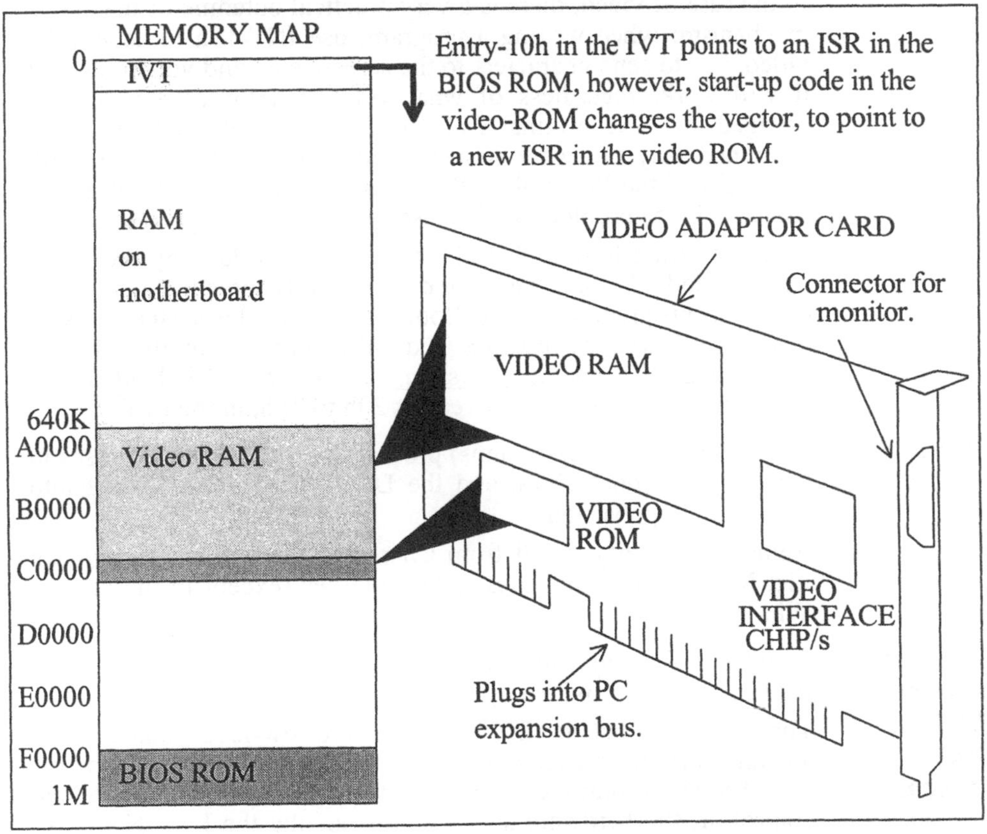

A little bit later in the start-up sequence, the memory address range C0000h to C8000h is scanned, in 2K increments, looking for any code that may be present on plug-in cards. Usually, an adaptor card has switches that set the address range of the on-board ROM to a vacant place in the PCs memory map. Note though, that *Plug and Play* is replacing switches with programmable configuration.

It is normally expected that a video adaptor will have video-ROM in the C0000h to C8000h region, in which case it executes. When the start-up code of the video-ROM executes, it changes the contents of entry-10h in the IVT to point to its own video-ISR, contained in its own ROM.

Note also that a little later, the start-up sequence scans the address range C8000h to F4000h looking for more ROMs, which will also be executed. Incidentally, valid code is identified by 55AAh at the first two memory locations, with offset-2 holding the size of the ROM module, expressed in 512-byte blocks. Execution will commence at offset-3 of the ROM.

In the case of video, there is a very practical outcome of the above mechanism: when writing a program, use INT-10h to access the video, i.e., to send characters to the screen, etc., and you know that it will work, regardless of what video adaptor card you have plugged in. The original INT-10h ISR in BIOS-ROM is fairly basic and may not work properly with your video adaptor card, especially if the PC is old. The redirection of INT-10h to a new ISR avoids the problem of obsolesence.

It is interesting to note that all of the above is done by the BIOS start-up code before the system disk is accessed. Later, the bootstrap program from the Boot Record on the system disk is loaded, followed by IO.SYS and MSDOS.SYS, in the case of loading the DOS operating system. When IO.SYS is loaded, and executed, it sets up interrupt vectors 20h to 3Fh, in the IVT.

BIOS & DOS vectors in IVT

BIOS-ROM (or the extensions) provides services with addresses in the IVT. So does DOS, and the DOS services are loaded into RAM during power-on.

Actually, you may recall from Chapter 1 that the hidden system file, MSDOS.SYS, has these DOS routines (except in the case of Windows 95).

Difference between BIOS and DOS services

So what is the major difference between the services provided by BIOS and those provided by DOS?

The answer is that the BIOS services are low-level, that is, they are for more basic access to, and control of, the hardware of the PC. The DOS routines provide mostly higher level access to, and control of, the hardware and resources of the the PC. Note also that some of the DOS routines are not actually for accessing hardware: rather they are operating system management functions.

Peripheral Connect Interface (PCI)

Figure 7.7 shows a typical configuration, though do note that there can be variations on this. For example, RAM memory could be interfaced to the PCI bus, rather than directly onto the CPU bus (or both).

Figure 7.7: PCI--CPU--ISA bridges.

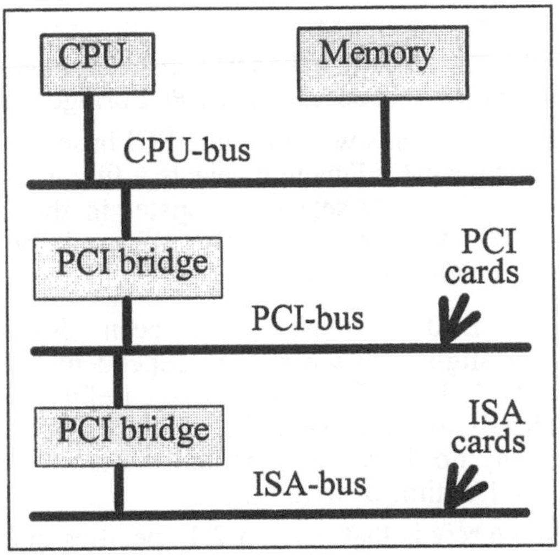

PCI bridge   The PCI *bridge* is a chip, and although it is not obvious from the figure, there are different kinds of chips for different bridges, such as between CPU-PCI and PCI-ISA. Also, the PCI plug-in cards themselves will have a PCI chip. One great advantage of having a special bridge chip between buses is that they allow address translation, so that a memory or I/O address on the CPU bus will be a different address on the PCI bus. In fact, the bridge chip is highly programmable and has its own configuration memory that, most importantly, is independent of the main memory and I/O map.

Configuration With PC systems, the standardized method of accessing the
memory   configuration memory of a PCI chip is by two reserved 32-bit I/O ports, 0CF8h and 0CFCh. The former is used for addressing a location in configuration memory and the latter for reading/writing it.

The former, 0CF8h, is called CONFIG_ADDRESS, and the latter, 0CFCh, is called CONFIG_DATA.

It is important to know that these two ports can allow you to access the configuration memory on any of the PCI interface chips (on any adaptor card). The 32-bit data that you write to CONFIG_ADDRESS, is formatted as in Figure 7.8.

Figure 7.8: CONFIG_ADDRESS write format.

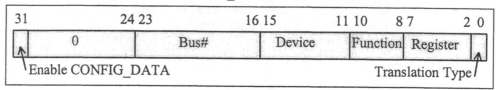

Bit-31 has to be set, otherwise the OUT instruction is treated like a normal I/O operation (not accessing the PCI bridge chip).

Bus# is for use in systems with multiple PCI buses, Device selects a particular adaptor card, Function selects a function that the card understands, and Register selects a register in the configuration memory. An OUT to CONFIG_ADDRESS would be followed by an IN or OUT to CONFIG_DATA.

PCI BIOS extension

Fortunately, a BIOS extension has been defined to give programmers a slightly less hardware-dependent mechanism for accessing the PCI chips. One point to be careful about, however, is that not all BIOSs implement the new specification fully, or, maybe, they may not have implemented the latest version of the specification (2.1 at time of writing).

Of particular interest is that version 2.1 specifies entry points for Real mode, 16-bit Protected mode, and 32-bit Protected mode. The 386 and later CPUs can operate in 16-bit Protected mode, which is what Windows 3.x applications run in, and they can also operate in 32-bit Protected mode, which is what "native" Windows NT and Windows 95 applications run in.

Reference source

A further source of information about this is *PCI System Architecture* (third edition) by Tom Shanley and Don Anderson, Addison-Wesley, USA, 1995.

Protected mode PCI BIOS

The normal BIOS that we have considered so far in this book is designed, at least originally, for an 8088 CPU, which only runs in Real mode. In a nutshell, Real mode uses the now-familiar segment:offset form of addressing, which has a 1M upper limit. The 286 and 386 CPUs are able to operate in Protected mode, which uses a different addressing mechanism and is able to address extended memory beyond 1M (as explained in Chapter 1).

One of the greatest criticisms of Windows 3.x, is its reliance on DOS and BIOS — to call any of these software interrupts, the CPU must switch back into Real mode (which takes time).

It is possible to write code that can execute in either Real or 16-bit Protected mode, and PCI BIOS has done this — via INT-1Ah, function B1h (Table 7.2). Great — you can call this from a Windows 3.x application, and the CPU will not have to switch back to Real mode.

The PCI BIOS requires an entirely different set of routines for 32-bit Protected mode.

Table 7.2: PCI BIOS access.

Real mode.	Use INT-1Ah, AH = B1h, like any other software interrupt
16-bit Protected mode	... ditto ...
Virtual-86 mode	... ditto ...
32-bit Protected mode	BIOS is scanned, for a signature, indicating presence of 32-bit BIOS, and an entry point is located. The services are accessed by a FAR CALL.

Note that, technically, it is possible, if you are writing a 32-bit application, to get it to call the Real mode/16-bit Protected mode PCI BIOS services, but this is starting to get too involved at this stage.

PCI summary

Here are some of the highlights of the PCI architecture:

- Multiple independent PCI buses in the one PC.
- 32-bit data bus at up to 132M/sec (megabytes/sec), and 64-bit at up to 264M/sec.
- Fully synchronous with CPU bus up to 33MHz.
- PCI connector can be mounted alongside an ISA/EISA connector, so either type can occupy that physical space on the motherboard/chassis.
- Processor independent
- Support for 64-bit addressing
- Support for 5V and/or 3.3V supply
- Full multi-master capability, allowing any PCI master peer-to-peer access to any other PCI master/target.
- Full auto-configuration (no dip switches on cards).

Postamble

I have introduced PC hardware, but so much remains to be explained. I covered the keyboard interface and expansion bus, but these are only "samplers". What about parallel and serial, disk drive, timer, real-time clock, and other interfaces? Some of these I do touch on in later chapters, however this book will grow into something enormous if I try to cover everything.

I could cover these in the next edition though. Let me know if you really like the idea.

Choice of keyboard interface and expansion bus serve as case studies, so that you can see how the principles earlier in the chapter are applied.

8

BIOS, DOS, &
Windows Low-Level
Services

Preamble

*What's
in this
chapter*

This chapter introduces the services available to the Windows programmer, but from a viewpoint that you would expect of a book on assembly language. I have covered two major aspects: the DOS services and the Windows low-level services.

This chapter gives an overview, and the next chapter provides practical code.

We haven't been so far away from the operating system in earlier chapters, but now is the time to delve in further.

*DOS/BIOS
INTs*

In this chapter I have particularly been concerned about the relationship between DOS and Windows. We have a new operating system running on top of DOS, with the CPU in Protected mode — how much of the old DOS can we still use?

Then there is the related issue of how DOS itself has been changed to handle the new CPUs and operating conditions. What are these changes?   For example, INT-16h, the keyboard handler under DOS, doesn't work under Windows.

I have already mentioned the problem of calling the old DOS interrupt services with the CPU running in Protected mode (page 33).

I introduced some of the first DOS services to utilize Protected mode (page 18).

Why use DOS/BIOS services?

Old habits die hard, and DOS programmers are going to be loath to give up their familiar DOS and BIOS services in favour of Windows functions, especially if some of the old services seem better suited to certain tasks or if the Windows functions don't seem to do anything equivalent, or do it poorly.

In many cases, the Windows solutions are painfully slow. If you are after performance, for certain kinds of applications it may be optimal to use certain DOS services.

An interesting example comes to mind — that of printing. Windows printing is designed for dumping a complete page at a time to the printer, but if all you want to do is output a line at a time to your faithful old dot matrix, perhaps to log some systems events, it is darned awkward. It is, of course, a pushover for DOS — you can use INT-21h to output a single character at a time, and when you send a carriage-return character, the line prints.

Since Windows uses its own special printer drivers for output, the question naturally arises about whether you can use the old DOS service. Will it work? Will there be a clash?

The answer is that it works fine, but yes clashes **are** possible. However for every problem there is a solution, including that of contention over resources.

Another qualification that needs to be made is that Microsoft has taken the opportunity with 32-bit applications to restrict BIOS/DOS and other low-level access. This will be explained as you read ahead.

DOS in the future

The advent of Windows 95 does not mean that DOS is dead. Even though Windows 95 does not identify DOS as a separate product, still, it is there. You can start the PC with the DOS prompt, or launch a DOS box from Windows, just as before. It's really more of the same thing, despite the Windows 95 publicity hype.

There are a number of issues with regard to how DOS lives alongside Windows, some of which I have gone into in Chapters 11 and 14.

BIOS and DOS Services

Overview   This is a mysterious gray area, very poorly documented by Microsoft. Although Windows runs in Standard or Enhanced Protected mode, most BIOS and DOS services still work, with various caveats.

DPMI   Apart from the standard services, Windows also supports a special group of DOS services, called the *DOS Protected Mode Interface* (DPMI). These consist of some INT-2Fh services and INT-31h services.

INT-2Fh has a range of sub-functions available under DOS, but Windows adds some extra functions. If you look in any DOS programming book you won't find anything on these extra functions, nor on INT-31h. Even Microsoft's own reference bible, *The Programmer's PC Sourcebook* (second edition) by Thom Hogan, USA, 1991, has nothing on these services.

Reference sources   You have to scratch around in strange places to find the information. This book brings much of it together, and where it does not, I give the appropriate reference. Microsoft's *Device Development Kit* (DDK) has reference material on DPMI, and I think their *Archive Library* CD-ROM has also. Obtaining these requires that you join the *Microsoft Developer's Network* (MSDN), and this is where Microsoft has us "over a barrel" — they want quite a lot of money for membership.

You can find a lot of information on the Internet. For example, a site with lots of links for developers is:

http://www.r2m.com/windev/

Another site with DPMI reference information is:

http://www.delorie.com/djgpp/doc/

DPMI overview   First, I will fit DPMI into its place in the overall scheme of things (the meaning of life and all that), before getting into a look at the standard BIOS and DOS services:

"DPMI enables DOS applications to access the extended memory of PC architecture computers while maintaining system protection. It also defines a new interface, via software interrupt 31h, that Protected mode applications use to do such things as allocate memory, modify descriptors, and call Real mode software (using segment:offset addressing and running within the 1M limit)."

This is a direct quote from some loose-leaf pages sold by Microsoft under the title *Windows Developer's Notes* (part number 050-030-313). It is extra material not found in the SDK[1] and has a couple of pages on DOS and DPMI — hardly anything, though, as it appears that Microsoft has the attitude that the less we know about how Windows works "under the hood", the better.

DPMI 0.9 and 1.0

Despite documentation to the contrary (see quote below), Windows 3.0, 3.1, and 95 only support DPMI version 0.9. The *Windows Developer's Notes* have the following warning:

> "Windows 3.0 running in 386 Enhanced mode supports DPMI version 0.9. Windows 3.0 running in Standard mode supports a subset of DPMI that enables applications to call TSR programs and device drivers running in real (or virtual-86) mode."

> "Windows applications should call *only* the following AX values for DPMI version 0.9 functions: 0200h, 0201h, 0300h, 0302h, 0303h, 0304h, 0305h. Windows applications should not use DPMI's MS-DOS memory management functions. The Windows 3.0 Kernel has two functions, GlobalDOSAlloc() and GlobalDOSFree(), that should be used by Windows applications and DLL's for allocating and freeing MS-DOS addressable memory. Other than those listed above, no DPMI functions are required for Windows applications since the Kernel provides functions for allocating memory, manipulating descriptors, and locking memory.
> Non-Windows applications running in 386 Enhanced mode can use all the DPMI version 0.9 functions, since they are not restricted by the Kernel."

However, to throw a spanner into the works, Microsoft has stated this in documentation supplied with the SDK v3.1:

> "Windows **3.0** and later in 386 Enhanced mode supports DPMI **version 1.0**. Windows 3.0 and later in Standard mode supports a subset of DPMI that enables applications to call terminate and stay resident (TSR) programs and device drivers running in Real (or virtual-86) mode."

---

[1] Much of the material from the *Developer's Notes* has found its way into the latest SDK for Windows version 3.1. This consists of about 12 books. DOS and DPMI notes are to be found in *Microsoft Windows Programmer's Reference, Volume 1: Overview*, the first of four volumes. This is now on CD-ROM supplied with the SDK, though in many cases Microsoft will sell printed versions.

If you think that the above two quotations are contradictory, join the club. What's it to be: 0.9 or 1.0? I received a clarification from Microsoft that Windows 3.0 and 3.1 (and now 95) only support DPMI 0.9. Their reply to me also had another interesting comment:

> "... Standard mode understands how to allocate memory from a DPMI provider ... Enhanced mode does not."

When to use DPMI services

There are Windows functions that overlap DPMI services, but most of the latter are undocumented, and in the light of the above comments from Microsoft, we are left between a "rock and a hard place". Andrew Schulman, *PC Magazine*, Jan. 28, 1992, page 323, puts it this way:

> "You're stuck with using either DPMI INT 31h functions ... which Intel documents but Microsoft doesn't sanction ... or Windows KERNEL functions, which Microsoft doesn't document. What a choice!"

Newly sanctioned functions

Windows 3.1 does make some of the previously undocumented functions "official", by documenting them in the SDK, and also introduces some new low-level functions, many of which cannot be used with Windows 3.0. Since there are going to be a some (?) users out there still using 3.0, I have been careful in this chapter to clarify which functions are not backwards compatible.

Microsoft has put some functions into a library, TOOLHELP.DLL, that you can bundle with your application for backwards compatibility with Windows 3.0.

Restrictions on using DPMI

A final note is that other programmers have commented in the press (and it is my own empirical experience) that the DPMI services work under Windows. I've tried most of them, but not all.

The main thing to be careful about is using those DPMI services that might conflict with Windows' management of the memory, such as allocation of memory blocks (see quotation on page 198).

In a virtual machine other than the system virtual machine (see page 274), there should not be any conflict with Windows' memory management, and you can use all the DPMI services (Microsoft sanction this statement).

I've done the right thing and printed Microsoft's discouragement for extensive use of DPMI. Code that you will see in subsequent chapters has been tested in both Standard and Enhanced modes, but with a book of this nature I do have to insist on a total disclaimer of any liability. You use the code with this understanding.

Windows 95   Most of my code has also been tested under Windows 95 and works. However, this statement is true of 16-bit applications running in Windows 95 — native 32-bit applications are somewhat more restricted. There are work-arounds. For example, many of the low-level API functions are available as 32-bit versions, but are not supported by the import library (during linking), nor are they documented. However, we can still use them (refer page 235).

One problem is that you can't just call 16-bit functions such as the 16-bit API functions from 32-bit code. Most of the interrupt routines also assume that the caller is 16-bit code.

Standard DOS Interrupts

Microsoft implies from their *Developer's Notes* that most of the DOS services will work ok when called from a Windows program running in (16-bit) Protected mode.

Those specifically not supported in Protected mode, and which will fail, are:

- INT-**20**h Terminate program
- INT-**25**h Absolute disk read
- INT-**26**h Absolute disk write
- INT-**27**h Terminate and stay resident
- INT-**21**h/AH =

00h	Terminate process
0Fh	Open file with FCB
10h	Close file with FCB
14h	Sequential read
15h	Sequential write
16h	Create file with FCB
21h	Random read
22h	Random write
23h	Get file size
24h	Set relative record
27h	Random block read
28h	Random block write

The following DOS INT-21h functions will work, but will behave differently from Real mode DOS versions:

Hooking Protected or Real mode interrupts

- AH = **25h and 35h**   **Set/Get interrupt vector**.

"These functions set and get the Protected mode interrupt vector. They can be used to hook hardware interrupts, such as the timer or keyboard interrupt, as well as to hook software interrupts. Except for INT-23h, INT-24h and INT-1Ch, software interrupts that are issued in Real mode[1] are not reflected to Protected mode interrupt handlers. However all hardware interrupts are reflected to Protected mode interrupt handlers before being reflected to Real mode."

- AH = **38h Get country data**.

"This function returns a 34-byte buffer containing a doubleword (DWORD) call address at offset 12h that is used for case mapping. The DWORD contains a Real mode address. If you want to call the case-mapping function, you need to use the DPMI translation function to simulate a Real mode FAR call."

- AH = **44h, subfunctions 02h, 03h, 04h, and 05h**.

"These I/O control (IOCTL) subfunctions are used to receive data from a device or send data to a device. Since it is not possible to break the transfers automatically into small pieces, the caller should assume that a transfer of greater than 4K will fail unless the address of the buffer is in the low 1 megabyte."

- AH = **44h, subfunction 0Ch**.

"Only the minor function codes 45h (Get Iteration Count) and 65h (Set Iteration Count) are supported from Protected mode. The extensions of this IOCTL subfunction that are used for code page switching (minor function codes 4Ah, 4Ch, 4Dh, 6Ah and 6Bh) are not supported for Protected mode programs. You must use the DPMI translation functions if you need to use this IOCTL subfunction to switch code pages."

- AH = **65**h, **Get extended country information**.

"This function is supported for Protected mode programs. However, all the DWORD parameters returned will contain Real mode addresses. This means that the case-conversion procedure address and all the pointers to tables will contain Real mode segment:offset addresses. You must use the DPMI translation functions to call the case-conversion procedure in Real mode."

---

[1]   This is a direct quotation from the *Developer's Notes*. The term "Real mode" in this publication is also taken to cover virtual-86 mode.

NetBIOS interrupts

Windows also supports the DOS NetBIOS interrupts. The *Developer's Notes* advise that all of the network control blocks (NCBs) and buffers must reside in fixed memory that is page locked. Also, all code that calls NetBIOS directly should reside in a DLL to ease the porting of the application to other operating environments. I haven't written anything more about NetBIOS support in this book. If you want more information, go to the *Device Driver Developer Kit* (DDK).

Accessing BIOS/DOS from Protected mode

Earlier in the book (pages 33+) I explained about the Interrupt Vector Table (IVT) used by Real mode interrupts and the Interrupt Descriptor Table (IDT) used in Protected mode. I explained that Windows has in some cases provided alternative services via the IDT where necessary, but in many cases the vector in the IDT points to a handler that changes the CPU to Real mode (virtual-86 actually) and calls the Real mode service as pointed to by the IVT. This mechanism is shown diagrammatically on page 268.

More vagueness

The *Developer's Notes* say that Windows provides support for "all MS-DOS interrupts" other than those specifically blacklisted above.

Despite the above comment, heed the warning from *Guide to Programming* (SDK 3.0 manual):

> "... you should use interrupts with extreme caution and only when necessary".

The SDK documentation leaves you hanging on the cliff at that point — there is virtually no further clarification about what you can and cannot use and under what conditions and circumstances. Furthermore, the SDK 3.1 documentation does **not** have this warning! The Windows 95 SDK just about ignores BIOS and DOS interrupts entirely.

I have already mentioned that INT-16h, the keyboard handler, works fine — **except** that you need to be aware that Windows hooks the INT-9 hardware vector that puts characters from the keyboard into the keyboard buffer. Windows has its own 128-character buffer and its own keyboard handler.

Leaving the standard BIOS and DOS services for now, I will focus on DPMI.

DOS Protected Mode Interface (DPMI)

Reference sources

The main sources of information for DPMI are the specification itself: *DPMI Specification*, version 1.0, DPMI Committee, 1991. This committee is hosted by Intel Corporation, and members include Microsoft, IBM, and Borland. Further information is in Microsoft's DDK and in *Writing Windows Device Drivers* by D. A. Norton, Addison Wesley, USA, 1991, and on-line at:

```
http://www.delorie.com/djgpp/doc/
```

I have summarized the major DPMI services in Appendix C, and you will find practical code with further explanation in subsequent chapters.

DPMI elsewhere in this book

What follows are some of the underlying principles of DPMI. If any of it doesn't make sense, don't worry, as it should be much clearer when actual code is shown in the next chapter. I have also provided more underlying detail in Chapters 11 and 14.

You can get a good overall idea of what the DPMI services do by examining Appendix C. They provide the kind of services that the old Real mode DOS services don't, that is, services connected with the descriptor tables, managing extended memory, going between Real and Protected modes, getting at real memory from Protected mode, and getting at the CPU control registers.

I introduced some DOS services back on page 18, but they are primitive. DPMI does a much more thorough job and is specially designed for the multitasking environment.

Host and client

Windows provides the DPMI services for our program to use, so the correct terminology is that Windows is the DPMI **host**, while our program is the **client**.

INT-31h, INT-2Fh

The DPMI services are available through INT-31h, which is only available in Protected mode. DPMI provides INT-2Fh services to obtain information about DPMI — these run in Protected or Real mode (see Appendix C and Chapter 9). A DPMI host must be running to provide INT-31h services, though note that Windows is not the only DPMI host. Other DOS extenders and memory managers are also DPMI hosts. For example, 386Max is a superb memory manager and DPMI 0.9 host from Qualitas Corp. that enables you to write DOS applications that can run in Protected mode.

INT-2Fh extensions provided by a DPMI host

The basic INT-2Fh services are:

- AX = **1680**h   Release current virtual machine's
time slice.
- AX = **1686**h   Get CPU mode.
- AX = **1687**h   Return Real-to-Protected mode
switch entry point.
- AX = **168A**h   Get vendor-specific API entry point.

Apart from these functions, DOS has a lot of other functions under INT-2Fh. Other software products provide enhancements to INT-2Fh. Windows provides extra services for device driver development (discussed a bit further on), and the new specification for FAX and modem communication adds further functions. INT-2Fh is a mixture of all sorts of stuff.

INT-31h
logical
groups

INT-31h has these major groups of services:

- **Extended memory management services**.
Works with blocks of linear memory above 1M and deals with *linear addresses* (refer back to pages 28+). These services allocate and release memory, but you still have the problem of accessing it, for which you need a descriptor — for that you need the descriptor management services.
- **LDT descriptor management services**.
These allocate, modify, inspect, and deallocate descriptors in the application's Local Descriptor Table (LDT).
- **Page management services**.
These will only work on a system with paging. They are used for locking and unlocking pages in memory.
- **Interrupt management services**.
These allow Protected mode applications to intercept Real mode interrupts and hook processor exceptions. Some also enable cooperation with the DPMI host in maintaining a *virtual interrupt flag* for the application.
- **Translation services**.
These enable Protected mode programs to call Real mode software directly. They also provide the reverse.
- **DOS memory management services**.
These work like the DOS INT-21h functions 48h, 49h, and 4Ah, but work from Protected mode. They automatically create and destroy descriptors, so that memory blocks can be accessed easily from Protected mode.
- **Debug support services**.
These set and clear watchpoints; used by debuggers.
- **Miscellaneous services**.
These provide information about DPMI, support for the

creation of TSRs, direct access to memory mapped peripheral devices, interrogation of the numeric coprocessor status, and emulation of the coprocessor.

INT-2Fh Extensions

Display driver services

Apart from the DPMI extensions to INT-2Fh provided as part of the DPMI, Windows also provides other extensions.

Functions 4000h to 4007h are for use with the display driver. Note that conceptually there are two different display drivers: the *virtual* driver (VDD) at the Windows end and the actual driver that does the dirty work:

- AX = **4000**h
  A program calls this function to determine how much work the Windows Virtual Display Driver (VDD) must do when it switches Windows between the foreground and the background. It also tells the VDD to allow the program to have direct access to the video hardware registers.
- AX = **4001**h
  Tells the display driver to save the current video state.
- AX = **4002**h
  Tells the display driver to restore the video hardware state saved by 4001h.
- AX = **4003**h
  Tells Windows Virtual Display Driver (VDD) that execution is currently in a critical section. This function appears to make the VDD pause until 4004h releases it.
- AX = **4004**h
  Tells VDD that critical section is finished.
- AX = **4005**h
  Similar to function 4001h
- AX = **4006**h
  Similar to function 4002h.
- AX = **4007**h
  A program tells the VDD that it has finished accessing the hardware registers. This is the complement of 4000h.

I think it unlikely that you will need to call 4000h and 4007h, unless you are designing your own display driver. 4000h is designed for use by a display driver to communicate with the VDD prior to the VDD calling 4005h. This sequence terminates when the VDD calls 4006h to let the display driver restore its state and continue functioning. After this the display driver calls 4007h to tell the VDD that it's all over. Chapter 9 has an example of usage, and Appendix D is an INT-2Fh reference.

Real and virtual driver interaction

Another group of INT-2Fh functions has to do with communication between DOS Real mode drivers and virtual mode drivers (VxDs).

I have noted below that some of the functions have been used in example programs, along with more detail on their usage. Also Chapter 11 discusses these functions in more depth.

Note that only functions 1605h and 1606h are available in Windows Standard mode.

Note also that these services, although designed for communication between device drivers, are quite general and can be used by any program. Chapters 11 and 14 develop a TSR that uses them.

- AX = **1600**h
  Obtains the version number of 386 Enhanced mode Windows.
- AX = **1605**h
  Windows calls this to tell DOS drivers that it is loading (example of usage Chapter 14).
- AX = **1606**h
  Windows calls this to tell DOS drivers that it is quitting (example of usage Chapter 14).
- AX = **1607**h
  A virtual driver calls a DOS driver.
- AX = **1608**h
  Windows calls this to tell DOS drivers that it has completed initialisation.
- AX = **1609**h
  Windows calls this to tell DOS drivers it is exiting Enhanced mode.
- AX = **1680**h
  Yields the current virtual machine's time slice.
- AX = **1681**h
  A driver calls this to tell Windows not to switch virtual machines.
- AX = **1682**h
  This is the complement of 1681h
- AX = **1683**h
  Returns the ID of the currently executing virtual machine.

- AX = **1684**h
  Allows a DOS mode driver to request services from a virtual driver.
- AX = **1685**h
  Allows a driver to switch virtual machines (examples of usage, Chapters 11 and 12).

INT-4Bh:
DMA
services

Windows drivers also make use of INT-4Bh for virtual *Direct Memory Access* (DMA), and I refer you to page 264.

Again, these are extensions that are not part of DOS but are provided by Windows. They are designed especially for the difficulty of using DMA controllers with a CPU running in Protected mode.

Windows Functions

There are some Windows functions that perform in a similar manner to DPMI services, so there is overlap.

Overview

What I have done in this section is not give exhaustive definitions of the functions, as that would require a complete book on its own. You need a lot of reference material for Windows development, and where appropriate I have given the reference.

There are two broad groups of functions: those available in USER, KERNEL, or GDI DLLs and those available within device drivers and other DLLs.

In the latter case, you will find functions of the same name. For example, enable() and disable() exist in all drivers. Obviously your program must be able to select which one it is to call, and that I have shown in the next chapter.

The Windows functions are all in files known as *Dynamic Link Libraries* (DLLs), and are loaded at run-time.

Low-level
function
summary

What follows is a collection of Windows functions that you may find useful for low-level work. The list immediately below all belong in either USER, KERNEL, or GDI DLLs.

Note that although many of the memory management functions could be considered low-level, I have only included those directly concerned with descriptors and selectors, with one exception: GLOBALPAGELOCK.

Functions are in Windows 3.0 and 3.1, unless stated otherwise, even if documented in one version and not the other. References to the "SDK" without specifically naming 3.0 or 3.1 apply to both.

I have used an asterisk if a function is not *directly* supported by 32-bit applications in Windows 95, optionally followed by a recommended 32-bit alternative. I have used a "$" if a function is unofficially available in the 32-bit Windows 95 API.

Low-level USER/GDI/ KERNEL function summary

- **ALLOCCSTODSALIAS**
  Not described in the SDK. Allocates a new data selector that aliases an existing code selector. *

- **ALLOCDSTOCSALIAS**
  Accepts a data segment selector and returns a code segment selector that can be used to execute code in a data segment. *

- **ALLOCSELECTOR**
  Allocates a new selector. *

- **ALLOCSELECTORARRAY**
  Not described in the SDK. Allocates an evenly spaced array of selectors. *

- **CALLMSGFILTER**
  Passes a message and other data to the current message filter function.

- **CATCH**
  Copies the current execution environment to a buffer. Complement is THROW. *

- **CHANGESELECTOR**
  Generates a temporary code selector that corresponds to a given data selector, or a temporary data selector that corresponds to a given code selector. Note that SDK 3.1 has renamed this PRESTOCHANGOSELECTOR! (both names will work). *

- **DEATH**
  Not documented in the SDK. Turns off the Windows display driver and changes screen to text mode. Used in Chapter 9. Complement is RESURRECTION. *

- **DEBUGBREAK**
  Not documented in the SDK. Forces a break to the debugger.

- **DEBUGOUTPUT**
  Available with Windows 3.1 only. Sends formatted messages to a debugging terminal.

- **DEFHOOKPROC**
  Calls the next filter function in a filter function chain. *CallnextHookEx()

- **DIRECTEDYIELD**
  Not documented in SDK 3.0. Forces execution to continue at a specified task. *

- **DISABLEOEMLAYER**
  Not documented in the SDK. Turns off Windows display, keyboard, and mouse and changes to text mode; restores DOS I/O. Complement is ENABLEOEMLAYER. *

- **DOS3CALL**
  Issues a DOS INT-21h interrupt (but doesn't use INT). *

- **ENABLEHARDWAREINPUT**
  Enables or disables keyboard and mouse input throughout the application. *

- **ENABLEOEMLAYER**
  Not documented in the SDK. See Chapter 9. Complement of DISABLEOEMLAYER. *

- **ENABLEWINDOW**
  Enables or disables keyboard and mouse input to a specified window or control.

- **FATALEXIT**
  Displays current state of Windows on debugger monitor and prompts on how to proceed.

- **FREESELECTOR**
  Frees a selector originally allocated by ALLOCSELECTOR(), ALLOCCSTODSALIAS(), or ALLOCDSTOCSALIAS() functions. *

- **GETASYNCKEYSTATE**
  Returns interrupt-level information about the key state.

- **GETCURRENTPDB**
  Returns the current DOS Program Segment Prefix (PSP). *GetCommandLine(),GetEnvironmentStrings()

- **GETCURRENTTIME**
  Returns the time elapsed since the system was booted.

- **GETDOSENVIRONMENT**
  Retrieves the environment string of the currently running task. *GetEnvironmentStrings()

- **GETFREESYSTEMRESOURCES**
  Only available in Windows 3.1. Returns the percentage of free system resource space. *

- **GETINPUTSTATE**
  Returns TRUE if there is mouse or keyboard input.

- **GETINSTANCEDATA**
  Copies data from a previous instance of the application to the data area of the current instance. *

- **GETKBCODEPAGE**
  Determines which OEM/ANSI code pages are loaded.

- **GETKEYBOARDSTATE**
  Copies an array that contains the state of keyboard keys.
- **GETKEYNAMETEXT**
  Retrieves a string containing the name of a key from a list maintained by the keyboard driver.
- **GETKEYSTATE**
  Retrieves the state of a virtual key.
- **GETNUMTASKS**
  Returns the number of tasks currently executing in the system. *
- **GETSELECTORBASE**
  Not described in SDK 3.0. Gets the linear base address of the specified selector from the descriptor table. *
- **GETSELECTORLIMIT**
  Not described in SDK 3.0. Gets the limit of the specified selector from the descriptor table. *
- **GETSYSTEMDEBUGSTATE**
  Only available in Windows 3.1. Returns system status information to a debugger. *
- **GETWINDEBUGINFO**
  Available in Windows 3.1 only. Queries current system debugging information. *
- **GLOBALDOSALLOC**
  Recommended by Microsoft instead of equivalent DPMI service. Allocates a block below 1M linear address space. Returns both a selector and segment. Complement is GLOBALDOSFREE. *
- **GLOBALFIX**
  Prevents the memory block from moving in linear memory. You would use this in Standard mode to lock a block in place. Complement is GLOBALFREE. $ *WOWGetVDMPointerFix
- **GLOBALHANDLE**
  Supplies a selector and returns a handle to the memory block.
- **GLOBALPAGELOCK**
  Prevents a segment from being paged out or moved. You can use this in Enhanced mode to guarantee a segment will be present at all times. Locks the segment at a physical address. Complement is GLOBALPAGEUNLOCK. *VirtualLock()
- **GLOBALWIRE**
  I'm not sure what this one does — it seems to be similar to GLOBALFIX. $ *

- **HARDWARE_EVENT**
  Available in Windows 3.1 only. Places a hardware-related message into the system queue. *

- **HMEMCOPY**
  Available with Windows 3.1 only. Copies a block of data from one address to another. *

- **LOADMODULE**
  Executes a separate application. *Supported but recommend CreateProcess().

- **LOCKINPUT**
  Available in Windows 3.1 only. Locks (and unlocks) input to all tasks except the current one. *

- **LOCKSEGMENT**
  Locks a segment in memory. Its complement is UNLOCKSEGMENT(). *

- **NETBIOSCALL**
  Issues a NetBIOS INT-5Ch interrupt. *

- **OUTPUTDEBUGSTRING**
  Sends a debugging message to the debugger if present, or to the AUX device if the debugger not present.

- **PEEKMESSAGE**
  Checks the application message queue without waiting.

- **PRESTOCHANGOSELECTOR**
  Described in the SDK 3.1. Same as CHANGESELECTOR documented in SDK 3.0. Obtains an alias to a code or data selector. *

- **REPAINTSCREEN**
  Not described in SDK. Tells the GDI to repaint the entire display. *

- **RESURRECTION**
  Not documented in SDK. Turns on Windows display driver. See the example, Chapter 9. Complement is DEATH. *

- **SELECTORACCESSRIGHTS**
  Not described in the SDK. Sets the attributes of the specified selector in the descriptor table. *

- **SETSELECTORBASE**
  Not described in SDK 3.0. Sets the linear base address of the specified selector in the descriptor table. *

- **SETSELECTORLIMIT**
  Not described in SDK 3.0. Sets the limit of the specified selector in the descriptor table. *

- **SETWINDOWSHOOK**
  Installs a system and/or application filter function. Applications specific to Windows 3.1 should use SETWINDOWSHOOKEX. *

- **THROW**
  Restores the execution environment to the specified values. Complement is CATCH. *

- **UNHOOKWINDOWSHOOK**
  Removes a Windows filter function from a filter function chain. Complement is SETWINDOWSHOOK. Applications specific to Windows 3.1 should use UNHOOKWINDOWSHOOKEX. *Supported but recommend UnHookWindowsHookEx().

- **SETWINDEBUGINFO**
  Only available with Windows 3.1. Sets current system debugging information. *

- **WINEXEC**
  Executes a separate application. *Supported but recommend CreateProcess()

- **YIELD**
  Halts the current task and starts any waiting task. *

Low-level GDI functions

There is a group of low-level GDI functions apart from REPAINTSCREEN() listed above and apart from those inside the display and printer drivers. They are:

ADVANCEDSETUPDIALOG, BITBLT, CHECKCURSOR, COLORINFO, CONTROL, DEVICEBITMAP, DEVICEBITMAPBITS, DEVICEMODE, DISABLE, ENABLE, ENUMDFONTS, ENUMOBJ, EXTDEVICEMODE, EXTTEXTOUT, FASTBORDER, GETCHARWIDTH, GETDRIVERRESOURCEID, GETPALETTE, GETPALTRANS, INQUIRE, MOVECURSOR, OUTPUT, PIXEL, QUERYDEVICENAMES, REALIZEOBJECT, SAVESCREENBITMAP, SCANLR, SELECTBITMAP, SETATTRIBUTE, SETCURSOR, SETDIBITSTODEVICE, SETPALETTE, SETPALTRANS, STRETCHBLT, STRETCHDIBITS, UPDATECOLORS, USERREPAINTDISABLE, WEP.

Low-level Comm functions

There is also a group of low-level communication functions:

BUILDCOMMDCB, CLEARCOMMBREAK, CLOSECOMM, ESCAPECOMMFUNCTION, FLUSHCOMM, GETCOMMERROR, GETCOMMEVENTMASK, GETCOMMSTATE,

OPENCOMM, READCOMM, SETCOMMBREAK, SETCOMMEVENTMASK, SETCOMMSTATE, TRANSMITCOMMCHAR, WRITECOMM.

In addition, there is a group of sound functions, utility macros and functions, file I/O functions, and debugging functions. For lists of these groups refer to *Microsoft SDK Reference Volume 1*.

TOOLHELP low-level functions

A special group of low-level functions have been provided with Windows 3.1 and documented in the SDK 3.1. They are supplied in TOOLHELP.DLL, and are backwards compatible with Windows 3.0, but you must bundle TOOLHELP.DLL with your program. The equivalent Win95 functions follow this list. The TOOLHELP functions are:

- **CLASSFIRST**
  Retrieves information about the first class in the class list.
- **CLASSNEXT**
  Retrieves information about the next class in the class list.
- **GLOBALENTRYHANDLE**
  Retrieves information about a global memory object.
- **GLOBALENTRYMODULE**
  Retrieves information about a specific memory object.
- **GLOBALFIRST**
  Retrieves information about the first global memory object.
- **GLOBALHANDLETOSEL**
  Converts a global handle to a selector.
- **GLOBALINFO**
  Retrieves information about the global heap.
- **GLOBALNEXT**
  Retrieves information about the next global memory object.
- **INTERRUPTREGISTER**
  Installs a function to handle system interrupts.
- **INTERRUPTUNREGISTER**
  Removes the function that processes system interrupts.
- **LOCALFIRST**
  Retrieves information about the first local memory object.
- **LOCALINFO**
  Fills a structure with information about the local heap.
- **LOCALNEXT**
  Retrieves information about the next local memory object.
- **MEMMANINFO**
  Retrieves information about the memory manager.

- **MEMORYREAD**
  Reads memory from an arbitrary global heap object.
- **MEMORYWRITE**
  Writes memory to an arbitrary global heap object.
- **MODULEFINDHANDLE**
  Retrieves information about a module.
- **MODULEFINDNAME**
  Retrieves information about a module.
- **MODULEFIRST**
  Retrieves information about the first module.
- **MODULENEXT**
  Retrieves information about the next module.
- **NOTIFYREGISTER**
  Installs a notification callback function.
- **NOTIFYUNREGISTER**
  Removes a notification callback function.
- **STACKTRACECSIPFIRST**
  Retrieves information about a stack frame.
- **STACKTRACEFIRST**
  Retrieves information about the first stack frame.
- **STACKTRACENEXT**
  Retrieves information about the next stack frame.
- **SYSTEMHEAPINFO**
  Retrieves information about the USER heap.
- **TASKFINDHANDLE**
  Retrieves information about a task.
- **TASKFIRST**
  Retrieves information about the first task in the task queue.
- **TASKGETCSIP**
  Returns the next CS:IP value of a task.
- **TASKNEXT**
  Retrieves information about the next task in the task queue.
- **TASKSETCSIP**
  Sets the CS:IP of a sleeping task.
- **TASKSWITCH**
  Switches to a specific address within a new task.
- **TERMINATEAPP**
  Terminates an application.
- **TIMERCOUNT**
  Retrieves execution times.

Windows 95 replaces all of the above with the following:

- **CreateToolhelp32Snapshot**
  Takes a snapshot of the Win32 processes, heaps, modules, and threads used by the Win32 processes.
- **Heap32First**
  Retrieves information about the first block of a heap that has been allocated by a Win32 process.
- **Heap32ListFirst**
  Retrieves information about the first heap that has been allocated by a specified Win32 process.
- **Heap32ListNext**
  Retrieves information about the next heap that has been allocated by a Win32 process.
- **Heap32Next**
  Retrieves information about the next block of a heap that has been allocated by a Win32 process.
- **Module32First**
  Retrieves information about the first module associated with a Win32 process.
- **Module32Next**
  Retrieves information about the next module associated with a Win32 process or thread.
- **Process32First**
  Retrieves information about the first Win32 process encountered in a system snapshot.
- **Process32Next**
  Retrieves information about the next Win32 process recorded in a system snapshot.
- **Thread32First**
  Retrieves information about the first thread of any Win32 process encountered in a system snapshot.
- **Thread32Next**
  Retrieves information about the next thread of any Win32 process encountered in the system memory snapshot.
- **Toolhelp32ReadProcessMemory**
  Copies memory allocated to another process into an application-supplied buffer.

Driver functions   What follows are functions available inside the drivers. They cannot be called directly as you would a normal Windows function, but require an extra step. See the practical code in the Chapter 9. Also, they are not documented in the SDK.

Mouse driver functions

- **INITIALIZATION**
  Initialises the mouse device driver.
- **DISABLE**
  Suspends interrupt callbacks from the mouse device.
- **ENABLE**
  Enables calls to the Windows mouse event procedure.
- **INQUIRE**
  Gets information about the mouse characteristics.
- **MOUSEGETINTVECT**
  Gets the interrupt level used by the mouse hardware.
- **WEP**
  Performs cleanup when the Windows session ends.

COMM driver functions

- **CCLRBRK**
  Clears the Comm line break state.
- **CEVT**
  Returns the address of the Comm event word.
- **CEVTGET**
  Clears and gets specified events in the Comm event word.
- **CEXTFCN**
  Performs an extended driver function.
- **CFLUSH**
  Discards the contents of a receive or transmit buffer.
- **COMMWRITESTRING**
  Transmits a block of data over the serial port.
- **CSETBRK**
  Initiates a Comm line break state.
- **CTX**
  Transmits a single byte before all others in the transmit queue.
- **GETDCB**
  Returns the address of the DCB structure for the specified port.
- **INICOM**
  Initializes the specified Comm port.
- **REACTIVATEOPENCOMMPORTS**
  Re-enables Comm ports disabled by SUSPENDOPENCOMMPORTS().
- **READCOMMSTRING**
  Reads bytes from the Comm receive buffer.
- **RECCOM**
  Reads a byte from the Comm receive buffer.

- **SETCOM**
  Sets the device configuration and state.
- **SETQUE**
  Specifies the memory input/output buffers.
- **SNDCOM**
  Places a character in the transmit queue.
- **STACOM**
  Gets the hardware and buffer status of the specified port.
- **SUSPENDOPENCOMMPORTS**
  Temporarily disables all Comm port activity.
- **TRMCOM**
  Closes the specified port.

Keyboard driver functions

- **DISABLE**
  Suspends interrupt callbacks and removes hooks.
- **ENABLE**
  Enables calls to the Windows keyboard event procedure.
- **ENABLEKBSYSREQ**
  Enables or disables SysRq key processing.
- **GETBIOSKEYPROC**
  Gets the address of the BIOS interrupt service routine.
- **INQUIRE**
  Returns the keyboard configuration structure that contains the DBCS ranges.
- **NEWTABLE**
  Loads the keyboard translation tables.

System driver functions

- **CREATESYSTEMTIMER**
  Allocates a system timer to be used by a device driver.
- **GETSYSTEMMSECCOUNT**
  Gets the amount of elapsed time.
- **INQUIRESYSTEM**
  Gets various system configuration parameters.
- **KILLSYSTEMTIMER**
  Frees a timer to be used by a device driver.

Grabber functions

Earlier I described mechanisms for Windows to save and restore its video hardware state, if an application wants to do something with the video. From the application's point of view, after getting control of the video, it can call some functions to manipulate the display driver. Calling these functions is not straight forward; refer to *Writing Windows Device Drivers* by D. A. Norton, Addison Wesley, 1991, page 79. This reference also has more detail on these functions in its appendix D, page 247. They are summarised here:

- **DISABLESAVE**
  Disables switching between Windows and DOS sessions.
- **ENABLESAVE**
  Enables switching between Windows and DOS sessions.
- **GETBLOCK**
  Copies the specified rectangular portion of the screen to a buffer.
- **GETINFO**
  Gets the grabber's GRABINFO structure.
- **GETVERSION**
  Returns the grabber version number.
- **INITSCREEN**
  Initializes the screen to text mode.
- **INQUIREGRAB**
  Gets the size of the text or graphics grab buffer.
- **INQUIRESAVE**
  Gets the size of the text or graphics save buffer.
- **RESTORESCREEN**
  Restores the state and contents of the display.
- **SAVESCREEN**
  Saves the state and contents of the display.
- **SETSWAPDRIVE**
  Specifies the drive and path of the grabber swap file.

The above group only work in Real and Standard modes. The 386 Enhanced mode has a different set of functions.

Undocumented functions Many functions available in Windows 3.0, 3.1, and 95 are not described in the SDKs, nor anywhere for that matter. These are "undocumented" functions, which means that Microsoft doesn't want us to know about them (see also page 235).

Reference books There are various chaps who have dug up the dirt, and written books.

Undocumented Windows: A Programmer's Guide to the Reserved Microsoft Windows API Functions by A. Schulman, D. Maxey, and M. Pietrek, Addison Wesley, USA, 1992.

Unauthorized Windows 95 by Andrew Schulman, IDG Books, USA, 1994.

Windows 95 Systems Programming Secrets by Matt Pietrek, IDG Books, USA, 1995.

Thunking

The mismatch between 16- and 32-bit code is a major headache. Windows internally is also a mixture, including Windows 95 (especially Windows 95!). I have shown in this chapter that some functions available to 16-bit applications are not available to 32-bit applications and vice versa. This is because each has its own set of API DLLs (see page 235).

However, we *can* "mix and match" — with caution of course.

Generic thunking
The process of translating between 32- and 16-bit code is known as *thunking*, and Windows 95 provides two mechanisms: *Generic* thunking and *Flat* thunking.

Flat thunking
Flat thunking is specific to Windows 95 — it is not portable to Windows NT. It allows 16- to 32-bit and 32- to 16-bit function calls, so it is most flexible.

Generic thunking works on both Windows 95 and NT but only allows a 16-bit application to call 32-bit functions, not the other way around.

Universal thunking
Universal thunking is for Windows 3.1 applications to access the win32s API.

Reference sources
A good explanation of Flat thunking is to found in *Inside Windows 95* by Adrian King, Microsoft Press, USA, 1994. Also look at the Win95 SDK CD-ROM.

Generic thunking is also explained in the Win95 SDK CD-ROM, in file DOC\MISC\GENTHUNK.TXT. The following information is based on this and other documents on the SDK CD-ROM.

Another excellent document that covers both Generic and Flat thunking and has detailed descriptions of all the Generic API functions is *Programmer's Guide to Microsoft Windows 95* by the Microsoft Windows Development Team, Microsoft Press, USA, 1995.

Generic Thunking

Windows on Win32 (WOW) presents 16-bit APIs that allow you to load the Win32 DLL, get the address of the DLL routine, call the routine (passing it up to thirty-two 32-bit arguments), convert 16:16 (WOW) addresses to 0:32 addresses (useful if you need to build a 32-bit structure that contains pointers and pass a pointer to it), and free the Win32 DLL.

Function prototypes
I hope you can read C code. I have taken these examples straight from the SDK documentation.

The following prototypes should be used:

```
DWORD FAR PASCAL LoadLibraryEx32W( LPCSTR, DWORD, DWORD );
DWORD FAR PASCAL GetProcAddress32W( DWORD, LPCSTR );
DWORD FAR PASCAL CallProc32W(DWORD,...,LPVOID,DWORD,DWORD );
DWORD FAR PASCAL GetVDMPointer32W( LPVOID, UINT );
BOOL  FAR PASCAL FreeLibrary32W( DWORD );
```

Note that although these functions are called in 16-bit code, they need to be provided with 32-bit handles, and they return 32-bit handles. Do not forget that the 32-bit functions must be called with the STDCALL convention.

CallProc32W() CallProc32W() follows the PASCAL calling convention. It is designed to take a variable number of arguments, a Proc address, a mask, and the number of parameters. The mask is used to specific which arguments should be treated as being passed by value and which parameters should be translated from 16:16 pointers to Flat pointers. Note that the low-order bit of the mask represents the last parameter, the next lowest bit represents the next to the last parameter, and so forth.

Code examples I didn't really want to put actual code into this chapter, but a little sample of Generic thunking is useful while I'm on the topic.

Assume that the Win32 DLL is named DLL32. First you need to load the 32-bit library:

```
ghLib = LoadLibraryEx32W( "dll32.dll", NULL, 0 )
```

Then you need to get the address of the 32-bit function, in this case MyPrint():

```
hProc = GetProcAddress32W( ghLib, "MyPrint" )
```

Then call MyPrint(), passing it the required parameters TestString and hWnd:

```
CallProc32W((DWORD)TestString,(DWORD) hWnd|0xffff0000,hProc,2,2);
```

The hWnd is OR'd with 0xffff0000, because this is the way to convert a 16-bit window handle to a 32-bit window handle in Windows NT and 95. If you want to convert a 32-bit window handle to a 16-bit window handle, simply truncate the upper word. Note that this only works for window handles, not for other types of handles. You should use the following functions exported by WOW32.DLL: WOWHandle32() and WOWHandle16(), in all cases, rather than relying on this relationship. These functions are discussed in the SDK.

A mask of 2 (0x10) is given because we want to pass TestString by reference (WOW translates the pointer), and we want to pass the handle by value.

Finally, we must free the 32-bit library:

```
FreeLibrary32W( ghLib );
```

NOTE: When linking the Windows-based application, you need to put the following statements in the .DEF file, indicating that the functions will be imported from the WOW kernel:

```
IMPORTS
    kernel.LoadLibraryEx32W
    kernel.FreeLibrary32W
    kernel.GetProcAddress32W
    kernel.CallProc32W
```

The use of the 16-bit versions LOADLIBRARY() and GETPROCADDRESS() is described in Chapter 9. The principles apply to the 32-bit versions also.

WOW functions called from 16-bit code

Very briefly, here they are:

- **CallProc32W, CallProcEx32W**
  Used by 16-bit code to call an entry point function in a 32-bit DLL.

- **FreeLibrary32W**
  Allows 16-bit code to free a 32-bit thunk DLL that it had previously loaded by using the LoadLibraryEx32W() function.

- **GetProcAddress32W**
  Allows 16-bit code to retrieve a value that corresponds to a 32-bit routine.

- **GetVDMPointer32W**
  Allows 16-bit code to translate a 16-bit FAR pointer into a 32-bit FLAT pointer for use by a 32-bit DLL.

- **LoadLibraryEx32W**
  Allows 16-bit code to load a 32-bit DLL.

WOW functions called from 32-bit code

These are a different group of WOW functions:

- **WOWCallback16, WOWCallback16Ex**
  Used in 32-bit code called from 16-bit code to call back to the 16-bit side.

- **WOWGetVDMPointer**
  Converts a 16:16 address to the equivalent FLAT address.

- **WOWGetVDMPointerFix**
  Converts a 16:16 address to the equivalent FLAT address. Unlike the WOWGetVDMPointer() function, this calls the GlobalFix() function before returning.
- **WOWGetVDMPointerUnfix**
  Uses the GlobalUnfix() function to unfix the pointer returned by WOWGetVDMPointerFix().
- **WOWGlobalAlloc16**
  Thunks to the 16-bit version, GLOBALALLOC().
- **WOWGlobalAllocLock16**
  Combines the functionality of WOWGlobalAlloc16() and WOWGlobalLock16().
- **WOWGlobalFree16**
  Thunks to the 16-bit version of GlobalFree().
- **WOWGlobalLock16**
  Thunks to the 16-bit GlobalLock()
- **WOWGlobalLockSize16**
  Combines the functionality of WOWGlobalLock16() and GlobalSize().
- **WOWGlobalUnlock16**
  Thunks to 16-bit GlobalUnlock().
- **WOWGlobalUnlockFree16**
  Combines the functionality of WOWGlobalUnlock16() and WOWGlobalFree16().
- **WOWHandle16**
  Maps a 32-bit handle to a 16-bit handle.
- **WOWHandle32**
  Maps a 16-bit handle to a 32-bit handle.

More Win95 "Improvements"

Device I/O Control

Windows 95 introduced DeviceIoControl() as a standardized channel for performing I/O, that is, to communicate directly with virtual device drivers. This is also the preferred way to access INT-21h services, though very few are supported.

Software interrupts will crash a 32-bit application, so Microsoft is trying to force you to do most low-level and direct access to the hardware through device drivers.

Much wider usage of the file I/O functions is found in Windows 95. DOS programmers will know that the INT-21h file handling functions can also operate on device drivers. That is, a device driver can be opened, a handle obtained, the "file" read from and written to, and then closed. The concept is alive and well in Windows 95. CreateFile() is used to open a virtual device driver prior to using DeviceIoControl(), and CloseHandle() is used to close the driver.

Dynamically Loadable Drivers

The ability to open a virtual device driver at any time is related to the new capability of Windows 95 to support dynamic loading. CreateFile() loads a driver and CloseHandle() unloads it.

Threads

A Windows *process* is an application, be it a Windows application or a DOS Virtual Machine (VM). However, 32-bit Windows 95 applications can also have multiple *threads* of execution, and the thread becomes the basic unit that can be scheduled by the operating system.

With Windows 3.x, the System VM (running all the Windows applications) and the DOS VMs (each running a DOS application) are preemptively scheduled, while the Windows applications themselves are cooperatively scheduled (i.e., amongst themselves).

Windows 95 adds to this picture with 32-bit applications that have one or more threads that can be preemptively scheduled. Because scheduling is thread-based, the term process is awkward — the 16-bit applications become one thread and each DOS VM is one thread.

Here are all the Windows 95 thread- and process-related functions:

AttachThreadInput, CommandLineToArgvW, CreateProcess, CreateRemoteThread, CreateThread, ExitProcess, ExitThread, FreeEnvironmentStrings, GetCommandLine, GetCurrentProcess, GetCurrentProcessId, GetCurrentThread, GetCurrentThreadId, GetEnvironmentStrings, GetEnvironmentVariable, GetExitCodeProcess, GetExitCodeThread, GetPriorityClass, GetProcessAffinityMask, GetProcessShutdownParameters, GetProcessTimes, GetProcessVersion, GetProcessWorkingSetSize, GetStartupInfo, GetThreadPriority, GetThreadTimes, OpenProcess, ResumeThread, SetEnvironmentVariable, SetPriorityClass, SetProcessShutdownParameters, SetThreadAffinityMask,

SetThreadPriority, Sleep, SleepEx, SetProcessWorkingSetSize, SuspendThread, TerminateProcess, TerminateThread, TlsAlloc, TlsFree, TlsGetValue, TlsSetValue, WaitForInputIdle, WinExec.

Memory Mapped Files

Windows 95 introduces memory mapped file functions for sharing data between applications. **CreateFileMapping()** creates such a file, while **MapViewofFile()** maps it. **OpenFileMapping()** and **DuplicateHandle()** can be used by processes to access the file.

Despite the name "file", it does not have to be on a disk — the file can reside entirely in memory. Such a global file is visible to all applications.

Postamble

This chapter is notable more for what it doesn't say than what it does! Various functions, interrupts, and concepts introduced here are developed in the chapters ahead.

9

Direct Hardware Access

Preamble

This chapter contains practical code to "get behind the scenes". The first part of the chapter focuses on the issues of direct reading from and writing to memory, particularly video-RAM, and the second part focuses on I/O.

I have shown the use of DPMI INT-31h services and of the INT-2Fh extensions, plus the use of low-level Windows functions. I have pointed out overlap between the two where it occurs.

You will be amazed to learn that it is possible to have an application running in a window, yet the application can write directly to the video hardware, at breathtaking speed, without all of the Windows rigmarole. This is the kind of practical code developed in this chapter.

You will also learn about I/O aspects, such as use of the IN and OUT instructions.

Mostly I view the material of this chapter as educational. It pokes around doing fun things that may be viewed as "hacking". It may be that you will never use some of the less orthodox material in professional applications, but what will be formed now is a good solid foundation of understanding of the fundamentals.

Initialisation

First I'll address the question of initialisation. Since your program is running in Protected mode, alongside other programs, you can't simply go reading and writing all over memory and I/O. There have to be rules to prevent contention. Initialisation is code that clears the way for you to get directly at the hardware.

The code below is a good way to start. For the moment, don't worry about the red tape of PROC — ENDP; etc. You'll put it together later.

Before you can use DPMI services, you need to check out a few things:

```
.DATA
dpmiflag            DB  0    ;=1 dpmi running ok
dpmiversion         DW  0    ;ah=major, al=minor.
386modeflag         DB  0    ;=1 386 dpmi type.
realmodeintsflag    DB  0    ;=1 Real mode interr.
virtualmemflag      DB  0    ;=1 virt. mem support.
cputype             DB  0    ;=2,3,4  286,386,486
;...
.CODE
   mov  ax,1686h             ;test if dpmi running.
   int  2Fh
   or   ax,ax
   jnz  nodpmi
   mov  dpmiflag,1           ;set flag, dpmi ok.
   mov  ax,0400h             ;get dpmi version.
   int  31h
   mov  dpmiversion,ax
   mov  al,bl
   and  al,01                ;bit-0 =1 if 386 dpmi
   mov  386modeflag,al
   mov  al,bl
   shr  al,1
   and  al,01     ;bit-1 =1 if not virtual86 int handling
   mov  realmodeintsflag,al
   shr  bl,2
   and  bl,01               ;bit-2 =1 if virtual mem. supported.
   mov  virtualmemflag,bl
   mov  cputype,cl          ;cl=2,3,4 if 286, 386, or 486.
```

Refer to Appendix C for a full description of all DPMI services. INT-2Fh/AX = 1686h is used to check if the CPU is running DPMI and is in Protected mode. This service returns a false flag if Windows is loaded in Real mode or DPMI is not running.

You would only need to perform this check if you were running Windows in Real mode, but version 3.1 of Windows won't even run in Real mode, so these days this test isn't required.

Note that most of the INT-2Fh services work in Real and Protected mode, with or without DPMI, but INT-31h will only work with DPMI and in Protected mode.

Which version of DPMI is running?

The next service is INT-31h/AX = 0400h, which returns the DPMI version number, plus other status information. Since version 1.0 of DPMI has more features than version 0.9, this test is necessary if you want to use the extra features of v1.0. Note that Windows 3.x and 95 only support DPMI v0.9 (refer to page 198). I have written the code in this chapter for v0.9. So, again, this test is not really required.

I have stored all of the flags as static data, to be used as needed by the rest of the program.

Addressing Segments

Direct access to memory and video -RAM

Assuming that DPMI is up and running, which it should be under Windows, you are ready to start doing interesting things. One of your objectives is to access real memory directly. That is, you hunger for the good old days when you could write directly to the video-RAM, not via some tortuous method using GETDC(), TEXTOUT(), and RELEASEDC(), with a hundred messages to worry about. You want **control** (slobber, slobber), and you want **speed**!

You may even be totally retrograde and want to run your Windows program with the screen in **text mode** (horror!). Remember good old text mode? It was good enough for most things, and even did quite a good job at graphics, using the IBM graphics character set. The MDA (Monochrome Display Adaptor) only has a 4K video-RAM, with the result that screen redrawing is instantaneous. Forget about delays with text mode.

Text-mode Windows applications

This text mode topic raises an interesting side issue. There are a lot of other "retros" like me out there, and there is even a special product available for those who want to write Windows programs but don't want to give up the advantages of Real mode and of text-mode video. The product is called Mewel,[1] and it is a complete library for writing Windows applications that run without Windows, under DOS, in Real mode (or Protected mode), with the screen in text mode (or graphics mode). It's a lovely product and works well. The only major deficiency is that there is no multitasking. Mewel even allows source code to specify

---

[1]   Magma Systems, 15 Bodwell Terrace, Millburn, New Jersey 07041, USA. http://www.uno.com/magma.html

standard Windows screen coordinates, so a stock-standard Windows program will compile and run under Mewel. Mewel applications are stand-alone, as the library files are linked statically. But it does mean that a simple "Hullo world" program is about 100K. Mewel even manages to represent icons in text mode!

Addressing memory below 1M in Protected mode

Back to the main topic. Let your first challenge be to directly access memory. No problem. DPMI has INT-31h/AX = 0002h:

```
.DATA
0000selector    DW    0    ;selector, addr-0.
B000selector    DW    0    ;selector video-RAM
.CODE
  mov   ax,0002¹ ;supply a segment, returns a selector.
  mov   bx,B000h         ;segment addr of video RAM.
  int   31h
  mov   B000selector,ax   ;save selector.
  mov   ax,2            ;get selector for segment addr 0000.
  mov   bx,0000h         ;start of physical memory.
  int   31h
 mov 0000selector,ax ;save.(label cannot start with 0-9).
```

LDT

What this service does is create an entry in your application's Local Descriptor Table,[2] and returns the index to that entry, that is, the selector. The way selectors work is that you can treat them just like the old segment values. Something like:

```
  mov   ax,B000selector
  mov   es,ax
  mov   bx,0
  mov   es:[bx],"x"
```

Linear address

This code will write the ASCII character "x" directly to the video-RAM at address B000:0000. From the theory in Chapter 1, that will be a physical and/or linear address of 000B0000h. I made the complete linear[3]/physical address up to 32 bits, since that's

---

[1] Intel's DPMI specification places a few caveats upon the 0002 function.

The descriptor's limit will be set to 64K.

Multiple calls to this function with the same segment address will return the same selector.

Descriptors created by this function can never be modified or freed. For this reason, the function should be used sparingly. Clients which need to examine various Real mode addresses using the same selector should allocate a descriptor with INT-31h/AX = 0000h and change the base address in the descriptor as necessary, using function 0007h.

[2] Note that all WinApps share a single LDT. The system VM maintains one each LDT, GDT, and IDT.

[3] Notice above that I used the word "linear" address. This is explained in Chapter 1. Basically, in

what the 386 actually puts out. In the case of the 286 it will only be 24 bits.

Look carefully at that above code fragment. See that I treated the selector as the exact equivalent of the segment (paragraph) address it represents. Behind the scenes, the CPU will use the selector value in ES to lookup the LDT and get the physical address.

This service is wonderful, because it gives you direct access to all memory below 1M. It also gives you enormous potential to "stuff up" the system.

... now, the same thing, but using a Windows function

Pardon the crudeness, but "there's more than one way to skin a cat". Ditto with DPMI services and low-level Windows functions. If the two overlap, which ones do you use? Interestingly, some of the Windows functions internally call the DPMI services!

In the above case, the Windows function equivalent is — well, there are choices here, just as there are some different avenues with DPMI. SETSELECTORBASE()[1] is appropriate: it creates a new entry in the LDT and will set the "base address" (linear address) field in the descriptor. You provide a selector value as a parameter to this function, the descriptor of which is used as the model for the new descriptor. So, if you want to treat the new memory block as data, use DS as the model. The SDK 3.1 documentation does not explain any of these vital details.

Note that SETSELECTORBASE() is available in Windows 3.0 but was undocumented until Windows 3.1 made it official.

Direct Video

You don't want to "stuff up" the system, of course, so you need to take whatever precautions are necessary. If your Windows application is going to do something drastic, like change the screen to text mode, then obviously it will not be outputting to a pretty little Windows box. The Windows screen will no longer be there. This may be ok for what we want, but if our program is to work with other Windows programs, and with Windows itself, then our program must be able to restore the original screen.

Saving and restoring the Windows screen: INT-2Fh/ AX=4001/2h

Microsoft does have a very suitable service: INT-2Fh/AX = **4001h**. It is summarised in Appendix D, described in Microsoft's

---

Windows Standard mode (286 mode), the linear and physical addresses are one and the same. In Enhanced (386) mode, an address goes through an extra paging step, so the physical address is renamed as the linear address, and is no longer the actual physical address.

[1]   SETSELECTORBASE() is passed two parameters: the selector (16 bits), and the starting linear address (32 bits). It returns a new selector value in AX, or AX = 0 if an error.

Device Development Kit (DDK), and rather briefly touched on in *Writing Windows Device Drivers* by D. A. Norton, Addison Wesley, USA, 1992. Windows Virtual Display Driver (VDD) uses it to control the actual (non-virtual) display driver. Basically, it saves the adaptor registers: it works and is darn useful, so I have used it in this example code.

The complement of the above is available also; INT-2Fh/AX = **4002h**, which tells the display driver to take back control.[1]

Directly overwriting the Windows screen

Mighty handy. So far you know how to write directly to the video-RAM and you have a selector to do it with. It may be that you don't want to save the screen at all — you just want to scribble all over what is already there. Perhaps if you want some little message to appear on the screen, independently of everything else that Windows is doing on the screen, then yes, go ahead (see page 239).

Running the screen in text mode

At the moment, I'm thinking more along the lines of taking over the screen directly for very fast video output, such as games or where text-mode is good enough or preferred.

To save the current video state:

```
mov  ax,4001h ;Note that undocumented DEATH() does this
int  2Fh      ;also, plus switches screen to text mode.
```

The next obvious step is to change the video mode. There are some interesting thoughts here. Won't Windows and other applications expect to be able to output to the screen also?

Yes they will, but always remember that Windows' 16-bit task management is non-preemptive. This means that once Windows has passed control to your program, you can keep control for as long as you like. You can lock out other applications and do whatever you want.[2]

Normally, when Windows sends a message to the callback function, your callback processes it, then has nothing more to do so returns to Windows. If control stays in the callback, for

---

[1]   There are two other services, AX = 4005h and 4006h, that are similar to 4000h and 4001h, respectively. The description of 4005h is "The Windows VDD calls this function to tell the display driver to save the video hardware state." And for 4006h "The Windows VDD calls this function to tell the display driver to restore the video hardware state that was saved by the last call to function 4005h" (*Writing Windows Device Drivers*, page 78). The 4000h and 4007h services are used in conjunction with 4005h and 4006h. 4000h gives the display driver direct access to the video hardware registers, while 4007h disables register access and tells the VDD that the display driver has finished accessing the video hardware.

[2]   Though the DPMI host does perform preemptive time slicing between VMs (see Chapter 11). Even this can be disabled by a DPMI service.

whatever reason, then no more messages are sent to it; therefore, your callback is not receiving anything from Windows. At least not in the normal way — just register that as an interesting point for now.

Restore Video

For now, I'll just say that you can save the screen upon entry to the callback, and you can write directly to the screen. But before going back to Windows, you must restore things to how they were. This means that whatever you displayed in text mode (or whatever) will be lost, unless you save it in a buffer.

This is some video cleanup code prior to returning to Windows:

```
mov  ax,4002h     ;Note undocumented RESURRECTION()
int  2Fh          ;is similar.
call REPAINTSCREEN PASCAL
```

REPAINT-SCREEN() function

REPAINTSCREEN() is a Windows function, but you won't find it mentioned in Microsoft's *Software Development Kit* (SDK), nor in most other places. It is described in the *Device Driver Development Kit* (DDK) (3.x versions), from Microsoft, and is another one that Microsoft seems to want to maintain a low profile on. In the latest set of MSDN CD-ROMs (January 1997), REPAINTSCREEN() is mentioned only in the *Library Archive* CD-ROM.

Although it is in the Windows library file USER.EXE (the other two are GDI.EXE and KERNEL.EXE, located in \WINDOWS\ SYSTEM directory), you may not be able to simply call it as I've shown above.[1] Later on, when you see the whole program together, you'll see what I did to call it.

32-bit applications

I am referring through most of this chapter to the 16-bit API DLLs. Thirty-two-bit applications can, by indirect means such as thunking (see Chapter 8), or some kind of separate 16- and 32-bit programs that cooperate (see Chapters 12 and 14) access the 16-bit API. However, many of the low-level functions have been ported to the 32-bit DLLs, except that linkage information is not provided in the IMPORT library (see footnote below and page 235).

---

[1] The Windows library file supplied by your software vendor, such as LIBW.LIB (Microsoft) or IMPORT.LIB (Borland), provide your program with access to the DLL functions. Whether or not you can access REPAINTSCREEN() directly from your program is determined by the inclusion of the linkage information in these link files. You will find that later versions may provide the linkage, even to undocumented functions; however, I have shown how to do it the hard way here, in case you have to do it for any functions, including those in other DLLs.

REPAINTSCREEN() redraws the screen, and is redundant here actually. INT-2Fh/AX = 4002h restores VDD (Windows) access to the display driver and also causes the screen to be redrawn. REPAINTSCREEN() *is* required after RESURRECTION().

Change Video Mode

BIOS
INT-10h

So in between having saved the screen and cleaning up prior to going back to Windows, how do you change the video mode?

If you are familiar with DOS and BIOS INTs, you'll know it is INT-10h — well, it still is!

Since an INT causes the CPU to look in the IDT (Interrupt Descriptor Table) for the location of the routine and not in the old IVT (Interrupt Vector Table) (see pages 33+), any of the routines can be replaced as required or the CPU redirected to the Real mode routine with appropriate translations. Thus INT-10h stills works, even though it is called from a Protected mode Windows program.

Here is how to go to text mode 7:

```
mov  ax,0007h ;Note that DEATH() will have got us to
int  10h      ;the text mode prior to Windows loading.
```

A Direct-Video Text-Mode Routine

I'll put it all together. I have named this routine directvideo(). You can call it from wherever in your program you want and modify it as required — some suggestions and possibilities follow after the listing. If you want to test it, you could take one of the earlier programs and perhaps call it from the WM_CHAR case, so whenever a key is pushed the routine will execute. There is code for this section on the Companion Disk.

Text-mode
direct-video
listing

Here is the listing:

```
EXTRN   GETMODULEHANDLE:FAR
EXTRN   GETPROCADDRESS:FAR
;.................
.DATA
dpmiflag            DB  0     ;=1 dpmi running ok
dpmiversion         DW  0     ;ah=major, al=minor.
mode386flag         DB  0     ;=1 386 dpmi type.
realmodeintsflag    DB  0     ;=1 Real mode interr.
virtualmemflag      DB  0     ;=1 virt. mem support.
cputype             DB  0     ;=2,3,4  286,386,486
```

```
BOOOselector            DW   0              ;selector video-RAM
szmodulename            DB   "USER.EXE",0
lprepaintscreen         DD   0
;.................................................
.CODE
directvideo PROC  PASCAL NEAR
   LOCAL          winvideomode:BYTE
   USES           ax,bx,cx,dx,si,di
;...
   call  GETMODULEHANDLE PASCAL, ds,OFFSET szmodulename
   mov   si,ax                  ;gets a handle for user.exe.
   or    si,si                  ;Returns handle in AX.
   jne   userexists             ;user.exe doesn't exist.
   jmp   nomodule
userexists:
;...
   call FAR PTR GETPROCADDRESS  PASCAL,si, 0,275
                    ;275=ordinal value of REPAINTSCREEN(), in
   mov   WORD PTR lprepaintscreen,ax          ;USER.EXE.
   mov   WORD PTR lprepaintscreen+2,dx  ;Returns far addr
                                        ;DX:AX.
;.....
   mov   ax,1686h               ;test if dpmi running.
   int   2Fh
   or    ax,ax
   jz    yesitis
   jmp   nodpmi
yesitis:
   mov   dpmiflag,1             ;set flag, dpmi ok.
   mov   ax,0400h               ;get dpmi version.
   int   31h
   mov   dpmiversion,ax
   mov   al,bl
   and   al,01                  ;bit-0 =1 if 386 dpmi
   mov   mode386flag,al
   mov   al,bl
   shr   al,1
   and   al,01   ;bit-1 =1 if not virtual86 int handling
   mov   realmodeintsflag,al
   shr   bl,2
   and   bl,01         ;bit-2 =1 if virtual mem. supported.
   mov   virtualmemflag,bl
   mov   cputype,cl          ;cl=2,3,4 if 286, 386, or 486.
;...
   mov   ax,0002
   mov   bx,0B000h                  ;segment addr of video RAM.
   int   31h                   ;Note that although DEATH() is
   mov   BOOOselector,ax       ;undocumented, I figured out
   mov   ax,2                  ;how to use it ...
;...                             call GETDC PASCAL,hwnd
   mov   ax,4001h               mov  hdc,ax
   int   2Fh                    call DEATH PASCAL,hdc
;...                           ;...Windows display driver is
   mov   ah,0Fh                ;now turned off and scrn in
   int   10h                   ;text mode.
   mov   winvideomode,al       ;Note that DEATH() leaves the
   mov   ax,0007h ;mode 7      ;CPU in Protected mode.
```

```
    int  10h
;...
    mov  ax,B000selector
    mov  es,ax
    mov  bx,0
mm:
    mov  cx,0FFFFh
mmm: nop
    loop mmm                    ;delay
    mov  BYTE PTR es:[bx],"X"
    mov  BYTE PTR es:[bx+1],10001111b        ;attribute
    inc  bx
    inc  bx
    cmp  bx,1998                ;put 1000 X's on screen.
    jbe  mm
;...                           |;Undocumented RESURRECTION()
    mov  ah,00                 |;will change back to graphics
    mov  al,winvideomode       |;mode and restore Windows
    int  10h                   |;display driver...
    mov  ax,4002h              |  call RESURRECTION PASCAL\
    int  2Fh                   |        ,hdc,0,0,0,0,0,0
;...                           |  call RELEASEDC PASCAL,hwnd\
  call lprepaintscreen \       |        ,hdc
            PASCAL             |;(Thanks to Undocumented Windows
;...                           |; for showing me how many params
nodpmi:                        |; to feed RESURRECTION()!)
nomodule:
    ret
directvideo ENDP
```

There are a host of things I can say about this routine. I have itemized major points below.

Call REPAINTSCREEN()

Calling a function in a DLL

I mentioned earlier that I have used REPAINTSCREEN() as an example to show how to get at a DLL function at run-time, which is one option if linkage information is not provided in the library file. The standard technique is to call **GETMODULEHANDLE()** to get a handle for USER.EXE (a file is a *module* in Windows parlance, *but* the file name can be different from the module name), then call **GETPROCADDRESS()** to get the FAR address of the function within that module. If you would like to see another example of accessing a function in this way, *Microsoft's Programmer's Reference, Volume 2: Functions*, provided with the SDK 3.1 (and available separately), gives an example of **LOADLIBRARY()** (instead of GETMODULEHANDLE()), **GETPROCADDRESS()**, and **FREELIBRARY()** to access a function in TOOLHELP.DLL.

Thity-two-bit applications are somewhat more constrained — see notes on page 235.

Ordinal Coordinates

*EXE
header
extraction
utilities*

USER.EXE is a Dynamic Link Library and is a standard feature of Windows. It has a heap of useful functions, and the question naturally arises: what are the other functions in USER.EXE? Furthermore, where did I get that *ordinal coordinate* of 275?

Each function in USER.EXE, or any DLL for that matter, can be referenced by a unique ordinal coordinate. You can find out all of the functions in a DLL and their ordinal coordinates, by use of a utility program supplied with Microsoft C/C++, called EXEHDR.EXE (or TDUMP.EXE from Borland C++). Since you may not have access to this utility, I have listed the output of EXEHDR.EXE for many of the Windows DLLs and drivers (see the Companion Disk). The file on the disk has a comprehensive alphabetical list of functions, with a short description, where it is documented, what DLL it belongs to, and its ordinal coordinate. Each device driver has built-in functions that can be called also.

*32-bit
applications*

Thirty-two-bit applications are a problem. Apart from crashing if you try to use a software interrupt, the low-level undocumented (and many previously documented) functions are not readily available. Matt Pietrek, arguably the Windows systems programming guru of gurus, covers this problem in *Dirty Little Secrets about Windows 95*, on-line at:

```
http://ftp.uni-mannheim.de/info/OReilly/windows/win95.update
                                                    /dirty.html
```

*Reference
book*

In this Web page, Matt is actually quoting from his book *Windows 95 Systems Programming Secrets*, IDG Books, USA, 1995:

> "In *Unauthorized Windows 95,*[1] Andrew Schulman made extensive use of undocumented functions in KERNEL32.DLL. Although there obviously weren't header files for these functions, the functions appeared in the import library for KERNEL32.DLL. Calling these functions was as simple as providing a prototype and linking with KERNEL32.LIB.
>
> In subsequent builds of Windows 95 after Andrew's book came out, these functions disappeared from the import library for KERNEL32.DLL. (Surprise! Surprise!) At the same time, these function names disappeared from the exported names of KERNEL32.DLL. These undocumented functions were still exported, however. The difference is that they were exported by ordinal only.

---

[1]   IDG Books, USA, 1994.

Now, normally this would have been only a small nuisance to work around. You should be able to simply call GetProcAddress and pass in the desired function ordinal as the function name (0 in the HIWORD, the ordinal in the LOWORD) and get back the address. In a normal, sane world, this would work. However, at some point during the beta, Microsoft added code to GetProcAddress to see if it's being called with the ordinal form of the function. If so, and if the HMODULE passed to GetProcAddress is that of KERNEL32.DLL, GetProcAddress fails the call. In the debugging version of KERNEL32.DLL, the code emits a trace diagnostic: "GetProcAddress: kernel32 by id not supported."

Now, let's think about this. Since the undocumented functions aren't exported by name, you can't pass the name of a KERNEL32 function to GetProcAddress to get its entry point. And GetProcAddress specifically refuses to let you pass it an ordinal value. The Microsoft coder responsible for this abomination really didn't want people (Andrew Schulman?, myself?) from calling these undocumented KERNEL32 functions. Apparently, the only way you can call these functions is if you have the magic KERNEL32 import library that Microsoft isn't supplying with the Win32 SDK.

Never fear. As you'll see later in the book, I make extensive use of the KERNEL32 undocumented functions (for good, not evil). With a little bit of work, I was able to coerce the Visual C++ tools to create a KERNEL32 import library that contains these "documentation-challenged" functions.

Appendix A contains information about these functions and an import library for them."

Page 208 lists some of these functions

To and From Text Mode

RESURREC-
TION()

If you choose to use RESURRECTION() to come back from text mode, the screen will stay black, and bits will be redrawn as you use Windows. If you do want the entire Windows screen to be redrawn, then REPAINTSCREEN() is necessary.

Video
mode 7

There are various options for going to and fro between text and graphics modes, apart from INT-10h. You could try the C run-time library, or DEATH/RESURRECTION. The latter, although undocumented, is probably the best supported and cleanest method.

Notice from the above listing that I used INT-10h/AH = 0Fh to obtain the current video mode before changing to mode 7. After doing my thing, I used INT-10h/AL = 00 to change the mode back to what it was. Mode 7 is the original text mode for the MDA card, giving monochrome 80 columns by 25 rows. It works on EGA and VGA adaptors, but not on CGA. The reason for this is that CGA does not have high enough resolution. CGA text mode is mode 2 or 3, and is only 640 x 200 pixels, while mode 7 is 720 x 350 pixels. The old mono MDA screen gives a nice sharp image.

This problem is a point in favour of DEATH().

Video Output Issues

Overwriting the current Windows screen

You do not necessarily have to change the video mode. A typical application might be to leave the Windows screen as-is and overwrite it. Think about this — there are Windows functions to obtain coordinates of your application's window, or you could call functions to set your window to certain coordinates. Then you will know exactly where it is, so when you go into direct-video-access mode, you will be able to write to the portion of the screen that is within your window.

This means that you can have your program running as a window, but you are still employing super-fast direct access to the video-RAM. Yes, you can have your cake and eat it too!

Virtual vs physical video-RAM

By getting a selector to the video-RAM, you can write directly to it. But what about "virtual" video-RAM? Since we are running in virtual machines, shouldn't output to the video-RAM be to a virtual video-RAM, that does not necessarily correspond with the physical video-RAM? This is potentially true, but all WinApps run in the system VM, and the virtual video-RAM does correspond to the physical.

I am perhaps getting a bit ahead here. Even though the concept of a VM was introduced in Chapter 1, I haven't fully developed it until Chapters 10, 11, and 12. A DOSApp running in another VM does write to a virtual video-RAM, which Windows can map directly to full screen or into a window (depending upon the settings of the .PIF file).

Suppressing redrawing

The problem with the above (overwriting the current Windows screen) is that when you exit your callback and return control to Windows, the screen will be redrawn. Of course you may not want to return to Windows until you have finished running your game or whatever, but suppose you do. A return to Windows without redrawing the screen can be done by not executing INT-31h/AX = 4002h, or REPAINTSCREEN().

These can be executed later, when the time is right, or not at all.

Message Input

Dumping the queue

One thing to bear in mind is that although Windows 3.x is non-preemptive, the device drivers are still working asynchronously, as indeed is the case in Windows 95. Key presses and mouse activity can still generate messages, which will be placed into your application's queue.

So, your program may have saved the Windows video state and gone to mode 7, or whatever, and done its thing. When finished, and after the clean-up of restoring the video state and maybe calling REPAINTSCREEN(), your program would normally continue on in the normal fashion — if execution is within a callback, control will continue on and return to Windows, and a message waiting in the queue will then be sent to the message loop in WinMain().

PEEK-MESSAGE

If, perchance, you don't want to respond to messages received during the direct-video period, you can use PEEKMESSAGE() to see what is there, and discard it.

Note that PEEKMESSAGE() can be used at any time within your callback to interrogate the queue. It gives you the options of checking the queue with or without removing messages, checking for a range of messages only, and of not yielding to Windows.

The main advantage of PEEKMESSAGE() is that it doesn't wait if there are no messages on the queue; it returns immediately — great for getting keyboard or mouse input in a non-event-driven manner (a bit like old times!). The next advantage is that if you are doing some kind of direct access and don't want any other application to run, you can call PEEKMESSAGE() with the "no yield" option.

DOS keyboard/ mouse input

Windows undocumented functions can be a bonanza, if we can figure out what they do. DEATH(), RESURRECTION(), DISABLEOEMLAYER(), and ENABLEOEMLAYER() are extremely interesting. The latter two go the whole way, turning off Windows screen, mouse, and keyboard and restoring all normal DOS I/O, with the vital exception that we are still in Protected mode.

Experimenting

Writing pixels to the video -RAM

One thing that you might like to do as an exercise is modify my code so that the mode is not changed. Leave it as it was, and change the segment address from B000h to A000h, then you will have a selector to the graphics video buffer. The EGA and VGA physical video buffers are at segment address A000h. If the program sends ASCII "X"s to the screen, you won't see "X"s, because the screen is in graphics mode.

ASCII codes are only appropriate when the screen is in text mode. In graphics mode you write pixels to the buffer, and to know how to do that you need a good EGA/VGA programming book. In this simple example, the "X"s will produce an interesting pattern on the screen. You might like to experiment with commenting out the INT-31h/4002h and the REPAINTSCREEN().

A Direct-Video Window Program

Overwriting the screen using BIOS and DOS services: a renegade window

I introduced the idea of overwriting the existing Windows screen back on page 237, and above I suggested the relatively complicated approach of writing directly to the video RAM (which will also involve manipulation of I/O registers on the adaptor card). However there is another way: the BIOS and DOS services. These services will do whatever you want.

What I have listed below is a complete program that is an extension of the OO program with a control button, developed in Chapter 6. Clicking on the button causes another window to appear — but this window is different! It is a pseudo-text-mode window, that uses the IBM graphics character set (not ANSI characters. See Appendix B). Furthermore, this window always remains visible — no matter what you do, this window will always appear, until the application is terminated.

The most fascinating aspect of this program is that it uses the plain-vanilla BIOS and DOS video services to generate this window, without the least complaint from Windows. This window is your own personal possession: as far as Windows is concerned it doesn't exist. Yet I have arranged the program so that Windows can never overwrite it (unless you want it to).

I have listed this program here (and it is also on the Companion Disk in directory \DPMI0) as a source of ideas — possible building blocks for other more wondrous programs.

Ok, here it is:

```
;This demo program is written in TASM v3.0.
;It uses the WINASMOO.INC OO-file developed in Chapter 7.
```

```
;This program makes use of BIOS/DOS & low-level Windows
;functions.
;remember that Windows funcs only preserve SI,DI,BP & DS.

INCLUDE WINDOWS.INC
INCLUDE WINASMOO.INC
IDM_QUIT          EQU    100
IDM_ABOUT         EQU    101

.DATA
window1  WINDOW  { szclassname="DPMI",sztitlename= \
   "DPMI DEMO", paint=w1paint, create=w1create, command= \
   w1command,  createstylehi= WS_OVERLAPPEDWINDOW+ \
   WS_CLIPCHILDREN, char=w1char, sziconname="icon_1", \
   y_coord= 10,timer= w1timer,destroy=w1destroy }
control1 CONTROL { \
      szclassname="BUTTON",sztitlename="OK",\
      x_coord=20,y_coord=40,wwidth=30,wheight=20, \
      hmenu=IDOK,createstylehi=WS_CHILD+WS_VISIBLE,\
      createstylelo=BS_PUSHBUTTON }

.CODE
kickstart:
   lea  si,window1              ;addr of window object.
   call [si].make PASCAL,si        ;make the window.
   lea  si,control1
   call [si].make PASCAL,si        ;make child window
   ret
;.........................
w1paint PROC  PASCAL
   LOCAL          hdc:WORD
   LOCAL          paintstructa:PAINTSTRUCT
   lea  di,paintstructa
   call BEGINPAINT PASCAL, [si].hwnd, ss,di
   mov  hdc,ax
   call SELECTOBJECT PASCAL,ax, [si].hfont
   call TEXTOUT PASCAL,hdc,10,20, cs,OFFSET outstring,29
   call ENDPAINT PASCAL,[si].hwnd, ss,di
   ret
outstring       DB "Click button for direct video "
w1paint ENDP
;..................................
w1create:
   call GETSTOCKOBJECT  PASCAL,OEM_FIXED_FONT
   mov  [si].hfont,ax
   ret
;..................................
w1command:
   cmp  WORD PTR [si].lparam,0
                   ;lo half=0 if a menu selection.
   jne  notmenu
   ret
notmenu:
   cmp  [si].wparam,IDOK  ;button child window selected?
   ;note that lo-word of lparam has handle of control
   ;window, hi-word of lparam has notification code.
   jne  notbutton
```

```
      lea  si,control1              ;since si points to window1.
      call DESTROYWINDOW PASCAL,[si].hwnd  ;kill the button
      mov  [si].hwnd,0            ;must clear hwnd, if want to
                                  ;make() later.

;what we will do now is make the new window always stay
;visible....
      lea  si,window1
      call SETTIMER PASCAL,[si].hwnd,1,200, 0,0  ;1=timer id.
                  ;post WM_TIMER to window every 200mS.
notbutton:
      ret
szmsg   DB       "Created by Barry Kauler, 1992",0
szhdg   DB       "Message Box",0
;.............................................
w1char:
;let's bring back the button if any key pressed...
      lea  si,control1              ;since si points to window1.
      call [si].make PASCAL,si
      ret
;.............................................
w1destroy:
      call KILLTIMER  PASCAL,[si].hwnd,1    ;kill the timer.
      call POSTQUITMESSAGE  PASCAL,0     ;normal exit.
      ret
;.............................................
w1timer:
;comes this way if a WM_TIMER message....
;this WinApp keeps on posting a WM_TIMER message to
;itself, thus this section is in a continuous loop...
      call dpmidemo
      ret
;.............................................
dpmidemo:
;comes here if button selected.  now we will do some direct
video...
      mov  ah,0Fh
;get current video state
      int  10h
;-->al=mode,ah=width,bh=page
      mov  mode,al       ;save
      mov  columns,ah    ;     /
      mov  vpage,bh      ;     /
      mov  ah,3    ;get current cursor position
      mov  bh,vpage      ;video page
      int  10h     ;-->dh=row,dl=col,cx=cur.size
      mov  curpos,dx            ;save.
;all of this below, writes the pseudo text mode window on
;to the scrn...
      mov  ah,2                 ;set cursor position
      mov  dh,5                 ;row=5
      mov  dl,columns
      shr  dl,1                 ;centre cursor on screen
      mov  bh,vpage             ;video page
      push dx                   ;save
      int  10h
      mov  dx,OFFSET sdirect
```

> Note that this only works for standard VGA. Modification is required for SVGA — see \DPMI0\README.TXT on the Companion Disk.

```
        mov   ah,9                          ;write a string to scrn
        int   21h
        pop   dx                            ;restore
        inc   dh                            ;next row
        mov   bh,vpage
        mov   ah,2
        push  dx                            ;save
        int   10h                           ;set cursor
        mov   ah,9                          ;write string
        mov   dx,OFFSET sdir2
        int   21h
        pop   dx                            ;restore
        inc   dh                            ;next row on scrn
        mov   bh,vpage
        mov   ah,2
        int   10h                           ;set cursor
        mov   ah,9                          ;write string
        mov   dx,OFFSET sdir3
        int   21h
        mov   ah,2                          ;restore cursor pos.
        mov   dx,curpos
        mov   bh,vpage
        int   10h
        ret
.DATA
mode DB 0
columns DB        0
vpage   DB        0
curpos  DW        0
sdirect           DB  "  ┌──────────┐  $"
sdir2             DB  "  │ BIOS/DOS O/P │  $"
sdir3             DB  "  └──────────┘  $"
;. . . . . . . . . . . . . . . . . . . . . . . . . . . . . . . . . . . . .
        END
```

Reference sources   Refer to a good DOS/BIOS programming book for details on the video services. A person by the name of Ralph Brown has compiled a detailed document on all of the interrupts and this can be located at various places on the Internet, such as:

http://www.cs.cmu.edu/afs/cs/user/ralf/pub/WWW/files.html

The above code for drawing the box isn't particularly elegant (there are a hundred possible ways) but shows the idea.

VGA and SVGA   One lovely feature of the BIOS/DOS services is that the cursor and text I/O services treat row and column just like text-mode, even if the screen is in graphics mode. The number of rows and columns for each graphics mode can be found from a table, such as in Thom Hogan's *PC Sourcebook* (see page 82). I used INT-10h to obtain the current mode and number of columns, but do read \DPMI0\README.TXT for special information on Super-VGA.

Before you run this program, change Windows to standard VGA, 640 x 480 x 16, that is, 16 colors, and restart Windows. Yes, it works on Windows 95!

The not-entirely-appropriately-named dpmidemo() is called every 200ms, which is how the window always manages to stay on top. You can see that I called SETTIMER() to create a Windows software timer, and note also that I killed it before exiting the program. The reason for this is that Windows timers are a limited resource.

Figure 9.1 shows the result of this program on-screen.

Figure 9.1: BIOS/DOS O/P to screen.

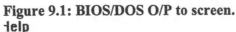

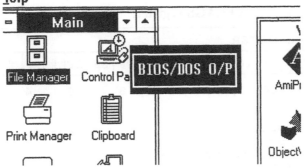

It's a bit like trying to mix water and oil!

Ghosting

If you try the program, one "feature" that you will observe is that "ghosting" can occur in windows moved underneath, so an improvement would be to hook all WM_MOVE messages and append a WM_PAINT message. The problem is that whenever you move (drag) a window on the screen, Windows simply performs a shift of the window image, and does not tell the window callback to repaint the window. Thus, shifting a window under our "special" window can result in the underlying window picking up a ghost of our special window. I have toyed with various ideas for telling the window to repaint its client area, but did not put any code into this example, for the sake of simplicity.

Anyway, I see this more as a learning exercise, and I don't think you should put these techniques into that professional office business suite you're working on! On the other hand, you never know when low-level knowledge like this will come in handy.

I/O Ports

DOS assembly language programmers will be accustomed to using the IN and OUT instructions to talk with I/O ports.

Of course, with DOS it was very straightforward. Execute "OUT 28h,AL" to send a byte of data from the AL register to port (address) 28h, and it happens immediately, without question.

EFLAGS register

However, with the CPU running in Protected mode, there is some extra rigmarole. Since more than one task can be executing, there has to be a mechanism to prevent contention. First, look at the flags register inside the CPU (Figure 9.2):

Figure 9.2: EFLAGS register.

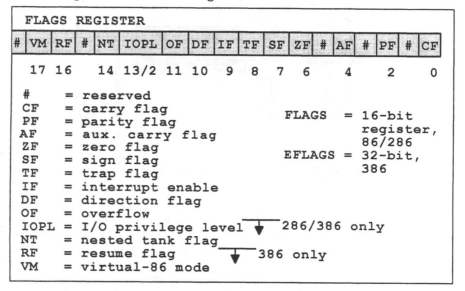

```
 FLAGS  REGISTER

 # VM RF # NT IOPL OF DF IF TF SF ZF # AF # PF # CF

    17 16   14 13/2 11 10  9  8  7  6    4    2    0

 #    = reserved
 CF   = carry flag
 PF   = parity flag              FLAGS  = 16-bit
 AF   = aux. carry flag                   register,
 ZF   = zero flag                         86/286
 SF   = sign flag                EFLAGS = 32-bit,
 TF   = trap flag                         386
 IF   = interrupt enable
 DF   = direction flag
 OF   = overflow
 IOPL = I/O privilege level      286/386 only
 NT   = nested tank flag
 RF   = resume flag              386 only
 VM   = virtual-86 mode
```

IOPL field

The field of immediate interest to us is IOPL, which means Input/Output Privilege Level. Referring back to page 31, privilege level varies from 0 to 3, with 0 being most privileged. IOPL is set by the operating system, and an application must have a privilege level at least as privileged as IOPL for it to be allowed to perform I/O.

With Windows, the IOPL is set to 0, while applications and DLLs run down at 3, so it appears that they can't do I/O. But this is only how it appears, for the protection mechanism is only used by Windows as a control mechanism, and in *some* cases I/O is allowed — clarification is needed here!

I/O and related instructions: CLI, STI, PUSHF, POPF

First however, we should be clear about what we mean by "allowing I/O". Hardware interrupts are, of course, part of I/O, as is control of the interrupt flag, *IF*.

IF is a mask that determines whether external interrupts are allowed to occur. It is 0 if interrupts are disabled and 1 if enabled. There are two instructions that control this flag: *CLI*, meaning CLear Interrupt, and *STI*, meaning SeT Interrupt.

There is another means of controlling this flag: *PUSHF* and *POPF*, which push and pop the flags register respectively. If POPF is executed, whatever value is on top of the stack will be popped into the flags register, thus affecting IF.

I/O exception handler

IN, OUT, CLI, STI, PUSHF, and POPF all work without question under DOS in Real mode. However, in Protected mode, since the application doesn't have permission to do I/O, execution of any of the above causes an exception, which is a special reserved interrupt that causes a Windows exception-handler to execute. It is the hardware in the CPU that does this detection.

The Windows exception handler *may* allow the IN, OUT, CLI, STI, and PUSHF to go ahead, with due regard for contention with other applications, but it modifies the POPF instruction so that it does not change the interrupt flag IF. The moral of this story is never use POPF to change the interrupt flag.

Windows 95 I/O

There are many caveats to I/O under Windows. Yes, it's fine in Real mode. It may also be fine in virtual-86 mode, i.e., running in a DOS box inside Windows. Note that some references call virtual-86 mode Real mode, though it is really a Protected mode simulating the "real" Real mode! Interrupts and IN/OUT to most of the ports is allowed for WinApps in Windows 3.0 and 3.1.

It works for 16-bit WinApps in Windows 95 also.

The problem is 32-bit WinApps. I recently read this nice submission to the comp.os.ms-windows.programmer.win32 newsgroup from Eugine Nechamkin:

```
> I require to be able to
> intercept an interrupt (that being int 0x78) performed by
> a dos application, and respond to it from my windows 95
> application.

if((don't care about interrupt latency time) && !(must write VxD))
   {
    // Make Win16 DLL controlling your interupt vector and
    // processsing interrupts in usual DOS - like manner
    // (setintvect(), getintvector()) !!!
    if (you are happy with Win16 app. under Windows 95)
      {
       //Make some front end Win16 app communicating with your Win16
```

```
        // DLL ;
      }
    else
      {
       // Make Flat-Thunk Win32 DLL to communicate with Win16 DLL;
       // Make some front end Win32 app using Flat-Thunk Win32 DLL ;
      }
  }
else
  {
   // Spend $$$$$ on Win95 MSDN, DDK, Nu-Mega's "Soft-Ice";
   // Write a VxD for Win95;
   // Write your Win32 (or Win16) front end;
  }
```

There is a cheap option for developing VxDs, known as VxD-Lite (see Chapter 14). Chapters 12 and 14 explore transitions between DOS, 16-bit and 32-bit WinApps, and VxDs. There are many options for getting at the low-level (hardware and underlying OS management) from 32-bits, though it's all awkward.

Example program: loudspeaker control
Now for some example code. What I have here is a simple routine to emit a tone from the loudspeaker. Nothing startling, but it is significant because it is done by programming the I/O circuitry directly. The PC has three hardware timers, the first dedicated to producing an interrupt every 55 milliseconds (ms): the INT-8 hardware interrupt. The second generates continuous pulses that are used by the dynamic RAM refresh circuitry. The third is general purpose, and is most often used to produce tones on the loudspeaker, since its output is connected physically to the loudspeaker.[1]

```
;to initialise and start hardware timer ...
    mov   al,0B6h      ;setup the mode of timer-2.
    out   43h,al       ;       /
    mov   bx,0700h     ;load the counter.
    mov   al,bl        ;       /
    out   42h,al       ;       /
    mov   al,bh        ;       /
    out   42h,al       ;       /
    in    al,61h       ;set bit-0 and 1 of port 61h.
    or    al,3         ;       /
    out   61h,al       ;       /
```

---

[1]   The hardware timers used in the PC are 16-bit down-counters that can be loaded with a value and then started. Once started, they count to zero and then either stop or automatically reload and count down again. They can be programmed to produce a pulse at the output pin upon reaching zero or simply flip state (which gives a square wave output). In continuous mode, their frequency obviously depends upon the value first loaded; this has to be programmed to port 42h in two steps. Bit-0 of port 61h starts and stops the timer. It is important to note that once started, the timer is completely independent of the CPU and will keep on going until an OUT to port 61h stops it.

```
   . . .
;timer is now running.  To turn it off ...
   in   al,61h           ;clear bit-0 and 1 of port 61h.
   and  al,0FCh          ;    /
   out  61h,al           ;    /
```

You could arrange this code in your Windows application in whatever way is appropriate. It illustrates the use of both IN and OUT instructions and shows that they work fine from Protected mode (subject to the above-mentioned caveats).

Real-time events

While we are on the subject of timers, another issue arises; that of synchronisation, or response, with or to real-time events. An interrupt from a hardware timer is a real-time event. Any hardware interrupt is a real-time event. "Real time" need not necessarily imply external hardware: if one application wants to signal another and if the other is to respond immediately, it will be a real-time response.

It may be that an external device has to signal a Windows application, and it may be important that the application respond in a very predictable way, within a predictable time frame. Unfortunately Windows' response is anything but real time. This is a very important issue, and worthy of a new chapter.

Windows timers

A little note before I launch into the real-time section — Windows does have "software" timers that can be programmed to time out at regular periods, just like the hardware timers discussed above. See the usage of SETTIMER() on pages 239+. However, upon time out they send a message to the application over the standard message queue, so its arrival time at the application is highly unpredictable. It is even possible for the timer to time out a few times, and queue the messages, before the application gets them — suddenly the application will get three or more timer timeout messages at once! Hardly useful if you want your application to be triggered at precise intervals.

As a final thought, Windows has an undocumented function, **CREATESYSTEMTIMER()**, that is documented in the DDK, Daniel Norton's book (see page 203), and in *Undocumented Windows* (see page 218). It bypasses the message queue and calls the callback directly. Thus, it is possible to make code execute at precise intervals (though the callback has the major restriction that it can only call certain Windows functions, just like an ISR).

Threads

Windows 95 has made timers less important, with the introduction of threads. These introduce an execution overhead though. Threads are only supported in 32-bit applications, with the Win32 API. Even though a 16-bit application can call 32-bit API functions (see thunking section in Chapter 8), it can't use the thread functions.   Threads can synchronise their execution also (see Chapter 8 page 223).

10

Real-Time Events

Preamble

Windows
preemptive
aspects
So you think 16-bit Windows applications are non-preemptive? Think again!

Just about everything you read will tell you that a disadvantage of Windows 3.x is non-preemption. That is, once control is passed to an application, Windows cannot regain control until the application has passed control back, by a RET. One of the touted advantages of 32-bit applications under Windows 95 is preemption.

Actually, whether it be Windows 1.0 or 95, interrupt-driven device drivers, including keyboard input, must always be working in the background. When a key is pressed, a hardware interrupt is generated, which invokes the keyboard device driver.

The immediate response to a key press is preemption, nothing else, and contrary to common knowledge, Windows 3.x applications can make use of similar mechanisms.

Also, the DPMI host maintains preemptive time-sliced switching between VMs on Windows 3.x and 95.

Application
preemption
A Windows application can respond immediately to an external hardware event, or a timer interrupt for that matter (refer back to page 246 for a brief introduction to the PCs hardware timers, and to pages 239+ for an introduction to the Windows "software" timers).

You can also signal between Windows applications, immediately, without going through the messaging mechanism.

Preemption by interrupts

Just as device drivers can be interrupt driven, so too can your own application to provide predictable real-time response.

It is not all peaches and cream however.

The chapter starts with code for software interrupts, because it is the easier case. The interrupt mechanism is particularly useful for signalling and passing data between Windows programs.

The chapter then progresses to hardware interrupts, with example code.

TSRs

What originally started me thinking about this topic was a problem some colleagues of mine at Edith Cowan University were having. They wanted a Windows 3.0 application to sit in memory, like a TSR (Terminate and Stay Resident) program, logging external real-time events, while Windows was running other applications. In other words, they were asking for preemption. Windows, they concluded, was not suitable, so they chose OS/2.

Hooking an interrupt

After some experimentation, I discovered that it is very simple to create a Windows application that behaves just like a DOS TSR and hook an interrupt vector, yet be operating in Protected mode and be in every respect a normal Windows application.

"Hooking an interrupt vector" means to change the entry in the interrupt table (refer back to page 33) to point to the new TSR. In DOS it was very common for a TSR to hook INT-16h — the code that follows also hooks this vector, but note that Windows doesn't use INT-16h for keyboard input, so it doesn't matter what damage we do to this vector!

Hardware vs software interrupts

An interrupt can be either a hardware or a software interrupt — a good DOS programming book will clarify the distinction, but basically a hardware interrupt occurs as the result of an external event, via the Interrupt Controller chip, and maps to various reserved entries in the interrupt table.

Software interrupts are invoked from a program by the instruction "INT n", where "n" is any number from 0 to 255. Note that some of those numbers will also correspond to hardware interrupts, which means that such interrupts can be called either by a hardware event or from a program.

Exceptions   Yet another class of interrupts is *exceptions*, generated by the CPU.

Invocation of the hooked interrupt, either software or hardware, will result in transfer of execution to the TSR. This happens "immediately". The TSR terminates with an IRET instruction, which sends control back to whatever was running before the interrupt.

Hooking a Vector

What I have done in the first part of this chapter is put together a program that hooks INT-16h. The new INT-16h service routine uses the music code from page 246, so there is audible feedback of it executing.

Once the service routine is installed, INT-16h can then be executed from anywhere, including another program, and the service routine will be invoked.

TSR installation routine   The program can be any basic skeleton to which you patch the following code. The "install" portion could be wherever you want it; in WinMain(), in kickstart: (OO program), or in the callback. You could start the program up as an icon (or invisible) and immediately execute the install code.

This is what the install code would have to be:

```
.DATA
offsetint       DW    0           ;old int-vector
selectorint     DW    0           ;   /
.CODE
install  PROC   PASCAL NEAR   ;no params
   USES ax,bx,cx,dx,si,di,es
   mov  al,16h                 ;get vector in idt
   mov  ah,35h
   int  21h                    ;returns vector in es:bx
   mov  offsetint,bx           ;save old vector.
   mov  selectorint,es         ;   /
   mov  dx,OFFSET runtime      ;new vector.
   push ds                     ;save ds.
   push cs
   pop  ds                     ;new vector in ds:dx
   mov  al,16h                 ;int to be hooked.
   mov  ah,25h                 ;set vector
   int  21h
   pop  ds                     ;restore ds.
   ret
install  ENDP
```

INT-21h/AH = 35/25h   Some interesting points arise from this code. INT-21h/AH = 35h or 25h are functions for getting the interrupt vector and for setting it. Look back to the special note on how these work with Windows on page 200.

IDT vs IVT   It is most important to know that they work on the IDT, not the IVT. When the CPU is running in Protected mode, an interrupt will cause the CPU to look in the IDT to find the selector:offset of the interrupt routine.

In the code above, I have not hooked the old INT-16h routine in the IVT. I have only hooked INT-16h in the IDT, which for normal Windows programs isn't used.

The rest of the WinApp   Notice in the above code that I saved the old vector. This is in case I want to call it or jump to it, possibly from within the new interrupt service routine.

Having done that, all that remains is to go into the usual message loop, as per a normal program, which returns control to Windows. There is one little complication with this — since the vector has been hooked, don't close the application, because executing that interrupt from some other application will cause the CPU to try to execute a service routine that is no longer there. In fact, it will crash rather rudely. It is possible to create a window for the program but keep it invisible,[1] to prevent accidental closure, or unhook it before closing. See an example of unhooking on page 260.

Fixed vs moveable segments   Is that all there is to it? Yes. Even the old .DEF file can be used, and you can have MOVEABLE and DISCARDABLE segments. It is not necessary for the CODE and DATA segment statements in the .DEF file to have FIXED qualifiers. FIXED forces Windows to leave the segments at a fixed place in memory, rather than moving them around as it normally does. You would think, from the way TSRs are designed under DOS, that a resident interrupt handler should be FIXED, but not with Windows.

If the operating system determines that the segment referenced via the IDT is not actually in memory, then it will get it back, and update the descriptor. If you want, modify the .DEF file as follows:

```
DATA    PRELOAD FIXED
CODE    PRELOAD FIXED
```

[1]   Note that it possible to have an application without any window at all. Since all messages usually are posted to a window, this requires special consideration. For example, POSTAPPMESSAGE() will post a message to an application without a window and leave the message's hWnd parameter NULL.

Related
issues

Specifying FIXED is not a bottleneck itself, from the point of view of memory management, as some books will have you believe: I discuss this issue on page 324.

Perhaps I am getting ahead of myself, since I haven't even discussed the service routine itself. The above points do tie-in with the service routine however. We may want to store writable data in the code segment of the ISR, which will cause problems. Also, hardware interrupts are a special case. In practise you may have to do more than just specify FIXED: I have gone into this in more detail on page 323. Also some relevant Windows functions (GLOBALHANDLE, GLOBALFIX, and GLOBALPAGELOCK) were introduced on page 210.

While I'm referring you all over the place for extra information, I might as well do it some more. The above install routine works for hardware or software interrupts, that is, any entries in the IDT (or IVT if the CPU is running in Real mode). There are DPMI equivalents: see the Appendices. What about exceptions? These have to be treated as a special case: see page 258.

Service Routine (ISR)

Accessing
data in the
ISR

No, an ISR doesn't have to be a DLL[1] or some other separate program. It can simply be a procedure in the same program that has the install code. It will not be called from the program however.

There is a problem with addressing data upon entry to the service routine, because DS will be an unknown value.[2] Look back to page 33 for a review of the steps that the CPU goes through upon an interrupt occurring. It pushes CS, IP, and flags on to the stack, and gets the new CS:IP from the IDT. The other registers are as they were before the interrupt.

Thus, upon entry to the service routine, only CS is set to the code segment of the service routine. How do you access data in the service routine? One solution is to put data into the code segment.

Normally this is not allowed, or rather it is but you can't write to it, because code descriptors have their access-field set to read-only — however DPMI has a service that gets around this very nicely. What you can do is obtain an "alias"; that is, a data selector that

---

[1] Implementation as a DLL does have some advantages, however. If a DLL segment is declared FIXED in the .DEF file, it loads below 1M, and is also guaranteed to be in contiguous memory. These features allow the DLL to have Real mode code as well as Protected mode code. The DLL runs at privilege level 3 (level 1 in Windows 3.0), so I/O still causes an exception.

[2] MAKEPROCINSTANCE() can be used to attach prolog code that binds data to code, though I have not used it here, for certain reasons. See further notes in the Companion Disk.

points to the same code segment. This will allow you to write to the code segment.

Windows has various functions for segment manipulation, though many of them were unofficial until 3.1 was released. Of most interest is CHANGESELECTOR(), which is official for both 3.0 and 3.1 (see page 208). ALLOCCSTODSALIAS() is an unofficial alternative. With Win95 they all go back to being unofficial.

There is another interesting, related function introduced with TOOLHELP.DLL, and so is backwards compatible with 3.0: MEMORYWRITE(). This will copy a block of memory from one segment to another, regardless of their attributes. Thus it will write to a code segment.

Actually, it is quite easy to get data segment addressability from within an ISR, but I'll leave that one for now.

Data alias to code   Before I show you the actual ISR, I'll provide a little bit of extra setup code using the abovementioned DPMI service:

```
;will create alias in LDT of CS ...
  mov  ax,000Ah
  push cs
  pop  bx                  ;selector to be aliased
  int  31h                 ;returns alias selector in ax
  push ax
  pop  es
  mov  es:dsselector,ax   ;save the alias in the code seg.
```

Normally I would perform the above aliasing in the install code and save the alias selector in the code segment. The ISR can then read it and use it. This works, as long as the ISR doesn't move in memory. The same principle can be used to obtain addressability to the WinApp's data segment.

Having got into the service routine and established data addressability, all that remains is to do something. I have used the code from page 246 to produce a tone on the loudspeaker. Here it is:

```
.CODE
;I've put this data in the code segment ...
dsselector      DW    0              ;data alias to code seg
musicflag       DB    0              ;turn music on/off
;
runtime:
  pusha                              ;save all regs.
  push ds
  push es
  mov  es,cs:dsselector      ;get alias
  push es                    ;can also set ds to alias.
  pop  ds ;(so seg.override isn't needed to access data).
```

```
     sti                          ;enable interrupts.
       ;(STI and reentrancy issues discussed on page 323).
     cmp  musicflag,20  ;musicflag is used as a counter, for
     jb   jumpout3       ;turning the tone on or off on each
     mov  musicflag,0          ;10th entry to the routine.
     jmp  turnoff
jumpout3:
     inc  musicflag
     cmp  musicflag,10
     jne  jumpout2
timeron:
     mov  al,0b6h                ;turn on the hardware timer.
     out  43h,al
     mov  bx,07c5h                  ;frequency 600Hz.
     mov  al,bl
     out  42h,al
     mov  al,bh
     out  42h,al
     in   al,61h
     or   al,03
     out  61h,al
     jmp  SHORT jumpout2
turnoff:
     in   al,61h                ;turn off the hardware timer.
     and  al,0fch
     out  61h,al
jumpout2:
     pop  es
     pop  ds                         ;restore all regs.
     popa
     iret
```

Testing

Stick this service routine somewhere in your program, then assemble and link as per normal. To test it, you will have to modify some other program, by inserting an "INT 16h" instruction into it. Perhaps you could put this instruction into the other program's WM_CHAR case, so whenever you press a key and the other program's window is active, the program will execute "INT 16h", which will call the service routine.

Don't be confused here. A key press has nothing to do with INT-16h under Windows, at least as far as normal code is concerned. I have just arbitrarily suggested that you use the WM_CHAR message as a convenient means of invoking the service routine.

Having modified another program, start both it and the "TSR" program. With the "other" program active, try key presses, at least ten, and you should be able to toggle the tone on and off.

What you are doing here is accessing a global variable! Other applications can also access that same variable, which raises interesting possibilities for interprocess communication.

*Only one
LDT and
IDT*

If you know much about LDTs and GDTs, you might be puzzled as to how the above code can work. The classical theory states that each application has its own LDT (see Chapter 1), so modifying the TSR's LDT has nothing to do with any other application's LDT. Not so with Windows! As is explained in more detail in the next chapter, all WinApps share the same LDT. Ditto for the IDT.

The IDT is a very grey area. It is another case of Microsoft hiding the truth. The classical model for the IDT would be that there is only one, but Windows does maintain copies, as far as I know, for each VM. So maybe there is just one "main" IDT that interrupts "go to" but the interrupt handler references the copy in the current VM. This is a very very grey area, but you can get by with just thinking that there is only one IDT. Certainly, as all WinApps are in the same VM, this assumption is safe.

Hardware Interrupts

You will notice that my example code earlier in this chapter dealt only with software interrupts. Hardware interrupts can work, but there are some complications. The problems are associated with how interrupts are mapped and the difference in treatment of interrupts in Protected and Real modes.

The issue is very complicated and it behooves us to start with the handling of hardware interrupts from the point of view of the XT; that is, with an 8088 or 8086 CPU.

XT Hardware Interrupts

IRQ 0-7

The PC model XTs are based upon the 8086 CPU and have a hardware interrupt controller chip that allows eight devices to interrupt the CPU. That is, the chip has eight inputs, labelled IRQ0 to IRQ7 and one output labelled IRQ (Interrupt ReQuest) that feeds into the maskable interrupt pin of the CPU.

A flag named IF (Interrupt Flag) enables this IRQ input with the STI instruction or disables it with the CLI instruction (see page 33).

The interrupt controller chip can be, and is, programmed to map IRQ0 to IRQ7 to any group of eight entries in the IVT or IDT (look ahead to page 268 for the relationship between the IVT and IDT) (see page 185 for an introduction to the interrupt controller chip).

The XT maps IRQ0 through 7 to entries 8 to 0Fh in the IVT. Thus if you were to access these by software interrupt, you would execute "INT 8" to "INT 0Fh".

AT Hardware Interrupts

The IBM model AT, based upon the 80286 CPU, introduced more hardware interrupts, by cascading a second interrupt controller chip, as shown in Figure 10.1.

Figure 10.1: AT hardware interrupts.

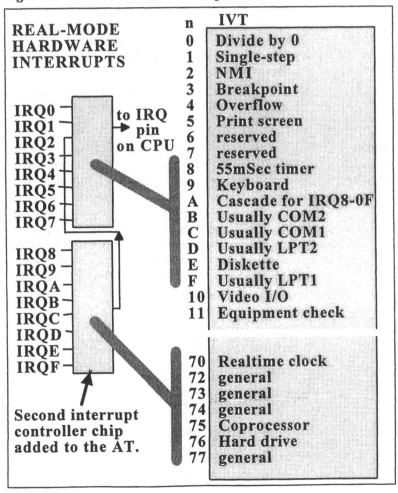

At power-on, the interrupt controller chips are programmed to map to certain entries in the IVT. When an interrupt arrives, IRQ is forwarded to the CPU, and the CPU then interrogates the controller chip, which passes the interrupt number "n" to the CPU over the data bus. The CPU then looks up that entry in the IVT and goes to the interrupt service routine.

Exception handling conflict

When the CPU is operating in Real mode, INT-0 is what is called a processor *exception*; that is, an interrupt generated by the CPU itself, not by the program or by external hardware. Ditto for INT-1.

I have shown INT-6 and -7 as reserved, which is the case for the XT. However on the AT, the 286 CPU uses these for "invalid op-code" and "device not available", respectively. Again, these are exceptions.

There is a very serious problem with this arrangement. With the 286 and 386, Intel uses the first 16 entries of the interrupt table — and now we must refer to the IDT — as exceptions when the CPU is operating in Protected mode.

However, the hardware interrupts IRQ0 through 7 are mapped into INT-8 to -Fh. Quandary — how is this conflict resolved?

Windows remapping of vectors

Windows (and OS/2) map IRQ0 through Fh elsewhere in the IDT, at INT-50h to -5Fh. Obviously, these entries would point to the same routines as before, but even so, there is room here for trouble.

You might deduce from this that if you wanted to hook the original INT-8, you should instead hook INT-50h. This is valid, but only to a certain extent. Windows can be in Protected or V86 mode at the time of interrupt, and in the latter case we have to go back to the IVT in the V86 virtual machine currently active.[1]

Therefore, we (may) actually have to hook **two** (or more) vectors. Headache!

Windows' Standard Mode Hardware Interrupts

Somewhere earlier in the book I promised not to mention Standard mode again, as it's history — almost. Maybe in some third world

---

[1]   I don't want to be misleading here. In Windows Standard mode when a DOS program is running, the CPU will be in the one-and-only Real mode and interrupts vector via the one-and-only IVT. However, in Enhanced mode with a DOS VM active, it is still a *Protected mode,* and hence, hardware interrupts still go to the IDT in the **current** VM (note the emphasis on current). Because the DOS VM is supposed to behave in all respects like an XT-model PC, the interrupt will eventually come down to the IVT.

countries it's all the rage. However, the following is extremely interesting, and I've put it in for the education it gives us about the warp and weave of interrupt handling.

INT-51h

What I have for you here is a useful program that hooks INT-9, the keyboard hardware interrupt — except to illustrate how Standard mode works, I have hooked INT-51h!

This keyboard hook can be very useful for filtering whatever comes from the keyboard before Windows has a chance to see it. Note that INT-51h is invoked every time a key is pressed or released, with bit-7 of the scancode distinguishing which.

POSTMES-SAGE()

Hardware interrupts are somewhat more delicate than their software cousins — for example you can't call Windows functions from them — with one exception: **POSTMESSAGE()**. Microsoft especially made sure that this would work from the hardware interrupt level, so that a hardware interrupt service routine can signal a Windows application.

There is a particular problem with these hardware interrupts, due to the way they are mapped. With Windows in Standard mode[1] I have shown on the previous pages that the keyboard interrupt maps to INT-51h in the IDT, with certain qualifications, and this example code hooks that vector. This point is elaborated on a little later.

Incidentally, if you need to know which mode Windows is running in, there is a function that will do that for you; **GETWINFLAGS()**. I haven't shown the call to GETWINFLAGS() in the example below, but in a practical program you could include it.

A skeleton hardware interrupt handler

What follows is just an extraction of the bare essentials to get a hardware interrupt working — the flesh can go on later.

Ok, now for some code:

```
;add this extra function to the external declarations...
EXTRN   POSTMESSAGE:FAR
;
.CODE
;put in the usual WINMAIN() function ...

;followed by a callback ...
   PUBLIC      DPMICALLBACK
DPMICALLBACK PROC WINDOWS PASCAL FAR \
      hWnd:WORD,msgtype:WORD,wParam:WORD,  lParam:DWORD
;put in the usual CASE structure to process messages,
;but with some additions ...
```

---

[1] Put Windows in Standard mode by typing "WIN /S" when loading it. That is what it will be anyway if the CPU is a 286, or a 386 with insufficient RAM (usually less than 2M).

```
    mov   ax,msgtype
    cmp   ax,WM_CREATE
    je    xcreate
    cmp   ax,WM_DESTROY
    je    xdestroy
    cmp   ax,WM_USER
    je    xuser
    ...etc ...
    ...
;here is the handling of the WM_CREATE case ...
xcreate:                           ;whatever you want, plus ...
    call installint                ;hooks the vector.
    jmp  xexit
;here is the handling of the WM_USER case ...
xuser:
    push ax
    push dx
    mov  ah,2                      ;write char to scrn
    mov  dl,07                     ;beep
    int  21h
    pop  dx
    pop  ax
    jmp  xexit
xdestroy:
    call POSTQUITMESSAGE    PASCAL,0
;unhook the int-51h...
      push dx
      push ds
      mov dx,offsetint             ;this is the old INT-51 vector
    mov  ds,selectorint        ;      /
    mov  ax,2551h ;before quitting, we are restoring it.
    int  21h
    pop  ds
    pop  dx
    jmp  xexit
;whatever else you want here ...
    ...
xexit:
    sub  ax,ax                     ;returns 0 in DX:AX.
    cwd                            ;return a 32-bit (long) value).
    ret
DPMICALLBACK    ENDP
```

Hooking/
unhooking
the vector

A WM_CREATE message is sent when the window is first created, so this is a convenient time to hook the vector. Therefore a call to INSTALLINT() is included.

Similarly, upon exit it is necessary to unhook the vector, otherwise Windows will crash. Unhooking on receipt of the WM_DESTROY message is most appropriate. This code simply uses INT-21h/AH = 25h to restore the old vector, which has previously been saved in "offsetint" and "selectorint" by the installint() procedure.

Interrupt Handler Code

***POSTMES-
SAGE()***

The interrupt service routine will be entered every time a key is pressed or released, and all that I have done inside it is call **POSTMESSAGE()** to send a **WM_USER** message to the window's callback function DPMICALLBACK().

WM_USER

WM_USER equates to a message number that is not used by Windows as a message, so it is free for an application to use. A range of such numbers is available for an application to use: look in WINDOWS.INC.

Ok, now for the installint() function:

```
.DATA
descrbuffer    DB    8 DUP(0)
offsetint      DW    0              ;old int. vector
selectorint    DW    0              ;            /
.CODE
dsselector     DW    0                     ;data alias to code seg
hwndcs         DW    0    ;save window handle for use in isr
;
installint  PROC                   ;no params
  pusha
  push es
  push ds
;will create alias in ldt of current task...
  mov  ax,000Ah ;create alias data descr. for code seg.
  push cs
  pop  bx                          ;selector to be aliased
  int  31h                         ;returns ax
     push ax
     pop  es
     mov  ax,hwnd
     mov  es:hwndcs,ax             ;handle of window
     mov  ax,es
     mov  es:dsselector,ax         ;alias
;now to get the old INT-51h vector, and save it ...
  mov  al,51h                      ;get vector in idt
  mov  ah,35h                      ;-->ES:BX
  int  21h
  push es
  pop  ax
  mov  offsetint,bx                ;save the old vector.
  mov  selectorint,ax              ;   /
  mov  dx,OFFSET runtime           ;get the new vector
  push cs                          ;   /
  pop  ds                          ;new vector in ds:dx
  mov  al,51h
  mov  ah,25h                      ;set vector
  int  21h
;
  pop  ds                          ;restore ds.
  push ds                          ;save it again
;let's hook int60, to use as old vector...
  mov  dx,offsetint
```

```
   mov  ax,selectorint
   mov  ds,ax
   mov  ax,2560h
   int  21h
;installation now finished ....
   pop  ds
   pop  es
   popa
   ret
```

I can put the interrupt service routine in the same procedure as the install code, if I wish, but before listing it, I want to comment on the above code.

To be able to get at data in the service routine (I'll call it an **ISR** from now on), I had to create a data alias; that is, a data selector that points to the code segment. This enables me to write to the code segment.

Into the code segment I saved the handle (hwnd) of the application's window. The reason for this is that within the ISR I called POSTMESSAGE(), which needs the handle as a parameter.

Calling the old handler   You can see that I hooked the vector and saved the old vector, but I also put the old vector into INT-60h. That is, I hooked INT-60h so that it now points to the Windows keyboard handler. This is convenient, because from within the ISR I wanted to be able to call the old ISR, for proper handling of the keyboard input.

Note that there are other ways of doing this, such as by use of a CALL instruction.

Now for the ISR:

```
runtime:
   int  60h                    ;call the old INT-51h
   pusha                       ;save all registers.
   push ds
   push es
   push ss
   mov  ax,cs:hwndcs           ;get window handle
;    call POSTMESSAGE PASCAL,ax,WM_USER,0, 0,0
;no, will do it this way, as PASCAL qualifier very
;inefficient ...
   push        ax
   push        WM_USER
   push        0
   push        0
   push        0
   call        POSTMESSAGE     ;put message on queue.
   pop  ss                     ;now restore and get out.
   pop  es
   pop  ds
   popa
   iret
```

```
installint ENDP
    END
```

See how simple the ISR is! I was able to call the original keyboard handler for proper handling of the key press/release, though note that I could have put the "INT 60h" at the end of the ISR if required.

I accessed "hwndcs", the handle of the window passed as data in the code segment, and then called POSTMESSAGE().

Note that I did not make use of aliasing in this simple skeleton.

I chose to explicitly push the parameters onto the stack prior to the CALL, rather than use the PASCAL qualifier — TASM's generation of code with the PASCAL qualifier is horribly inefficient, so I felt better about doing it this way.

Enhanced Mode Hardware Interrupts

INT-9 keyboard handler

So what about Windows in Enhanced mode? Remember that Windows 95 can *only* run in Enhanced mode.

I mentioned earlier that Windows gets up to some tricky business, and for both Standard and Enhanced modes reflects the INT-51h to INT-9.

However, this mechanism is different in each case, as Enhanced mode is able to make use of virtual machines, with the result that hooking **either** INT-51h **or** INT-9 will work in Standard mode, but in Enhanced mode **only** INT-9 will work.

Real mode keyboard handler

So the earlier example code that I wrote to hook INT-51h for illustration purposes simply needs to be modified to hook INT-9, and it will work in both Standard **and** Enhanced modes. Unfortunately there is is still one complication — DOS.

I keep hoping it will go away — but it won't. The hardware interrupt handler developed in this chapter will work with any number of Windows applications multitasking, but not when a DOS program is running. In the former case, it doesn't matter if the program containing the ISR is iconized and another WinApp has the active window — still, all key presses will in real time be routed to the ISR and be posted to the iconized program — and Windows will call the iconized program's callback function, giving it the message, even though it is iconized.

So you'll always get the beeps when pressing and releasing a key.

However, if you run the "DOS Prompt" program, the beeps will stop. Upon exiting back to Windows, the beeps will start once again.

If you really must have the ISR continuing to function when the CPU is running a V86 or Real mode program, refer to Chapter 11, as I decided to make the handling of Real mode a special chapter all on its own. See also the footnote on page 258.

What the program "does"

I suppose you do realise by now what the example program does —'it beeps the loudspeaker every time you press or release a key.

Because the ISR only posts a message to the main Windows program, it is what I would class as pseudo-real-time response. Don't forget, however, that the ISR shares the same code segment as the main program, and by way of a data alias, data can be passed to and fro. Or the actual WinApp data segment can be readily accessed.

For example, harking back to the problem that my colleagues had — they wanted to measure an external parameter at precise intervals and log it for internal analysis. The interrupt mechanism provided the precise intervals, and the ISR could have read the parameter from the input port and recorded it, then exited. Simple enough.

You will find the program on the Companion Disk in \ISR1.

Direct Memory Access

In this and the previous chapters I have covered the basic elements of hardware access, namely direct memory access, I/O port access, and interrupts, but there is another aspect that is worth introducing: DMA.

Reference sources

DMA is perhaps somewhat too esoteric for a book of this general nature; however, a few notes are in order and I can point you in the right direction.

The best reference would be Microsoft's *Virtual DMA Services Specification*, part number 098-10869.

Another introductory reference is "DMA Revealed" by Karen Hazzah, *Windows/DOS Developer's Journal*, April 1992, pages 5-20.

What is DMA?

Basically, DMA takes the job of data transfer away from the CPU for the sake of speedy transfer of blocks of data, usually between a hardware device and memory. It requires a DMA controller chip. Initialization involves telling the controller the address of the memory buffer and how many bytes to transfer.

Bolting a segment down

With Windows, there are complications, because the CPU can be in Real or Protected mode. In Protected mode the buffer should be constrained to be below 1M and should also be contiguous.

Paging normally will split a segment up all over the place, but there are mechanisms in Windows for keeping a segment together. The DMA controller is given the selector:offset and simply increments the offset without regard to paging — remember that the CPU is turned off at this time, and the DMA chip has complete control of the bus.

Another implication of this is that it is wise to keep memory buffers to no more than 64K.

I did note earlier that by declaring the DLL data segment FIXED, it will load below 1M and be contiguous. However there appears to be some doubt about the latter, as the recommendation is that to ensure that *pages* are contiguous, another service must be called: the INT-4Bh/Lock-DMA function.

INT-4Bh

INT-4Bh provides the extensions to DOS for DMA handling, and you will find these documented in the above Microsoft reference — not anywhere else, that I'm aware of.

The services, available from both Windows Standard and Enhanced modes, are:

- INT-4Bh/AX = 8103h   VDS_LOCK
- INT-4Bh/AX = 8104h   VDS_UNLOCK
- INT-4Bh/AX = 810Bh   VDS_ENABLE_TRANSLAT.
- INT-4Bh/AX = 810Ch   VDS_DISABLE_TRANSL.

Some of the discussion in this and earlier chapters has referred to Real mode. Although Windows normally runs in Protected mode, Real mode is still encountered, as is virtual-86 mode, and more specific treatment is provided on this topic in the next chapter.

11

Real Mode Access

Preamble

Why bother with Real mode?

The topic of Real mode has already been encountered at various earlier stages in the book. There is, however, a lot more to the issue of Real mode.

Windows 3.1 won't run in Real mode, only Standard or Enhanced, version 3.0 loads in any of the three, while 95 only loads in Enhanced mode. "Real mode" in this context means that the WinApps themselves run in Real mode, which just isn't practical. So, we load Windows in Standard or Enhanced mode — why bother with Real mode?

One need is to run a DOSApp. In the case of Standard mode, the CPU has to switch back to Real mode, effectively freezing Windows. However, Enhanced mode will create another VM (virtual machine) in which to run the DOSApp, and we still say that the DOSApp is running in Real mode (though it would be more correct to say virtual-86 mode).

Then there are DOS device drivers and TSRs. Most likely these will be running in Real mode. And there are the BIOS and DOS services that we may still want to use.

A lot of code is still being developed to run in a DOS box, maybe in Protected mode, but still involving transitions between virtual-86 ("Real mode") and Protected mode in the DOS VM.

Code in Protected and Real mode must be able to communicate, and interrupts occurring in both modes must be handled correctly. The former is the major topic of this chapter, with hardware interrupts focused on in Chapter 12.

This chapter is split into two major portions: getting at Real mode code from Protected mode in the first half, and vice versa in the second half.

Accessing Real Mode from Protected Mode

A typical problem with porting code from DOS to Windows

Recently someone came to me with a problem. They had ported a Pascal program from DOS to Windows, which was quite easy using the excellent Borland tools, but the program didn't work.

The problem was traced to a section of code that looked at a certain interrupt vector, which was a pointer to an interrupt routine. But at a certain offset in this routine is some data that the program accessed. The code used INT21h/AH = 35h to get the vector — but of course you and I know that the vector will come from the IDT not the IVT (running in Protected mode) (Figure 11.1):

Figure 11.1: Interrupt deflection to Real mode.

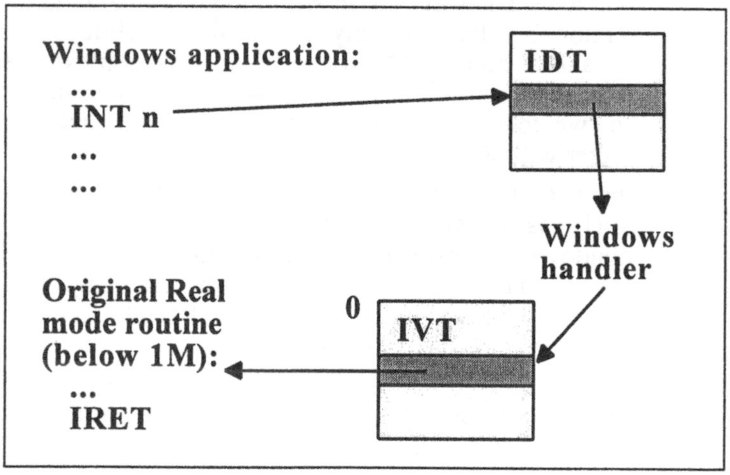

Interrupts reflected from IDT to IVT

The INT-21h/AH = 35h retrieves the vector from the IDT. When an interrupt occurs, the IDT points to a special handler that passes control to the Real mode DOS routine pointed to by the IVT (and remember that the IVT is located at Real mode address segment:offset of 0000:0000).

The routine terminates with IRET, which will bring it back to the Windows handler, which will change the CPU back to Protected mode and then return to your program.

Accessing Real Mode via the IVT

Solution to the above typical problem

So my advice to this person was: you have to look in the IVT, and for that you need DPMI service **0200**h: Get Real Mode Interrupt Vector. The vector obtained is in the form of segment:offset, which cannot be used while your program is in Protected mode. So then you need DPMI service **0002**h, Segment to Descriptor, which will create a descriptor for the segment:offset address and will return a selector (0002h was introduced on page 228).

Problem solved.

Real mode execution versus data access

Figure 11.1 relied upon a Windows handler to transfer control to the original Real mode routine, but this only works for the recognized BIOS and DOS services. Any other interrupt will most likely crash.

The question of an interrupt being reflected down to Real mode or not is a different question from the "typical problem" above, in which it was necessary to look at a certain offset inside the Real mode code.

I will not worry too much about the various scenarios that will require you to access Real mode software; just think for now what the solution is. I outlined above how to locate a Real mode routine for data access, but what if you want to call it?

DPMI to the rescue again!

Routine to call a Real mode ISR

There's an invaluable service, **0300**h, that does everything. Some code will illustrate:

```
.DATA
regstruc        STRUC ;Real mode register data structure
  edi1   DD     0
  esi1   DD     0
  ebp1   DD     0
  res1   DD     0
  ebx1   DD     0
  edx1   DD     0
  ecx1   DD     0
  eax1   DD     0
  flags1 DW     0
  es1    DW     0
```

```
ds1      DW       0
fs1      DW       0
gs1      DW       0
ip1      DW       0
cs1      DW       0
sp1      DW       0
ss1      DW       0
regstruc          ENDS
;  ...........................................
.CODE
callreal          PROC  PASCAL  NEAR
  LOCAL           reg1:regstruc
  USES            ax,bx,cx,dx,si,di
  push ss
  pop  es                 ;setup ES:DI point to data struc.
  lea  di,reg1            ;    /
  mov  WORD PTR [di].eax1,0500h       ;5 into ah.
  mov  WORD PTR [di].ecx1,0007h       ;07=beep
  mov  bx,0016h                       ;int to call
  mov  cx,0
  mov  ax,0300h                   ;simulate Real mode int
  int  31h                   ;    /
  jc error
  mov  ah,0                    ;get char from key buffer
  int 16h ;returns in ax ...will hang if no char in buff!
  mov  dl,al                   ;char in dl
  mov  ah,2                    ;display a char
   int            21h
  ret
```

INT-31h/   Intel's DPMI specification does place some caveats upon the
AX = 0300h   0300h function.[1]

---

[1]   The CS:IP in the Real mode register data structure is ignored by this function. The appropriate interrupt handler based upon the value passed in BL will be called.

If the SS:SP fields in the Real mode register data structure are zero, a Real mode stack will be provided by the DPMI host. Otherwise, the Real mode SS:SP will be set to the specified values before the interrupt handler is called.

The flags specified in the Real mode register data structure will be pushed on the Real mode stack's IRET frame. The interrupt handler will be called with the interrupt and trace flags clear.

Values placed in the segment register positions of the data structure must be valid for Real mode; i.e., the values must be paragraph addresses and not selectors.

All general register fields in the data structure are DWORDs, so that 32-bit registers can be passed to Real mode. Note, however, that 16-bit hosts are not required to pass the high word or 32-bit general registers or the FS and GS registers to Real mode.

The target Real mode handler must return with the stack in the same state as when it was called.

When this function returns, the Real mode register data structure will contain the values that were returned by the Real mode interrupt handler.

What the above program "does"

What I have done here is called INT-16h/AH = 5, which puts a character into the old DOS keyboard buffer. The character has to be provided in CX (as scancode:ascii).

All of the register values to be passed to Real mode have to be placed into an array pointed to by ES:DI.

That's it. The Real mode routine executes, then returns. To find out if the character really was placed in the buffer, I then called INT-16h/AH = 0, which gets a character from the buffer (and will hang if nothing is in the buffer!). Notice that I called this in the normal fashion — this will go via the IDT and IVT as per normal.

The previous INT-16h/AH = 5 would have worked in this way also, but I have used the DPMI service to show how to call code that is not necessarily a Standard BIOS or DOS service.

By this DPMI mechanism, you can call any code below 1M with the CPU running in Real mode — actually, this opens up some possibilities.

Staying on track for now, I used INT-16h/AH = 0 to get the character back off the buffer — and the character I chose was 07, the "beep" character. I sent it to the display, using INT-21h/AH = 2, supplying the ASCII code in DL.

The "beep" character doesn't go to the screen, however; it is treated as a control character (all characters below 32 decimal are) and in this case causes a beep on the loudspeaker.

Hence, there is immediate feedback that the code has worked.

Calling a DOSApp

The above code works fine, at least for calling a BIOS or DOS service, but if you want to call code or access data in a DOSApp, there are more complications.

A DOS program (DOSApp) running under Windows would be running in Real mode in what is sometimes called a "DOS compatibility box". Windows in Standard mode can only have one of these running at any one time, as Standard mode is based upon the capabilities of the 286 CPU (which cannot just flip between Real and Protected modes on a per-task basis). Windows in Enhanced mode is based upon the virtual-86 capability of the 386, which allows multiple "DOS boxes" or virtual machines.

Virtual Machines

Virtual Real mode

There is a section back on page 29 that introduces the concept of virtual 8086 machines. The 386 can happily multitask just about any number of these virtual machines, although Windows has a limit of 16. However, it does place a caveat on everything I've written so far about the so-called "Real mode".

You think of Real mode as using the segment:offset addressing method, without any of the memory management features and restriction to the first 1M.   That is quite true for Windows Standard mode, because to run a DOS program, Windows switches the CPU back to Real mode.

But in Windows Enhanced mode, to run a DOS program the CPU is not switched back to "real" Real mode. Instead it is switched to virtual-86 mode.

Virtual video -RAM

This can have unfortunate repercussions for those of us wanting direct access to hardware. I wrote in Chapter 9 about obtaining a selector to video-RAM and writing directly to it — you will have gained the impression that that is what really happens, and I didn't want you to think otherwise.   However, with Windows in Enhanced mode, what you are really doing is writing to a virtual video-RAM.[1]

In practise it worked, because Windows mapped the virtual video-RAM directly to the actual video-RAM, which is the normal situation for WinApps running in the *system VM*. However, the potential is there to cause trouble for you. Note however that it is possible to directly address the actual physical memory from within a VM — see page 344.

The idea of a virtual video-RAM and a virtual machine, in fact many of them, is awkward for many people to grasp, which is why I tended to delay this little detail until later in the book.

Mapping virtual address to physical address

So whenever we communicate with Real mode from a Windows program running in Enhanced mode, we are only communicating with a simulated Real mode, that is, a virtual-86 machine. The 1M address space of this machine will in reality be mapped, via paging, to anywhere in RAM that the operating system decides. The virtual addresses **may** map to the same physical addresses — see page 343 for more detail on this.

Multiple VMs means multiple IVTs and DOSApps

When Windows is running in Enhanced mode, and you load a DOS program, Windows will create a virtual machine just for it. You could in fact load any DOS program, including a TSR. Another way to do this is to go to the DOS prompt from within Windows and load the program from there.

A TSR loaded in this way will sit inside the virtual machine and will only be usable from within that virtual machine. This is a vital point.

---

[1]   It is this feature that enables Windows Enhanced mode to multitask DOS applications in Windows, not only full-screen as required by Standard mode.

Earlier, I described how to call a Real mode ISR directly by a DPMI service. I also explained that any BIOS or DOS service can be called by a software interrupt, "INT n", which is reached via the IDT and IVT.

Accessing a TSR via the IVT is a very convenient avenue for getting at Real mode code and data. Later on in the chapter I look at going the other way, and again the IVT is an excellent avenue.

However, I have been describing (above) the concept of multiple virtual machines, each with its own IVT, TSRs and DOSApps. The big question now is, how do we know **which** IVT and DOS program we are accessing from our WinApp? To answer this question, read on ...

DOS TSRs

Concept of TSRs

DOS TSR (Terminate and Stay Resident) programs, which also include device drivers, are covered in many DOS programming books. They load like any other program, but only have a short "install" procedure then exit back to DOS. The exit is via a special DOS service that leaves the program resident in memory, rather than freeing up that memory space, as with normal programs.

TSRs usually hook a vector, such as INT-8, -9, or -16h.

For example, by hooking INT-16h or -9 all DOS keyboard input can be filtered. Usually the TSR passes control to the old vector after doing whatever it wants.

Once a TSR is loaded and control returns to DOS, you can then load another program, so even under "single-user" and "single-tasking" DOS you have two (or more) programs sitting together in memory. The TSR will be executed, or rather its "run-time" portion will be executed, whenever the particular interrupt is called.

The Companion Disk has a useful TSR skeleton that hooks INT-16h with many of the tricks of the trade incorporated into it, fully commented for your convenience. Look in \DOSTSR.

A TSR sits inside a VM

If you load the TSR from within Windows or at a DOS prompt within Windows, the TSR will be inside a virtual machine. If the TSR hooks an interrupt vector in the IVT, it will only be hooking the vector in the virtual machine.

Whenever a DOS virtual machine is created, Windows copies everything from the actual 1M region into it, or rather, "maps" it in. The IVT is not the same IVT as the original IVT.

This is the crux of the problem. Perhaps Figure 11.2 will help:

Figure 11.2: A TSR is in a VM.

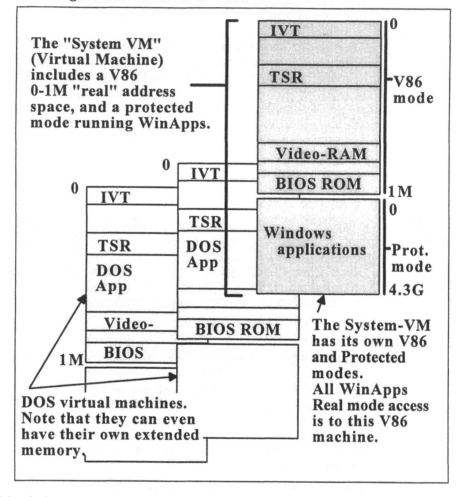

<table>
</table>

A TSR loaded By loading the TSR before loading Windows, for every virtual
before machine that Windows creates, it will also "copy" the hooked
Windows vector and the TSR. Thus by this method you ensure that the TSR
appears in is available to all applications.
every VM
Note that I put the word "copy" in quotes, as this is not always to
be taken literally. See ahead for clarification (page 343).

Each VM has Note also something most important: the descriptor tables. The
its own LDT system VM will have just one of each LDT and IVT. Despite the
fact that one of the fundamental concepts behind the LDT is that
there should be one per task, Windows maintains just one for the
entire VM. This is why obtaining a selector when installing a TSR

will work within the TSR's run-time code, no matter which WinApp is running at the time of the interrupt. However, an interrupt when in another VM will access a different LDT and IVT. This idea of one LDT per VM is in keeping with DPMI version 0.9. Version 1.0 has an LDT per client (task). Windows is one client only.

Accessing Real mode code in all VMs

The conclusion here is that accessing Real mode code (via the IVT) from a Protected mode WinApp accesses it in the system VM. If you want to get at code or data of a DOSApp or TSR in another VM you have to look into mechanisms for going between VMs — or, if you load a DOS TSR before Windows, it will be automatically in all VMs and thus its code and data will be global. Even its hooking of the IVT will be in every IVT.

Thus the DOS TSR is one convenient mechanism for communication between Protected and Real modes across all VMs and is developed further in this chapter. Also, a method for switching VMs is developed.

Accessing Protected Mode from Real Mode

Global data via a DOS TSR

Actually, using the global DOS TSR method by passing data between Real and Protected modes is very easy.

All that the DOS application has to do is execute a software interrupt to invoke the TSR or use the vector as a pointer to global data. Any data passed to the TSR's own data area will also be available to a Windows application that calls that same TSR. End of story.

This method works quite happily for Enhanced or Standard modes, but is awkward in that the TSR must be loaded before Windows. Also it takes up "valuable" space in that first 1M.

This simple technique for sharing data works across all VMs, because the same TSR is present in all. I develop this point as I go along. This idea of using a global DOS TSR is not the only approach but is quite powerful, and it is the basis for discussion in this chapter.

Reference source

The problem of different IVTs in each virtual machine is discussed by Thomas Olsen in "Making Windows & DOS Programs Talk", *Windows/DOS Developer's Journal*, May 1992, p 21.

He does not see any way around this problem except by loading the DOS TSR before loading Windows. Actually, if you only want the DOS TSR to load into the system VM, and not subsequent VMs, you can force this by naming it in a file called WINSTART.BAT, which Windows looks at to see what has to be done before loading itself (but after creating the system VM). Simply put the name of the TSR in it, as per a normal batch file.

Mapping the IVT and TSR across VMs

When I say "there's nothing to it", I'm being a bit flippant. A DOS TSR loaded before Windows can have a data area that a Windows program can get at, but there are certain extra considerations.

If the TSR is being copied to each V86 machine as it is created, won't each have its own code and data? Therefore, if a Windows program looks in the IVT to access the DOS TSR, which one will it see? Will it just see the copy in the system VM?

Yes, the WinApp will only see the IVT in the system VM and hence the TSR in the system VM, but Microsoft arranged things so that the subsequent copies of the TSR are not really "copies" as such — they all map back to the one physical TSR. So there only **appear** to be multiple copies of the TSR.[1] Thus the TSR is truly global.

I have elaborated upon this point with a supporting figure on page 343.

There is still another major problem. Yes, the WinApp can get at the DOS TSR, but what if a DOSApp in a VM, via the TSR (or whatever method), wants to asynchronously send a message to a WinApp in the system VM? I talked about signalling between applications back in Chapter 10, but that was between WinApps. Getting a DOSApp to signal a WinApp across VMs is a new ball game.

Signalling a WinApp from a DOSApp

A DOS TSR can be made to appear in all virtual machines or only in the system VM, so it is a ready means of providing the signalling.

---

[1] You can verify this by running COMMAND.COM in two different windows. Run the DOS "MEM" program to see where the DOS TSR is located, then go into DEBUG.COM and dump the start of the TSR (use the Dump command), then enter a new value somewhere (Enter command) ... and you will find the same new value showing up in the other DOS window. Note that DEBUG is a standard DOS program, and DRDOS also has a program (almost) equivalent to DEBUG. I do have a modified DEBUG that will run on any version of DOS, but at this stage I don't have permission from Microsoft to put it on the Companion Disk. You may be able to locate a similar modified DEBUG on the Internet. Usage of DEBUG is described in many DOS programming books.

A DOS application can call the DOS TSR by a software interrupt, but since the DOS TSR is running in Real mode, how does it communicate with a Protected mode WinApp?

Reference source

Walter Oney has solved this particular problem in "Using DPMI to Hook Interrupts in Windows 3", *Dr Dobb's Journal*, February 1992, page 16. He does not tackle hardware interrupts; his focus is purely on the issue of passing a message from a DOSApp to a WinApp across VMs.

Mechanism for forwarding up to a WinApp

A DOS TSR can be made to load into the system VM only, by specifying it in WINSTART.BAT; however, what we want is to hook an IVT vector that will appear in all VMs. The reason for this is that we want a mechanism for a DOSApp in any VM to be able to find out the address of a "forwarding" routine (in the DOS TSR) in the system VM.

Did I just say that we want the TSR to be in every VM? It will be, but the IVT hook's appearance in every VM is what matters: we want a DOSApp in another VM to pass control over to the "copy" of the TSR in the system VM, which can in turn pass control up to a WinApp. This may seem complicated, but hopefully I can explain it clearly.

First consider the DOS TSR. It will have to be loaded before Windows and will have to hook a vector in the IVT:

```
;DOSTSR.COM  Resident program to pass control up to a
;WinApp.
.286
DOSTSR SEGMENT  BYTE  PUBLIC  'CODE'
   ASSUME        cs:DOSTSR,ds:DOSTSR,es:DOSTSR
   ORG  100h
begin:
   jmp  start

; Put any local data in here.

;......
;This is the forwarder. It passes control up to Protected
;mode...
forwarder:
   push es                    ;save working registers
   push ds                    ;   /
   pusha                      ;        /
   sti  ;enable interrupts, unless you want a crash!
   push cs              ;routine entered with DS unknown.
   pop  ds              ;    / want to addr. local data.
;
;To pass control up to a WinApp, the WinApp has to
;provide its address (selector:offset) in the IVT.
;We must test if that has been done...
   xor  ax,ax                 ;get current int-60h vector.
```

```
      mov   es,ax         ;    / (don't use int-21/35, as under
      mov   si,60h*4       ;    /    certain circumstances DOS
      mov   bx,es:[si]     ;   /    may not be stable)
      mov   ax,es:[si+2]   ;   /
      or    ax,bx          ;    /   it will be 0:0 if not hooked.
      jz    done60        ;if not, don't forward to it!
      int   60h                ;issue int-60h to call WinApp.
;
done60:
   popa                      ;restore registers.
   pop ds                   ;    /
   pop es                   ;    /
   iret                      ;return to DOSApp (in another VM).
;..................................
endprog: ;transient portion below dumped after install.
start:
      mov   ax,2561h        ;hook int-61h  in ivt.
      lea   dx,forwarder ;  /  addr. of forwarder in TSR.
      int   21h             ;    /
;
   lea   dx,endprog+17 ;point past all code in this module.
   shr   dx,4                ;compute # paragraphs to keep.
   mov   ax,3100h           ;terminate and stay resident.
   int   21h             ;    /
;......
DOSTSR ENDS
   END  begin
```

So there you are, a complete DOS TSR! Note that this particular one has been written without the "simplified" directives, which is no big deal. Actually my own experience has been that it is difficult to write .COM programs using the simplified directives, and you are better off sticking with the "long hand" notation shown above. You can write a TSR using .EXE format and the simplified segment directives, which I have done for one of the examples of Chapter 14 (see also directory \TSR2WIN on the Companion Disk).

Install portion

Have a close look at what the "install" portion does. It hooks INT-61h in the IVT then exits.

Because this TSR is loaded before Windows, it will be in the system VM and will hook the vector in the system VM. But it will also be copied to every VM.

Thus, every time a DOS program is run within Windows Enhanced mode, the new VM will have that hooked vector.

But what you should note in particular is that INT-61h contains the address segment:offset of the "forwarder" code for the TSR.

Passing Control to the WinApp

A major problem is created if our code must work for both Standard and Enhanced modes. With Standard mode, the question of VMs doesn't arise.[1] This means that all access to the IVT from a WinApp is to the actual, original, real, physical, bona fide IVT!

That's not the problem: in fact that's good, because there's no need to jump VMs. However, Windows itself is in a strange state while a DOSApp is running. I have elaborated more upon this in Chapter 12.

Both Enhanced and Standard modes, however, can use the same mechanism for transferring up to Protected mode.

Installing a Real to Protected mode handler

There is a DPMI service that allows us to hook (from a WinApp) a vector in the IVT (function **0201h**:[2] Set Real Mode Interrupt Vector) and another that will redirect it up to Protected mode (function **0303h**:[3] Allocate Real Mode Callback Address). Actually, 0303h is called first, followed by 0201h.

Get the picture here — an interrupt occurs while the CPU is in Real mode, but the vector is to a DPMI routine that switches the CPU to Protected mode and passes control up to a WinApp.

The above may seem like a suitable method for a DOSApp to communicate with a WinApp, but executing 0303h and 0201h from the WinApp will only hook the vector in the IVT of the system VM in the case of Enhanced mode. However, in Standard mode, there's only one IVT anyway, so (in theory) this method works!

The obvious point here is that if Windows is loaded in Enhanced mode, then as well as installing the handler as outlined above, we will also have to perform a jump from the VM running the DOSApp into the system VM.

---

[1] Actually, this is a qualified statement. It is better to say that Standard mode cannot have V86 VMs, or DOS VMs, since it can, by the DPMI host, have multiple Protected mode VMs. Windows, however, only runs the one VM, in which all WinApps reside.

[2] Intel's DPMI specification places some caveats upon function 0201h:

The address placed in CX must be a Real mode segment address, not a selector. Consequently the interrupt handler must reside in DOS memory (below 1M) or the client must allocate a Real mode callback address. See functions 0100h and 0303h in Appendix C.

If the interrupt is a hardware interrupt, the memory that the interrupt handler uses must be locked.

[3] The Intel DPMI specification places these caveats upon function 0303h:

A descriptor may be allocated for each callback to hold the Real mode SS descriptor. Real mode callbacks are a limited resource. A client should use the Free Real Mode Callback Address function (0304h) to release a callback that is no longer required.

The contents of the Real mode register data structure are not valid after the function call, only at the time of the actual callback.

Handling hardware interrupts

A line of thought — If the CPU happens to be in Protected mode when a hardware or software interrupt occurs and if the interrupt is one of the BIOS/DOS services, Windows will redirect control down to Real mode and the routine pointed to in the IVT in the system VM.

Therefore, if you want your interrupt routine to work for the CPU in both Real or Protected mode, especially in the case of hardware interrupts, why not use functions 0303h and 0201h to hook **only** the IVT and have just one ISR?

This will work for all normal DOS interrupt services, which do get redirected from the IDT to the IVT. Unfortunately, the particular case of INT-9, which we have been using as a case study, does not get redirected in this way.

This deviates somewhat from my current line of thought. For more on handling hardware interrupts, refer to Chapter 12.

The DOSApp "Signaller"

Whenever a V86 machine is created, it will be in response to loading a DOSApp. This DOSApp may want to send a message to a WinApp, so it will need some code inside it to call the "forwarder" routine in the DOS TSR.

This is how the section of code would look:

```
;DOSAPP.ASM    DOS signaller program.
;
;   what follows is only a fragment of the whole DOSApp...
;
,DATA
ivt61off        DW    0 ;address of "forwarder" in DOS TSR.
ivt61seg        DW    0   ;    /
tsrloaded       DW    0   ;set if TSR has hooked int-61h.
;
.CODE
;
; Test if Windows was loaded in Enhanced or Standard
; mode.
; (the method for doing this is shown in Chapter 12...
; ... here I have just supplied a flag, "winmode",
; already set or cleared)

; ... it is only necessary to switch VM's if Enhanced
; mode.
  mov  al,winmode
  and  al,1                 ;set if Enhanced.
  jz   Enhanced
;
Standard:
; (see Companion Disk)
```

```
        jmp   doneit
;
Enhanced:
;see if forwarder TSR is present by checking interrupt
;vector 61h...
        mov   ax,3561h            ;get int-61h vector address
        int   21h                 ;    /  -->es:bx
        mov   ivt61off,bx         ;save it.
        mov   ivt61seg,es         ;    /
        mov   ax,es               ;be sure there is one.
        or    ax,bx               ;    /
        mov   tsrloaded,ax ;set if  TSR loaded.
        jz    cantcall            ;if not, complain and quit.
;
;use 2F/1685h to switch to system virtual machine and
;call forwarder program in the DOS TSR.
        cmp   tsrloaded,0
        jz    dontswitch
        mov   ax,1685h¹           ;switch VM's and execute.
        mov   di,bx   ;es:di = callback addr. (int 61 hndlr).
        mov   bx,1          ;bx = VM to switch to (system VM).
        mov   cx,3          ;cx = 3 wait until interrupts enabled
                           ;and critical section unowned.
        xor   dx,dx        ;dx:si = priority boost (0).
        xor   si,si        ;    /
        int   2Fh          ;switch to system VM and do INT-60.
dontswitch:
doneit:
;
; DOSApp continues...
;
```

This program, or any DOS application with this code in it, looks at the INT-61h vector to see if there is anything in it (there will be 0 if not hooked). If so, the program goes ahead and calls the "forwarder" portion of DOSTSR.COM, the DOS TSR.

However, this is where you need to think. If you loaded the TSR from the DOS prompt before loading Windows (in contrast to loading it from WINSTART.BAT), there will be a copy of the DOS TSR in the current VM where the DOSApp is running, but the TSR is useless. The reason is that its purpose is to call the WinApp, but it will try to call the Protected mode WinApp in the current VM, where it isn't.

I'll look at this diagrammatically in Figure 11.3:

---

[1] INT-2Fh/AX = 1685h is described in Appendix D. It is for switching VMs.

CX = bit-0 is set to indicate that Windows must wait until interrupts are enabled before calling the callback in the VM; bit-1 is set to indicate that Windows must wait until the critical section is unowned before calling the callback in the specified VM; the remaining bits must be zero.

DX:SI = the 32-bit amount by which to boost the target VM's priority before changing contexts.

ES:DI = the segment:offset of the routine to call in the target VM.

Figure 11.3: Execution in System VM from another VM.

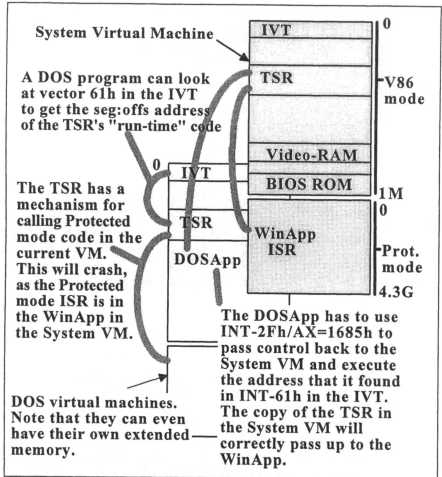

If you have a look at the DOSApp, you will see that it looks at vector 61h in the IVT to get the address of the "forwarder" routine in the TSR, and then it uses INT-2Fh/AX = 1685h to switch over to the system VM and also to execute the forwarder code in the copy of the TSR located in the system VM.

These little programs are two pieces of the puzzle, but there is a third. The WinApp has to hook INT-60h in the IVT of the system VM.

Hooking a Real Mode Interrupt from a WinApp

I have already written a little about this, back on page 279 and introduced two DPMI functions that will allow a WinApp to hook a vector in the IVT in the system VM and pass control up to a Protected mode ISR.

WinApp
ISR
install
routine

What your WinApp needs to do is call DPMI function **0201**h (Set Real Mode Interrupt Vector), and **0303**h (Allocate Real Mode Callback Address).

Here is an "install" portion of a WinApp:

```
.CODE
offsetrealint  DW     0        ;old ivt vector
segmentrealint DW     0        ;    /
dsselector     DW     0        ;data alias to code seg
hwndcs         DW     0 ;save window handle for use in isr
callbackbuffer REGSTRUC < > ;Real mode register structure
;
installint      PROC                    ;no params
install:
   pusha
   push es
   push ds
;will create alias in ldt of current task...
   mov  ax,000Ah ;create alias data descriptor for code.
   push cs
   pop  bx                    ;selector to be aliased
   int  31h                   ;returns ax
;
     push ax
     pop  es
     mov  ax,hwnd
     mov  es:hwndcs,ax ;save handle of window in code seg.
     mov  ax,es
     mov  es:dsselector,ax     ;save data alias in code
   . . .
     ;could put some code for hooking the IDT ...
   . . .
hookreal:
   pop  ds                    ;restore it again.
;OK, now to hook Real mode int....   hook 60....
   mov  ax,0200h              ;get Real mode vector
   mov  bl,60h
   int  31h                   ;-->cx:dx (seg:off)
   mov  es,cs:dsselector
   mov  es:offsetrealint,dx   ;save old vect
   mov  es:segmentrealint,cx  ;    /
;
;now must reflect the Real mode int up to prot mode
;code....
   push ds                              ;save
 mov es,cs:dsselector ;get alias. Addr of buffer in es:di
   mov  di,OFFSET callbackbuffer  ;   /
```

```
    mov  ax,0303h                   ;alloc Real mode callback
    push cs
    pop  ds                          ;addr of prot code
    mov  si,OFFSET runtime2          ;  /
    int  31h                         ;-->cx:dx (seg:off)
    pop  ds                          ;restore
;now hook the ivt....
    mov  ax,0201h                    ;set Real mode vector
    mov  bl,60h                 ;hook int in ivt
    int  31h
;
getout:
    pop  es
    popa
    ret
```

Real mode register structure   The data structure referred to as "callbackbuffer" is the same callback structure used to pass register values between Real and Protected modes, as discussed on page 269, where function 0300h is introduced (this is for calling a Real mode interrupt from Protected mode, which is going the other way).

WinApp ISR   Actually, the piece of the puzzle, consisting of the WinApp code, is in two parts: the "install" portion above, and a "run-time" portion. The latter is the ISR (Interrupt Service Routine) that is the end result. Wherever the interrupt originated, control should end up there. I want this ISR to behave much like the ISR introduced in the previous chapter; that is, to post a message to the main window.

A Protected mode ISR is shown back on page 262, illustrating how to post a message.

Because the WinApp has hooked INT-60h in the system VM, any software interrupt within the system VM while the CPU is in Real mode will cause execution of the Protected mode ISR "run-time" portion of the WinApp. You can see in the DOS TSR that this was very simply done by an "INT 60h" instruction.

Entry to the ISR   When control is "passed up" from Real to Protected mode, the ISR is entered with certain registers loaded:

> DS:SI = Real mode SS:SP
> ES:DI = Real mode call structure

The "call structure" is that same data structure containing the Real mode register values. Return from the ISR is by an IRET, but the data structure is modified as appropriate. At exit, the registers ES:DI must be pointing to the data structure, because the DPMI handler will put whatever is contained in the structure into the Real mode CPU registers.

***Exit from
the ISR***

For example, if we want the ISR to chain to the old ISR, we need
to get the old vector and put it into CS:IP in the data structure:

```
    . . .
;end of ISR ...
   mov   ax,cs:segmentreal
   mov   es:[di].cs1,ax
   mov   ax,cs:offsetreal
   mov   es:[di].ip1,ax
   iret
```

On the other hand, if the ISR is not to chain to the old vector but
instead is to return from whence it came, the return address on the
stack must be put in CS:IP in the data structure:

```
   cld
   lodsw                    ;get Real mode IP off stack.
   mov  es:[di].ip1,ax  ;put it into IP in data structure.
   lodsw                    ;get Real mode CS off stack.
   mov  es:[di].cs1,ax  ;put it into CS in data structure.
   lodsw                    ;get Real mode flags.
 mov es:[di].flags1,ax  ;put into flags1 in data structure
   add   es:[di].sp1,6     ;adjust SP on data structure.
   iret
```

The above mechanism is elaborated upon in Chapter 12.

***DPMI 1.0
global
memory***

This is all quite involved, just to post a message from a DOSApp
to a WinApp, but while I think of it, if your need is not to signal or
execute but just to share data, DPMI version 1.0 does have a neat
solution. Ok, this is academic, as no versions of Windows run
DPMI v1.0 — but maybe one day.

DPMI version 1.0 (not v0.9) has a function, **0D00h** (Allocate
Shared Memory), that creates and allocates a memory block that is
accessible across all VMs. Thus all Windows and DOSApps have
access to it.

There are also **0D01h** (Free Shared Memory), **0D02h** (Serialize on
Shared Memory), and **0D03h** (Free Serialization on Shared
Memory).

The latter two allow synchronization of access to the shared block.

12

32-Bit Ring 0

Preamble

Privilege levels

As explained in Chapter 1, the 286 and 386 have four privilege levels, numbered from 3 to 0. With Windows 3.0, the operating system kernel and device drivers run at the most privileged level, 0, while Windows applications and DLLs run at level 1. DOS applications, being the least trusted, run at level 3.

However, Microsoft changed its mind with Windows 3.1, and moved Windows applications and DLLs down to level 3 also. This includes all the DLLs of the Windows API.

When I upgraded from Windows 3.0, to 3.1, I had the distinct but subjective feeling that the new version was a tad slower. The changes in privilege could be the reason. Of course, Microsoft claimed just the opposite — that the new version was faster, which could have been true, taking into account the new 32-bit file and disk access (which I originally had turned off).

Then, when I upgraded to Windows for Workgroups 3.11, I again had the subjective feeling that everything had slowed down. I have never tried to quantify this. Version 3.11 seemed to take longer to load, which may have had something to do with the fact that when going from 3.1 to 3.11, I decided to network two PCs.

Then, when I upgraded to Windows 95 ...

Anyway, the current situation with Windows is that applications run at level 3, least privileged. Unfortunately, this seriously hampers my style, if I want to do my own I/O. If my requirement is direct access to memory and I/O ports or interrupt handling, invariably, the problem with Windows comes down to lack of speed and unpredictability of response times. A hardware interrupt will quite literally propogate through hundreds of instructions before it reaches your application.

Then there is the general issue of the protected environment: you may want to access a particular I/O port or memory location, but the operating system may prevent access. You may want to tweak the system hardware or operating system in some way but not be allowed to.

Device driver

The traditional way to obtain unrestricted access to everything is to write a device driver. Development of a device driver requires the *Device Development Kit* (DDK), and once developed, its name must be entered into the SYSTEM.INI file in the \WINDOWS directory. Device drivers are difficult to write, and it is a nuisance that the SYSTEM.INI file has to be altered. Though with Windows 95 you have the possibility of dynamically loading and unloading device drivers, on the fly.

This chapter, however, explores an alternative approach. It is a technique in which your application can switch up and down between rings 3 and 0 at will, without requiring a device driver. With this technique, you can get nearly all of the benefits of device drivers, with fewer hassles.

16- and 32-Bit Programming

Reference source

As far as I am aware, the first person to publish this technique was Matt Pietrek in an article titled "Run Privileged Code from Your Windows-based Program Using Call Gates", *Microsoft Systems Journal*, May 1993, pages 29-37.

Early in 1993, I was trying to figure out how to do this, but Matt had an advantage over me: "inside" information. His technique makes use of two undocumented features, which he thinks are likely to stay in future versions of Windows.

Basically, Matt was writing from the point of view of a Windows 3.1 application, which would normally be running in 16-bit mode. Now, let me clarify one point: this entire chapter assumes Enhanced mode Windows only, using a 386 or above. Windows 3.1 can run in Standard mode, but the 286 CPU has gone the same way as the 8088.

This confuses everybody, but Windows running in Enhanced mode can be running in 16-bit or 32-bit mode. Windows NT runs normal native applications in 32-bit mode, and Windows 95 encourages this. In Matt's article, his application was running in 16-bit mode, and when he switched up (or down!) to ring 0, he stayed in 16-bit mode.

Obviously, this is a point of great potential confusion, so this chapter commences by explaining the difference between 16-bit and 32-bit programming.

Chapter 1 shows the structure of a descriptor, however it is now time to examine it in more detail. Figure 12.1 shows the full detail:

Figure 12.1: Detail of the code descriptor.

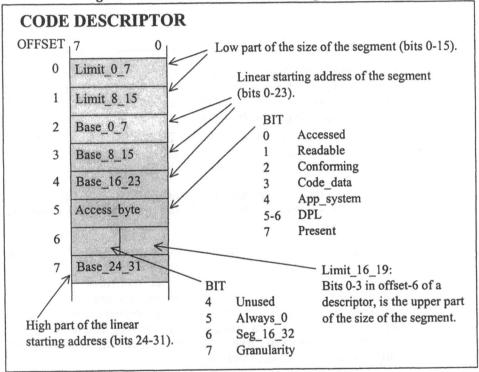

The vital bit in this descriptor is bit-6 in offset-6, labelled *Seg_16_32*. If this is set, the CPU is in 32-bit mode. When the *Granularity* bit is cleared, the limit value is the size of the segment expressed as number of 4K pages, or if set, the limit value expresses size in bytes. *DPL*, meaning *descriptor privilege level*, is the level of this segment. *Present* is set if the segment is physically present in memory. You likely will not need to modify the other fields of the *Access byte*.

The current mode of the CPU, whether 16-bit or 32-bit, is determined by Seg_16_32. If it is set, the 32-bit registers are enabled, and just about all operations become 32 bits. This includes PUSH and POP operations on the stack. However, this does not preclude you from using parts of the 32-bit registers, such as AL, AH, and AX in EAX.

Instruction size-prefix

Just to keep confusing you: even if the the segment is in 16-bit mode, you can *still* use the 32-bit registers!

For starters, I'll take the case of an "old fashioned" Windows application, running in 16-bit segments, and consider a very ordinary instruction that may appear in that program:

```
0907:0200      58         POP  AX
0907:0201      6658       POP  EAX
```

This example is a typical unassembly, showing address, machine code (in hex), and assembly language mnemonic. The first line contains no surprises: the code "58h" is the machine code for "POP AX"; a simple one-byte instruction.

You would logically expect "POP EAX" to have a different machine code, but note that the "58" is still there. All the assembler does is insert an *instruction prefix* of value 66h.

The 66h prefix is an *operand size prefix*, which tells the CPU to execute the following instruction in the opposite mode than it is currently in.

So the same "58h" is used for both pops, but the prefix determines the size of the pop. There is also another type of size prefix, the *address size prefix*, of value 67h, that overrides the current address-size mode. Therefore, even though you are programming in a 16-bit segment, you can use the 32-bit registers: the assembler will insert the prefix in front of any such instructions. **Note though, that the prefix overrides the default segment size, but only for the current instruction**.

32-bit default

When I first had to tackle this problem, I was using Microsoft's *Codeview* debugger, version 4.01. In my program, I had the instruction "POP EAX", but when the debugger unassembled my program, it showed "POP AX". This caused me enormous confusion, until I realised that Microsoft's own debugger can't even recognize what mode it is in.

The situation was, I was writing code in a 32-bit segment, i.e., with the *Seg_16_32* field in the descriptor set, as introduced in the previous Figure. In this situation, the *default* size is 32 bits.

Therefore, all instructions will reference 32-bit registers, operand size, and address size, without requiring an instruction prefix. Thus:

0907:0200	58	POP	EAX
0907:0201	6658	POP	AX

The situation is now reversed: the "58h" means "POP EAX", but if we write an instruction that only accesses a 16-bit register, it will have the prefix appended. It doesn't say much for Microsoft, but Codeview version 4.01, despite being fully operational in 32-bit mode, able to display the 32-bit registers, and able to trace, did not unassemble correctly. At the time of writing, 4.01 is my latest version — it came with MASM version 6.1 — and I'm sure that by the time you read this book, the bug will have disappeared.

32-bit Real mode

So what of Real mode and virtual-86 mode? In both of these modes, the default is 16 bits, but you may be very surprised to learn that in both modes, you can use the 32-bit registers. Of course, the prefix (or prefixes) will be in front of every 32-bit instruction.

This may come as a complete surprise, but use of 32-bit registers allows you to have segments greater than 64K — up to 4.3G — and thus break the 1M conventional memory limit for Real mode.

Of course, Real and virtual-86 modes have paragraph addresses in the segment registers, so these can only reference the first 1M: however, you are quite at liberty to use offsets to access code and data beyond 1M.

Reference book

A bit of setting-up is required to use Real and virtual-86 modes in this way, and I recommend a good book: Al Williams has worked it all out, and has an entire chapter dedicated to this, in his book *DOS5: A Developer's Guide; Advanced Programming Guide to DOS*, M&T Publishing Inc., USA, 1991. There is probably a more recent version of the book (probably with a new title!), but the chapter on 32-bit programming is still quite relevant, even in the 1991 book.

Ring Transition Mechanism

Say that for whatever reason, you want your program to have the unrestricted access, and the total control, of ring zero. Unfortunately, your program will be executing in ring 3 segments, which means that if you try to do an I/O operation, such as use the IN and OUT instructions, there will be a CPU exception. And if we want to hook an interrupt, we will be doing so at the "asse end" of the animal. What if we want to call some of the powerful functions in the Windows kernel and in virtual device drivers? Sorry, but even if you knew how to address them, you'd get a CPU exception, because they are ring-0 segments.

The 386 does provide a mechanism for going to a more privileged ring, called a *gate*, of which there are *call gates*, *interrupt gates*, *task gates*, and *trap gates*. However, only code in ring 0 is supposed to be able to create such gates.

Interrupt gate

I kind of glossed over this little detail in an earlier discussion (look back at Figure 11.1), but the interrupt services are at ring 0, so the entries in the interrupt descriptor table (IDT) of the form *selector:offset* reference an interrupt gate, not a descriptor.

An interrupt gate, or any gate for that matter, sits in the LDT or GDT as an 8-byte entry, just like any other descriptor (see Figure 12.1), but it has a different format. In the case of interrupt handling, if there is to be a ring transition, i.e., if the ISR is at a more privileged level than 3, then the entry in the IDT is not a descriptor: it is an interrupt gate. However, the code descriptor for the ISR is still there at another entry (also in the IDT, I presume).

Call gate

A call gate is the mechanism for a CALL instruction to call code at a more privileged level. It works just like the interrupt gate, in which the descriptor in the LDT or GDT, of the code to be called, is not called directly. Instead, you call a call gate, which in turn calls the more privileged code via its descriptor.

Call Gate Structure

I'll postulate that you want to call some ring-0 code at some address, say 0907:0000. How you would get the selector of some ring-0 code is another question, but I'll say you've got it. You try to perform a CALL to that address, but the CPU intervenes, since you are at lowly ring 3, and passes control to Windows, which informs you that there has been a general protection error.

The way around this problem is to create a call gate. Normally, only the operating system (ring 0) is supposed to be able to create a call gate, but we *can* do it from ring 3, using undocumented features. I wonder whether this loophole will be closed: the technique has been published in Microsoft's own *Microsoft Systems Journal*, which would tend to give it some authority (I suppose?), and the loophole remains in Windows 95.

A call gate is 8 bytes and can be an entry in the LDT or GDT, just like a descriptor. However, it has a different structure to a descriptor, as Figure 12.2 shows:

Figure 12.2: Detail of the call gate.

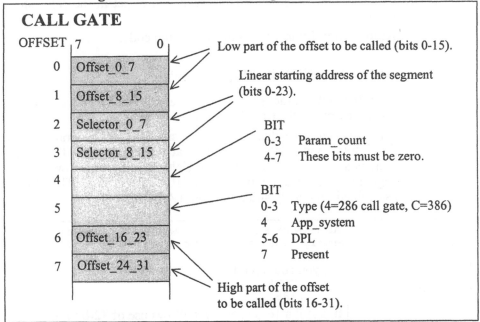

Actually, what distinguishes this as a call gate, and not some other kind of gate, is the *Type* field. The value Type = 4 means that it is a call gate to a 16-bit (286) segment, while a value of C (hex) means that it is a call gate to a 32-bit (386) segment. For the record, the other possible values are 5 = task gate, 6 = 286 interrupt gate, 7 = 286 trap gate, E = 386 interrupt gate, and F = 386 trap gate.

The *Selector* field is the ring-0 segment that we want to call, and *Offset* is where in the segment. Note that the code descriptor for the ring-0 selector still has to exist, and it will be elsewhere in the LDT or GDT.

App_system would normally be zero and *Present* set to 1. The *DPL* field is important: it specifies the least privileged code that is permitted to use this call gate. Therefore, we set it to 3.

Putting call gate & descriptor together

If we create a call gate, we can then put it in the LDT or GDT, and thus we will have a selector for it. Then, all we need to do in our program is call the selector: the CPU will recognize it to be a call gate, look inside it, and get out the selector:offset. The CPU will then use that selector to get the code descriptor and will call the code.

Note, however, one peculiar thing: if you perform a FAR call from your application to the call gate selector at some offset, any offset that you specify is ignored. Instead, the offset in the call gate is used.

At this point, I think it best to show some code.

Ring Transition Example Code

This first code extract shows just one example of how we could set up addressing of the ring-0 code and then call it.

The program is written as a 16-bit, small model program, hence the ".286" and ".MODEL SMALL" at the very start. The ring-0 code is a function, called RING0FUNC(), and is in its own 32-bit segment in a separate file called HEAVEN.ASM. This file is assembled separately and linked with ASMRING0.

```
;ASMRING0.ASM --> ASMRING0.EXE   Windows demo program.
;This skeleton assembly language program has been written
;for Microsoft
;MASM v6.1.           | Or, you could put this:
.286                 |      .MODEL SMALL
.MODEL SMALL         |      .386
                     | It is still a 16-bit seg., but allows use of 32-bit regs!
EXTERN RING0FUNC:NEAR      ;this is in HEAVEN.ASM.
                          ;It is the ring0 code.

 ....... stuff removed ........


;..........................................................
;callback starts .......
ASMDEMOPROC PROTO FAR PASCAL, :HWND, :WORD, :SWORD, \
                                               :SDWORD
ASMDEMOPROC PROC FAR PASCAL, ihWnd:HWND,\
       iMessage:WORD, iwParam:SWORD, ilParam:SDWORD
   LOCAL       dummy [5]:WORD
   LOCAL       @hDC:HDC
   LOCAL       s3:PAINTSTRUCT

 mov  ax,imessage              ;get message-type.
 .IF  ax==WM_CREATE       ;message received after
   call  xcreate  ;CreateWindow() function is called.
 .ELSEIF ax==WM_DESTROY ;message if window is closed.
   call  xquitmessage   ;posts WM_QUIT & does cleanup.
 .ELSEIF ax==WM_PAINT
   call  xpaint
 .ELSEIF ax==WM_COMMAND ;any selection of the menu will
   call  xmenu                    ;produce this message.
 .ELSEIF ax==WM_LBUTTONDOWN ;one of many mouse messages.
```

```
      call  xlbutton
   .ELSEIF ax==WM_CHAR          ;message that a key pressed.
      call  xchar
   .ELSE
      invoke DEFWINDOWPROC,ihWnd,imessage,iwParam, ilParam
      ret
   .ENDIF

   sub  ax,ax   ;returns 0 in DX:AX.  (callback functions
   cwd                          ;return a 32-bit (long) value).
   ret
ASMDEMOPROC    ENDP
;.....................................................
xcreate PROC
   call  makering0selector
   invoke GETSTOCKOBJECT,OEM_FIXED_FONT
   mov  hOemFont,ax          ;handle to font.
   ret
xcreate ENDP
;..............................
xmenu   PROC
   cmp  WORD PTR ilParam,0      ;low-half of lParam
   jne  zxcv                    ;test if a menu-message.
   cmp  iwParam,IDM_QUIT        ;wParam.
   jne  notquit
   call xquitmessage
   jmp  zxcv
notquit:
   cmp  iwParam,IDM_ABOUT
   jne  zxcv                            ;no other menu items.
   invoke MESSAGEBOX, ihWnd, ADDR szaboutstr, ADDR \
                            sztitlestr, MB_OK

   cli
   call RING0CALLGATE        ;will call ring 0 code
   sti

zxcv:   ret
xmenu   ENDP
;..........................................................
xquitmessage    PROC
   invoke POSTQUITMESSAGE,0
   call  freeourselectors
   ret
xquitmessage    ENDP
;..........................................................
errormsgproc    PROC
;entered with ds:si pointing to message....
   invoke MESSAGEBOX,handlemain, ds::si, ADDR szerror,\
                          MB_OK+MB_ICONEXCLAMATION
   ret
errormsgproc    ENDP
;..........................................................
   .... The rest of the message-handling code removed ....
```

Creation
of a call
gate

For this demo program, I chose to use the WM_CREATE message to call *makering0selector()*, which sets up the addressing to the ring-0 code.

Then, I arbitrarily chose to use a press of the "ok" button on the messagebox, which occurred in response to IDM_ABOUT, to call *RING0CALLGATE*, which is a pointer to the call gate, which takes execution to the ring-0 code.

Finally, before exiting from the program, it calls *freeourselectors()*, which removes the descriptor and call gate that we had created in the LDT.

Now for the part that does the real work:

```
.DATA
dpmiproc          DD    0         ;dpmi extensions entry point.
RING0CALLGATE LABEL DWORD  ;use this to call ring 0 code.
ring0_off         DW    0   ;callgate selector for RING0FUNC
ring0_cs          DW    0   ;       / (offset is ignored)
ms_dos_str        DB    "MS-DOS",0
ldt_selector      DW    0         ;for direct writing to ldt.
descriptor_selector DW 0          ;ring0, cannot be accessed
                                  ;directly.
ring0errormsg  DB "Error creating ring 0 access... \
                                  aborting program.",0

.CODE
makering0selector    PROC

    invoke GLOBALPAGELOCK,cs
    cmp  ax,0
    je   lockfailed

;find out where the LDT is ...
    lea  si,ms_dos_str
    mov  ax,168Ah           ;get dpmi extensions entry point.
    int  2Fh                ;-->es:di (undocumented)
;***        cmp      al,0                    ?????
;***        jne      extensionsnotfnd
    mov  WORD PTR[dpmiproc],di    ;save entry point
    mov  WORD PTR[dpmiproc+2],es  ;        /
    mov  ax,100h                  ;undocumented
    call dpmiproc                 ;-->ax=selector to ldt.
    jc   extensionsnotfnd
    mov  ldt_selector,ax
    mov  es,ax

;create a ring 0 32-bit descriptor...
    push es
    invoke  ALLOCSELECTOR,cs      ;-->ax=alias to cs.
    pop  es
    cmp  ax,0
    je   selectorerror
    and  ax,0FFF8h    ;get offset of descriptor in ldt.
    mov  bx,ax
```

```
    mov   al,es:[bx+5]              ;get access-rights byte.
    and   al,10011111b             ;clear dpl field.#
    mov   es:[bx+5],al
    mov   al,es:[bx+6]  ;get granularity & seg-size bits.
    or    al,01000000b             ;set bit7, for 32-bit.
    mov   es:[bx+6],al
    or    bx,0100b      ;set bit-2, selects ldt.leave dpl=0.#
    mov   di,bx                    ;temp save.
    mov   descriptor_selector,bx   ;save.
;create callgate, to above descriptor.....
    push  es
    invoke ALLOCSELECTOR,0         ;create a descriptor in ldt.
    pop   es
    cmp   ax,0
    je    selectorerror
    mov   ring0_cs,ax          ;save final selector.
    and   ax,0FFF8h        ;get offset of descriptor in ldt.
    mov   bx,ax
    mov   es:[bx],ring0func ;my ring0 code (declared EXTRN)
    mov   es:[bx+2],di                     ;ring0 alias.
    mov   BYTE PTR es:[bx+4],0   ;dwords copied to stack.***
    mov   BYTE PTR es:[bx+5],11101100b
;present=1,dpl=3,app=00,type
    mov   WORD PTR es:[bx+6],0   ; =C (386 callgate)
qwert:
    jmp   SHORT qwerty

lockfailed:
extensionsnotfnd:
selectorerror:
    lea   si,ring0errormsg
    call errormsgproc
    call xquitmessage                        ;quit program.
qwerty:
    ret
makering0selector ENDP
;...............................
freeourselectors       PROC
    invoke        FREESELECTOR,descriptor_selector
    invoke        FREESELECTOR,ring0_cs
    invoke        GLOBALPAGEUNLOCK,cs
    ret
freeourselectors              ENDP
```

INT-2Fh,
function
168Ah

The first thing that makering0selector() does is lock the segment in memory, as the ring-0 descriptor and call gate that are about to be created will have their present bit set, indicating that they are in physical memory.

The next problem is, where is the LDT? The exact location of the LDT is not something that a ring-3 program is supposed to know, but an undocumented feature of INT-2Fh, function 168Ah with address of string "MS-DOS" in the SI register, returns a selector to the start of the LDT.

Creation of a ring-0 code descriptor

The next job is to create a descriptor for the ring-0 code. This is a SMALL model program, which means that all code is in the same segment. ALLOCSELECTOR() creates a new descriptor in the LDT that is an alias to, in this case, CS. The code immediately after uses the selector to the LDT to directly access the LDT and modify the privilege level of the segment. Also, since the newly created descriptor is an alias to CS, it is a 16-bit segment: this example code requires the ring-0 code to be 32 bits by default. Therefore, the seg_16_32 bit is altered also.

16- & 32-bit code in same segment

Normally, an application cannot directly modify an entry in the LDT, for the simple reason that you don't know where it is. Now, having modifed it, you can't call it because it is a ring-0 descriptor whereas your code is running at ring 3.

Note that there is a trick being performed here, as there is only the one segment. I defined ASMRING0 as SMALL, and when the ring-0 file, HEAVEN, is linked, there will only be one code segment. CS is a ring-3, 16-bit descriptor, so that is how the code is treated when executed using CS. However, the newly created alias, *descriptor_selector*, is ring 0, 32 bits, but is referencing the same segment.

Call gate fields

The final step is to create the call gate. Again, an entry is made in the LDT, and it is directly written to, to make it into a call gate. The selector for this call gate is saved as *ring0_cs*. The call gate must contain the offset of the code to be called, which in this case is *ring0func*, defined as external, at the beginning of the code listing. You will see that descriptor_selector is also put into the call gate.

Offset-4 in the call gate, which I have marked in the listing with three asterisks, is where you can specify how many doublewords you have passed on the stack: the CPU will copy these from your ring-3 stack to the ring-0 stack. In this case, no parameters are copied.

Ring-0 stack

Now that the stack has been mentioned, this is an important issue that must be addressed. Windows maintains a separate stack for ring 0, and the call gate will automatically transfer to it. The CPU will copy the number of parameters specified from the ring-3 stack and will put the return address on top of the new stack.

Note that the ring-0 segment has also been defined (in this case) as a 32-bit segment, which means that the return address is two 32-bit values for selector:offset.

The default ring-0 stack is very small, which is why this program executes CLI (clear interrupt) before calling the ring-0 code. Have a look at the listing, and you will see *RING0CALLGATE*, which is the pointer that is called to get to the ring-0 code. Actually, RING0CALLGATE is an alias to ring0_cs:ring0_off.

32-bit ring-0 code
So the instruction "call RING0CALLGATE" will call the ring-0 code. What does the ring-0 code look like? Here is the listing for HEAVEN.ASM:

```
;this file is named HEAVEN.ASM, as it's as high as we can
;go...
.386P    ;masm is stupid... this proc has to be a separate
     ;file, to generate 32-bit code without the 66 prefix.
PUBLIC  RING0FUNC

_TEXT2 SEGMENT DWORD PUBLIC USE32 'CODE'
 ASSUME  CS:_TEXT2

RING0FUNC       PROC  FAR
  retf    ;NOTE must remove any params passed by callgate.
RING0FUNC       ENDP

_TEXT2 ENDS
 END
```

This example is doing absolutely nothing, just returning. You will know that it works if you don't get a "general protection error" message!

Structure of ring-0 segment
Note that I did not use ".MODEL" in this file, because it would create a code segment with the name "_TEXT", and the ".386P" at the very beginning of the file, if preceeding ".MODEL SMALL", would cause a 32-bit code segment. The linker would give the error message that two segments with the same name (_TEXT) cannot be combined if one is 16-bit and the other 32-bit.

Combining 16- & 32-bit segments
It's pretty stupid, but we are able to combine 16-bit and 32-bit segments, by giving them different names and placing them in the same "class". The _TEXT segment in ASMRING0.ASM has class "CODE", so putting "CODE" at the end of the SEGMENT declaration above will cause them to be combined. What I think is stupid is that I have to resort to the "old fashioned" SEGMENT directives to achieve this.

Anyway, note that I gave the ring-0 segment a different name, _TEXT2, but the choice is arbitrary. The qualifier "USE32" defines the segment as 32-bit, which means that the assembler will assemble 32-bit instructions without the prefix (and 16-bit instructions *with* the prefix).

The "P" on the end of .386P permits use of the ring-0 restricted instruction set; that is, the assembler will assemble them.

Finally, you can put a number after RETF to indicate the number of **bytes** to pop off the stack. Use this to remove parameters passed by the call gate, if calling in conformance with the Pascal convention.

So what can we do in this 32-bit ring-0 procedure?

FLAT Memory

You will find the program discussed so far on the Companion Disk in directory \ASMRING0.   This chapter also describes an enhancement to this program that is contained in \FLATASM0.

What you can do in ring 0

ASMRING0.EXE, as described so far, demonstrates how a 16-bit ring-3 program can make the transition to a 32-bit ring-0 code segment and come back. Once in ring 0, you can execute OUT, IN, CLI, STI, etc., without intervention by the CPU. You can also use the *privileged instructions* of the 386 that allow direct manipulation of LDT, GDT, and page tables.

However, one other thing you might want to do is call the functions in the *Virtual Machine Manager* (VMM), which you can think of as the "core" of Windows, and the functions in the *Virtual Device Drivers* (VxDs). Conceptually, you can view Windows as having two APIs — the ones you know about and that are described in all the Windows programming books (and in the SDK) and another set that can only be called by VxDs.

The latter functions are inside the VMM and the VxDs and are ring-0 code. The conventional wisdom is that you must write a VxD to be able to call them, but in fact our RING0FUNC can do so. The requirement simply is that you must be in ring 0 and you must be in the FLAT memory model. The program developed so far falls down on the latter point.

Fixing code & data at known linear addresses

One little note while I think of it — I used GLOBALPAGELOCK(), a Windows API function, to lock the entire code segment of ASMRING0, which means that it cannot be paged out and remains at the same linear and physical address. There are advantages to locking a segment, but one disadvantage is that Windows 3.1 tends to shift the segment down below 1M (physical) before locking it, which ties up some of that "valuable" conventional memory.

Fortunately, Windows 95 does *not* move it down below 1M.

If you write code that computes a certain linear address, you want to be sure that it stays at that linear address. Normally, when you use a selector, the linear address contained in the descriptor can change, and it is of no concern to you. GLOBALPAGELOCK() keeps it fixed and also ensures that the pages remain in memory — which would be optimal for interrupt handlers.

There is another API function that you might like to consider if your requirement is only that the linear address remains unchanged and paging is ok. If paging is left on, as per normal, the only repercussion is a possible access delay — unless you are doing coding that involves talking to specific physical locations.

Consider another alternative, GLOBALFIX(), which fixes a segment at a fixed linear address but allows paging-out.

Getting addressability to FLAT ring-0 code Back to the central argument. The objective now is for our ring-0, 32-bit procedure to be able to call VMM and VxD functions. The following code is a re-do of MAKERING0SELECTOR, which sets up addressability to ring 0:

```
makering0selector PROC
;get addressability of ring0, ring0func.....

    invoke GLOBALPAGELOCK,cs
    cmp  ax,0
    je   lockfailed

  lea  si,ms_dos_str
  mov  ax,168Ah      ;get dpmi extensions entry point.
  int  2Fh                 ;-->es:di (undocumented)
;***        cmp     al,0                     ?????
;***        jne     extensionsnotfnd
  mov  WORD PTR[dpmiproc],di     ;save entry point
  mov  WORD PTR[dpmiproc+2],es   ;         /
  mov  ax,100h                   ;undocumented
  call dpmiproc                  ;-->ax=selector to ldt.
    jc   extensionsnotfnd
    mov  ldt_selector,ax
    mov  es,ax

;find the linear address of CS...
  mov  bx,cs
  and  bx,0FFF8h              ;get offset in ldt
  mov  ax,es:[bx]            ;get size of segment.
    mov  cssize,ax
    mov  ax,es:[bx+2]         ;get lo-half of lin.addr.
    mov  WORD PTR flatlin,ax
    mov  al,es:[bx+4]         ;get hi-half of lin. addr.
  mov  ah,es:[bx+7]          ;         /
  mov  WORD PTR flatlin+2,ax
;calculate FLAT linear address of ring0func...
  mov  ax,WORD PTR flatlin
  add  ax,ring0func          ;note: "OFFSET" is optional
  jnc  moppi
    mov  bx,WORD PTR flatlin+2
```

```
      inc   bx
  mov   WORD PTR flatlin+2,bx
moppi:
  mov   WORD PTR flatlin,ax

;create callgate to ring0func.....
  push es
  invoke  ALLOCSELECTOR,0     ;create a descriptor in ldt.
    pop  es
    cmp  ax,0
    je   selectorerror
    mov  ring0_cs,ax                 ;save final selector.
    and  ax,0FFF8h   ;get offset of descriptor in ldt.
  mov  bx,ax
  mov  ax,WORD PTR flatlin    ;my ring0 linear address
  mov  es:[bx],ax             ;         /
    mov  ax,WORD PTR flatlin+2        ;       /
    mov  es:[bx+6],ax                 ;      /
  mov  WORD PTR es:[bx+2],28h;FLAT code selector(in gdt).
 mov BYTE PTR es:[bx+4],0 ;04;****?dwords copied to stack
  mov  BYTE PTR es:[bx+5],11101100b
    ;present=1,dpl=3,app=00,type=C (type=C: 386 callgate)

;find the FLAT linear address of this program's data
;segment...
  mov  bx,ds
  and  bx,0FFF8h            ;get offset in ldt
  mov  ax,es:[bx+2]         ;get lo-half of lin.addr.
  mov  WORD PTR flatdatalin,ax
  mov  al,es:[bx+4]         ;get hi-half of lin. addr.
  mov  ah,es:[bx+7]         ;        /
  mov  WORD PTR flatdatalin+2,ax

qwert:
  jmp  SHORT qwerty
lockfailed:                          ;... put in handlers ....
extensionsnotfnd:
selectorerror:
qwerty:
      ret
makering0selector ENDP
```

What you will notice in the above code is that I have not created code or data descriptors. What you do see above is the use of selector 28h. I have obtained the base addresses from ASMRING0's DS and CS descriptors, and to obtain the code FLAT linear address, I have added the offset of RING0FUNC to the base address of CS and saved the result in *flatlin*.

To obtain a FLAT linear address to the data segment, I extracted the base address from DS and saved it as *flatdatalin*.

Calling VMM and VxD services   Now, going up to ring 0 HEAVEN, by exactly the same method of "call RING0CALLGATE", will cause entry to RING0FUNC with CS = 28h, the FLAT selector.

HEAVEN,
enhanced

Here is some 32-bit ring-0 code for RING0FUNC that calls a VMM function:

```
.386P
INCLUDE vmm-tiny.inc      ;enables us to call vmm and vxd
          ;functions (derived from VMM.INC, in the DDK).
EXTERN   ring0stack:DWORD
EXTERN   default0esp:DWORD
EXTERN   default0ss:WORD

PUBLIC   RING0FUNC
_TEXT2 SEGMENT DWORD PUBLIC USE32 'CODE'
ASSUME CS:_TEXT2

RING0FUNC        PROC    FAR
;assuming that no parameters are passed, the ring 0 stack
;contains:
;return-EIP, return-CS, old-ESP, old-SS.
;the last two, deepest in the stack, reference the ring-3
;application stack.
;I think DS still points to old data segment, so can
;still use....

   cli                       ;make sure actual flag is clear.
   pushfd
   pushad
   push ds
   push es
   push fs
   push gs
   mov  default0esp,esp      ;save default ring 0 stack.
   mov  default0ss,ss        ;      /

;setup a new stack...
   mov  ax,30h
   mov  ss,ax
   lea  esp,ring0stack+1996
   add  esp,flatdatalin      ;calc Flat linear addr.
   sti

;ring-3-ds works here, but let's replace it with
;FLAT-ds...
   mov  ax,ds       ;use fs to access data in our prog.
   mov  fs,ax       ;              /
   mov  ax,30h                     ;Flat ds.
   mov  ds,ax       ;              /
   mov  es,ax
   mov  gs,ax

;example of calling a VMM service...
   int  20h
   DW   GET_CUR_VM_HANDLE      ;=1
   DW   VMM_DEVICE_ID          ;=1

;example of using a '386 privileged instruction...
   str  cx          ;get task (tss) register (selector)
```

```
;example of another call to the VMM...
;... parameters are passed by stack, so push them on...
  pushd 0                           ;get its descriptor out of gdt.
  pushd 0                           ;            /
  pushd ecx                         ;           /
  int   20h                         ;          /
  DW    GETDESCRIPTOR               ;         /
  DW    VMM_DEVICE_ID               ;        /
  add   esp,12                      ;       /
  mov   ecx,eax           ;zero in eax and edx if error.
  or    ecx,edx                     ;      /
  jz    error1                      ;     /
  jmp   bypass1

error1:  ;.... do something here ....

bypass1:
;restore default ring0 stack...
  cli                                       ;make sure.
  mov   ss,fs:default0ss
  mov   esp,fs:default0esp
  pop   gs
  pop   fs
  pop   es
  pop   ds
  popad
  popfd
  retf
RING0FUNC       ENDP
_TEXT2 ENDS
  END
```

Because the default stack is very small, I have replaced it with another that physically exists in the data segment of ASMRING0.

Execution enters RING0FUNC with DS still set to the data segment of ASMRING0, but I have moved it into FS and have put DS = 30h, the GDT FLAT data selector. There is no problem with accessing all the data in ASMRING0, using FS (ring-3 selector) or DS. In the latter case, we would also have to add "flatdatalin".

INT-20h

Notice the peculiar method for calling a VMM or VxD service by means of an "INT 20h" instruction, followed by a couple of parameters. Inserting data directly into the code may seem odd, but on the first execution-pass, Windows modifies these three lines and replaces them with a CALL. The first parameter specifies which service to call, and the second parameter specifies which VxD. These are simple equates defined in *VMM.INC* or in my cut-down version *VMM-TINY.INC* on the disk.

Note also that GETDESCRIPTOR() uses the standard C calling convention, which means that parameters are pushed right-to-left, and the stack must be cleaned up after return.

Moving On

To be able to go up to ring 0 from inside a ring-3 application is "real neat". This chapter also showed how to go from a 16-bit segment to a 32-bit segment, and actually have them overlap, that is, be the same segment, fitting the SMALL memory model.

The work done in this chapter can also be applied to Windows 95 native 32-bit applications, in which case the segment is already 32 bits, but the ring transition is still required.

It may be perverse, but I really like the idea of writing 16-bit applications that have 32-bit and/or ring 0 functions in them. These will run fine in both Windows 3.1 and 95.

A 32-bit application will run in Windows if it has the Win32s library installed, and it will run natively in Windows 95. So, I guess we need to move ahead into the pure 32-bit world. A lot of the material earlier in this book has focused on 16-bit code, although the principles are in most cases applicable to 32-bit code also.

We need a chapter that elaborates on the differences in coding for 32-bit segments and Win32, the 32-bit Windows API library. We also need to see a pure 32-bit application. The next chapter does this.

13

32-Bit Ring 3

Preamble

Other chapters

This book has been structured in a quasi-historical sequence, starting with 16-bit programming in the early chapters, gradually introducing 32-bit issues in latter chapters. I didn't want to dump 16-bit, as it is still relevant and will remain an issue for a long time. Even if a systems programmer wants to program entirely in 32-bit mode, Windows 95 internally is surprisingly 16-bit oriented. This means that a thorough knowledge of the 16-bit issues and the interaction between 16- and 32-bit modes is required. Therefore, the gradual progression of the chapters from a 16-bit foundation is most relevant.

Of course, many developers are still programming for Windows 3.x, and 16-bit applications run fine on Windows 95 and even have some advantages with regard to system privileges, compared with 32-bit applications. As described in the last chapter, putting 32-bit instructions into a 16-bit segment incurs only a small instruction prefix penalty. Putting 32-bit segments into a 16-bit application can also be done. Considering these points, many developers do not feel any urgency to go totally 32-bit.

However, if you want to move ahead and write a true native 32-bit application, this is the chapter.

TASM5 versus MASM6

Back in Chapter 5, I compared the features of the various versions of TASM and MASM, targeting 16-bit applications.

The two products have tended to leap-frog each other, but MASM has remained stuck on version 6.11 for some time now. Borland has recently released version 5.0, which does not leap-frog MASM: it only brings it to about even.

QUESTION: How many Microsoft Officials does it take to change a light bulb?

ANSWER: None. They will just declare darkness to be the new standard.

Not so far from the truth! Microsoft has put MASM "on the back burner" for some time, because it is a very "small fish" for them. At the time of writing, rumour is that they are selling it to another company.

Borland, to their credit, does not consider itself to be too big to ignore the lower-end of the market. That is, the relatively small-volume sellers like assemblers.

TASM 32-bit support
Both companies have moved toward less printed and more on-line documentation. My personal viewpoint is that you can't beat a good printed manual, which is why the supplementary printed books business is booming.

TASM5 supports 32-bit programming for Windows 95 and NT, but the documentation, both printed and on-line, is pitiful. The one example program is also pitiful, as it is written for TASM 4.

Porting code from MASM to TASM
So, I had to figure it out from scratch. I had a 32-bit program written for MASM, which I converted. Now, this is an interesting story, and there were nights spent working to 3:00 AM trying to figure it out.

TASM5 almost supports all of the features of MASM version 6.1. Therefore, the example program given in this chapter, though written for TASM5, should also be very easy to convert for MASM.

Itemising the differences between MASM6 and TASM5
• TASM5 has prototypes for procedures, except they are designated by the "PROCDESC" keyword, not "PROTO". Otherwise, the syntax is the same, and I was able to create an Include file, W32.INC, on the Companion Disk in directory \TASM32 that is very easy to convert for MASM.

- TASM5 does not use "INVOKE" for high-level procedures, just the plain old "CALL" keyword.
- TASM5 allows parameters passed to a procedure to be declared on the same line as the "PROC" declaration, but it is not quite so sophisticated as MASM. You cannot use MASM's "ADDR" prefix, or the "::" for composing two 16-bit registers into one 32-bit value. However, you can achieve the same results with different syntax.
- I'm not so sure about passing dynamic data parameters to high-level PROC declarations. If you specify a parameter "OFFSET S1", it means "pass the address of S1". However, that works if the data is declared statically, in the data segment. For data declared by the "LOCAL" directive, that is, dynamic, stack-based data, it seems to be necessary to load the data into a register first and pass the register as a parameter. MASM doesn't have this limitation with its "ADDR" directive.

Installing TASM5

All the TASM5 tools are DOS programs

TASM5 is designed to work from the command line in a DOS box. There is no editor or IDE. There is, though, the wonderful Turbo Debugger. I prefer to use the command line, though an IDE does have advantages, such as seeing where assemble errors occur in the source code. With the command line approach, the assembler spews out a list of errors and the developer must then find those lines in the source code, which is easy enough.

```
C:\> make -B -DDEBUG | more
```

If the assembler generates a huge error listing, this is what you do to make output fill the screen and pause. Simple enough. The "make" program will execute "makefile" if it exists, otherwise a filename needs to be entered on the above command line, after the switches. "-B" means to rebuild everything, "-DDEBUG" is interpreted inside the Make file to include debug information. The "more" postfix is what pauses the screen.

TASM32, TLINK32, BRC32 32-bit tools

For 32-bit development, you will be using TASM32.EXE, TLINK32.EXE, and BRC32.EXE. The latter is the resource compiler. There is also BRCC32.EXE, but the documentation does not mention anything about it. In fact, the documentation barely says anything about the resource compilers at all, and there is no demonstration 32-bit program that utilises them.   Never mind, Uncle Barry figured it all out.

Fine-tuning the installation for stability

After installation, when I first ran TASM32, it crashed. I fiddled around, and suddenly it started working. Later that night, for no apparent reason it crashed again. That is, it aborted on loading. I had no idea why. I made some changes, and everything has been ok since then.

I found a reference in the documentation that the WIN.INI file should have this entry:

```
;in WIN.INI file
[Windows]
spooler=yes
```

So, I put that in. Then I read that the install process puts these two lines into the SYSTEM.INI file:

```
;SYSTEM.INI file
[386Enh]
device=c:\tasm\bin\windpmi.386
device=c:\tasm\bin\tddebug.386
```

What is the purpose, I asked myself, of WINDPMI.386, when Windows already provides DPMI for DOS boxes? So, I erased that line.

Example Skeleton Program

Ok here it is. Thirty-two-bit coding has certain refinements, one of which is the prolog/epilog code: the simple use of the STDCALL language qualifier takes care of everything.

```
;By Barry Kauler 1997
;Companion Disk, "Windows Assembly Language & Systems
;Programming".
;W32DEMO.ASM --> W32DEMO.EXE Windows 95 demo program.
;This skeleton assembly language program has been written
;for TASM5.0.
;It has the startup code built-in, rather than as a
;separate object file.

.386
.MODEL FLAT,STDCALL
UNICODE = 0              ;this equate used by W32.INC.
INCLUDE W32.INC         ;equates, structures, prototypes.
```

```
IDM_QUIT     EQU 100          ;menu-identifiers -- must be
IDM_ABOUT    EQU 101          ;same as defined in .RC file.


.DATA
;-----------------------------------------------------------
hInst           DD 0
mainhwnd        DD 0
s1              WNDCLASS    <?>
s2              MSG         <?>
s3              PAINTSTRUCT <?>
szTitleName DB "Win32 Assembly Language Demo Program",0
szClassName     DB "W32DEMO",0
sziconname      DB "ICON_1",0    ;name of icon in .RC file.

g_hwnd      DWORD 0
g_message   DWORD 0
g_wparam    DWORD 0
g_lparam    DWORD 0

szaboutstr  DB "This is an about-box",0 ;messagebox
sztitlestr  DB "Barry Kauler 1997",0   ;/

.CODE
;-----------------------------------------------------------
start:

   call GetModuleHandle, NULL
   mov  hInst,eax

; initialise the WndClass structure
   mov  s1.w_style, CS_HREDRAW + CS_VREDRAW + CS_DBLCLKS
   mov  s1.w_lpfnWndProc, offset ASMWNDPROC
   mov  s1.w_cbClsExtra, 0
   mov  s1.w_cbWndExtra, 0

   mov  eax, hInst
   mov  s1.w_hInstance, eax

;call LoadIcon, NULL,IDI_APPLICATION ;loads default icon.
;No, let's load a custom icon....
   call LoadIcon, hInst, OFFSET sziconname
   mov  s1.w_hIcon, eax

   call LoadCursor,NULL, IDC_ARROW
   mov  s1.w_hCursor, eax

   mov  s1.w_hbrBackground, COLOR_WINDOW + 1
   mov  s1.w_lpszMenuName, OFFSET szClassName
   mov  s1.w_lpszClassName, OFFSET szClassName

   call RegisterClass, OFFSET s1

  call CreateWindowEx, 0,OFFSET szClassName, \
       OFFSET szTitleName,WS_OVERLAPPEDWINDOW, \
CW_USEDEFAULT,CW_USEDEFAULT,CW_USEDEFAULT,CW_USEDEFAULT,\
       0, 0, hInst, 0
```

```
   mov   mainhwnd, eax

   call ShowWindow, mainhwnd,SW_SHOWNORMAL
   call UpdateWindow, mainhwnd

msg_loop:
   call GetMessage, OFFSET s2, 0,0,0
   cmp  ax, 0
         je      end_loop
     call TranslateMessage, OFFSET s2
     call DispatchMessage, OFFSET s2
   jmp  msg_loop

end_loop:
   call ExitProcess, s2.ms_wParam

;-------------------------------------------------------------
   PUBLIC ASMWNDPROC
ASMWNDPROC proc STDCALL, hwnd:DWORD, wmsg:DWORD, \
                           wparam:DWORD, lparam:DWORD
   USES  ebx, edi, esi
   LOCAL hDC:DWORD

   mov eax,hwnd            ;useful to make these static.
   mov g_hwnd,eax          ; ... be cautious though, as
   mov eax,wmsg            ;sometimes Windows reenters
   mov g_message,eax       ;ASMWNDPROC. For example, it is
   mov eax,wparam          ;possible for Windows to call
   mov g_wparam,eax        ;ASMWNDPROC with a WM_PAINT
   mov eax,lparam          ;message even though execution
   mov g_lparam,eax        ;is currently inside ASMWNDPROC.
       ;...alternative is pass these via stack to functions.

   xor eax,eax
   mov ax,WORD PTR g_message
   .IF ax==WM_DESTROY
     call   wmdestroy
   .ELSEIF ax==WM_RBUTTONDOWN
     call   wmrbuttondown
   .ELSEIF ax==WM_SIZE
     call   wmsize
   .ELSEIF ax==WM_CREATE
     call   wmcreate
   .ELSEIF ax==WM_LBUTTONDOWN
     call   wmlbuttondown
   .ELSEIF ax==WM_PAINT
     call   wmpaint
   .ELSEIF ax==WM_COMMAND
     call   wmcommand
   .ELSE
     call DefWindowProc, hwnd,wmsg,wparam,lparam
     ret
   .ENDIF
   xor eax,eax
   ret
ASMWNDPROC ENDP
;-------------------------------------------------------------
```

```
wmcommand PROC
   mov ax,WORD PTR g_lparam
   .IF ax==0
     mov   ax,WORD PTR g_wparam
     .IF ax==IDM_QUIT
       call PostQuitMessage,0
     .ELSEIF ax==IDM_ABOUT
       call MessageBox, g_hwnd, OFFSET szaboutstr, OFFSET
sztitlestr, MB_OK
     .ENDIF
   .ENDIF
   ret
wmcommand ENDP
;---------------------------------------------------------------
wmpaint PROC
   call BeginPaint, hwnd,OFFSET s3
   mov  hDC, eax


   call EndPaint, hwnd,OFFSET s3
   ret
wmpaint ENDP

wmcreate PROC
   ret
wmcreate ENDP

wmdestroy PROC
   call PostQuitMessage,0
   ret
wmdestroy ENDP

wmlbuttondown PROC
   ret
wmlbuttondown ENDP

wmrbuttondown PROC
   call MessageBeep,0
   ret
wmrbuttondown ENDP

wmsize PROC
   ret
wmsize ENDP


;---------------------------------------------------------------
ENDS
END start
```

Elegant isn't it? You can refer to earlier chapters for explanations of how each part works. You might like to compare it with the 16-bit MASM6 program in Chapter 5.

The differences are small. Most importantly, you do everything in 32 bits.

The differences between 16- and 32-bit coding of Windows apps

- The fields of the structures mostly become 32 bits.
- FAR addresses become the same as NEAR addresses and are 32 bits. The OFFSET prefix in an instruction will load the 32-bit address of a static data item, and you do not need to worry about the segments.
- All stack pushes and pops are 32-bit.
- Values returned from functions are in EAX.
- Note that the Win32 API is a blend of C and Pascal calling convention. That is, stack cleanup is performed by the function, but parameters are pushed right-to-left. Please note that the 16-bit API pushes parameters left-to-right. However, using the high-level procedures, you do not need to worry about this. When using Turbo Debugger, you will need to be aware of this fact, though. For example, in GetMessage(), "OFFSET s2" gets pushed *last*.

Code conversion from MASM to TASM

When I first converted a MASM6 program for TASM5, it assembled and linked but crashed when execution got to CreateWindowEx(). I paid closer attention to the skeleton example supplied with TASM5, even though it is written for TASM4. I made a couple of changes, and it now works and is rock solid, though I'm not sure which change was the culprit.

Notice that there is an ENDS directive at the very end of the program. You could experiment and see what happens if that is left off. I never needed it for MASM programs.

The rest of the program looks very much like a MASM6 program, and TASM5 also accepts the same syntax for the high-level procedures, though it does not support ADDR and "::". "::" isn't needed in 32-bit programming, and ADDR can be replaced by OFFSET for static data.

You will notice that I have used correct case in all symbols. I used the "/ml" switch to turn on case sensitivity, which is a break from my past. I decided to invoke case sensitivity for all true 32-bit code, which is why I have shown correct case for all the 32-bit API functions.

Support Files

Resource files

There is nothing much to say about resource files. They work the same as before.

```
//W32DEMO.RC resource file.
//these (arbitrary) equates could have been in an include
//file...
#define IDM_QUIT      100
#define IDM_ABOUT     101


ICON_1 ICON GOOFEE.ico

W32DEMO  MENU
    BEGIN
      POPUP "File"
        BEGIN
            MENUITEM "Quit",   IDM_QUIT
            MENUITEM "About...", IDM_ABOUT
        END
    END
```

The program BRC32.EXE is required to compile a .RC file to .RES.

Make file   Now may be the best place to show the Make file:

```
#MAKEFILE.
#W32DEMO Win32 demo application.
#TASM32.EXE, TLINK32.EXE, BRC32.EXE, MAKE.EXE
#are from TASM v5.0. Make sure the path points to them.
#Path only needs to point to \bin subdirectory, TLINK32
#finds IMPORT32.LIB in the \lib subdirectory ok.

#You should be in a DOS box, by executing the PIF file
#B32TOOLS.PIF (make a shortcut on your desktop).

#TLINK32 switches: /Tpa = build 32-bit EXE, /aa = target
#Windows 32-bit application, /v = include debug info.
#TASM32 switches: /Zi = include debug info.
#the last parameter is the resource file to be bound to
#the executable.
#the 2nd last param. is the definition file.

# make -B          Will build .EXE
# make -B -DDEBUG  Will build the debug version.

FN = W32DEMO
OBJS = $(FN).obj
DEF  = $(FN).def

!if $d(DEBUG)
TASMDEBUG=/zi
LNKDBG=/v
!else
TASMDEBUG=
LNKDBG=
!endif
```

```
!if $d(MAKEDIR)
IMP=$(MAKEDIR)\..\lib\import32
!else
IMP=import32
!endif

 $(FN).EXE: $(OBJS) $(DEF)
tlink32 /Tpe /aa /c $(LNKDBG) $(OBJS),$(FN),,$(IMP),$(DEF),$(FN)

 .asm.obj:
    tasm32 $(TASMDEBUG) /ml $&.asm
    brc32 -r $(FN).rc
```

In the above Make file, you can see the invocation of BRC32.EXE. It is used with a "-r" switch to mean compile only, which is probably optional.

Binding resources to the executable

In earlier examples, I have run RC.EXE again after LINK, to bind the .RES file to the .EXE file. However, TLINK32 does this automatically if the name of the .RES file is appended onto the end of the command line. The last $(FN) achieves this.

Compatibility of Borland & Microsoft Make files

I have a lot of trouble with Borland Make files. Although there is a switch for setting compatibility with Microsoft's NMAKE.EXE, it is still *not* compatible. I have never been able to get a Make file I have created for NMAKE to work with Borland's MAKE.

I have to resort to taking an example Make file provided by Borland, which is what I have done above. It is not quite optimum, as the resource compiler executes every time, but at least it works. I recommend that you use the "-B" switch to force everything to build:

```
C:\> make -B -DDEBUG
```

There is something weird about Borland's MAKE.EXE and I personally use NMAKE.EXE mostly.

B32TOOLS.-PIF, to fine-tune the DOS box

Note also that Borland supplies B32TOOLS.PIF. I recommend that you put a shortcut to it from your Windows 95 desktop. It has the correct settings for the DOS box. You will find it in C:\TASM\BIN.

Also, place C:\TASM\BIN into the path statement of your AUTOEXEC.BAT file, so DOS can find the executables. TLINK32 finds the library file IMPORT32 without any help.

.DEF file

Finally, the definition file, W32DEMO.DEF:

```
NAME          W32DEMO
DESCRIPTION   'ASM program'
```

```
EXETYPE      WINDOWS
STUB         'WINSTUB.EXE'
CODE         PRELOAD MOVEABLE
DATA         PRELOAD MOVEABLE MULTIPLE
HEAPSIZE     8192
STACKSIZE    8192
EXPORTS      ASMWNDPROC
```

W32DEMO.DEF is referenced by the second-last parameter in the TLINK32 command line.

Turbo Debugger

If after assembling and linking, it doesn't work, it is time to use the debugger. Stay in the DOS box to use it, and type this:

```
C:\> TD32 W32DEMO.EXE
```

Turbo Debugger has been an old favourite of mine. It's really nice, and very easy to use.

One thing to bear in mind is that you are in a multitasking environment, so feel free to run Windows programs alongside the DOS box.

You can have File Manager (I mean, Windows Explorer) running for the purposes of testing the program.

A text editor/IDE for use with TASM

You can use a Windows-hosted text editor or a DOS-based text editor. Everybody has a favourite. I use Microsoft's Programmer's Workbench v4.01, which is DOS based.

There are a zillion editors on the Internet that can be downloaded. Arguably the best for assembly language development is ASM_EDIT, a complete IDE with extensive help files. It is quite interesting to see color-coded assembly code. ASM_EDIT is shareware, but the warning window comes up so frequently it is almost unusable — that is, if you have a low tolerance level!

The price in February 1997 was US$20. The main Internet page is:

```
http://www.skysurf.de/~asmedit/ae_whats.htm
```

WALK32, development suite for MASM

Various people have experimented with stand-alone Windows applications written entirely in assembly language. Sven Schreiber has developed WALK32, a complete package for MASM, even with its own linker. It is public domain and can be found at the site:

```
http://www.thepoint.net/~jkracht/pdnasm.htm
```

SKELETON.- A 32-bit skeleton program written by Wayne Radburn for MASM
ZIP, a v6.11 uses the latest features of MASM, much like the example
skeleton for given in this chapter. If you have MASM v6.1x, have a look at
MASM6 this package. It is on the Companion Disk in
\RADBURN\SKELETON.ZIP.

Wayne has produced an very nice help file that explains how the
program works. His Include file is very cut-down, without all the
equates, structures, etc. I took his file, Sven's Include file, plus
some extra stuff and put it together into one file, did a lot of
editing, and ended up with W32.INC.

He has a bit more code in the startup than my above skeleton, and
I suggest you examine it and maybe include the same code if you
want to use my skeleton for actual projects.

Postamble

Chapter 12 showed how a 16-bit application can move into 32-bit
ring-0 code. What about the 32-bit application of this chapter?
Another question: what if the 32-bit application wanted to call a
function in a 16-bit DLL? Or an interrupt? Or perform an IN or
OUT instruction?

It is a strange fact of the historical evolution of Windows that
16-bit applications have greater freedom getting into the insides of
Windows than 32-bit applications. DOS TSRs also have great
advantages. Because support for legacy applications is going to
continue for the forseeable future, it is sensible to use whatever
easy paths are available.

A 32-bit application cannot use the technique of Chapter 12. The
reason is that the interrupt handlers provided by Windows for
certain interrupts assume that it is 16-bit code executing the
interrupt. The most fundamental problem is that it is only a 16-bit
stack, so the interrupt handler will crash. Nor can a 32-bit
application call a 16-bit function.

The next chapter backtracks somewhat and looks at the transition
between DOS and Windows as Windows loads. Understanding
this can be very useful and will help with the above questions.

14

DOS-Win Transitions

Preamble

Integrating the code from previous Chapters

This chapter further develops many of the concepts introduced in the previous chapters and also discusses some overall and related issues.

In this chapter, I have built upon the issues of moving between various modes, such as between VMs and between Real and Protected modes. What happens to registers? What about the stack? What are the address mappings?

I have further developed the discussion of interrupt handling for Real and Protected modes.

I have also considered the issue of synchronizing between DOS and Windows. For example, how does a DOS driver know when Windows is loading? How do you get a virtual device driver to cooperate with a DOS device driver? Or to cooperate with a WinApp?

When writing the first edition of this book, I paid a lot of attention to Standard mode. In this edition, I have considered it to be "almost" history, so just about all of the code and description in this chapter is geared toward Enhanced mode, i.e., requiring at least a 386 CPU.

**API,
WinApps,
at ring 3**

Sometimes I feel quite disgusted with Miscrosoft, because the "playing field" keeps changing. For example, Windows 3.0 had WinApps and the API DLLs running at ring 1, while DOSApps ran down at ring 3. Then, in Windows 3.1, everything went down to ring 3, including the DLLs. Windows 95 also has everything at ring 3, except of course the "insides" of Windows, such as much of the VMM (Virtual Machine Manager) and the VxDs (virtual device drivers).

**16-bit
WinApps
inferior to
32-bit
WinApps?**

Actually, 16-bit Windows applications should not be viewed as inferior, as it may turn out that they will give better performance than equivalent 32-bit applications. As explained in Chapter 12, all that is meant by 32-bit is that instructions in a 32-bit segment default to address and size of 32-bits, and they no longer have the 64K segment-size limitation.

**System-level
access in
Windows 95**

Sixteen-bit WinApps actually have some advantages when it comes to global addressing and general messing around inside Windows and with the hardware. Microsoft has tried to "close the door" to low-level access for 32-bit WinApps, so there is no direct access to the interrupts or the low-level API functions. All the low-level facilities are still there, however, and will continue to be there — it is a matter of knowing how to get at them. Sixteen-bit WinApps running in Windows 95 have easy access to them, for backwards compatibility reasons.

Most of the development that ended up in the first edition of this book was on Windows 3.0, while for this edition I worked mostly on 3.1 and 95. Some descriptions in this book will be more appropriate to 3.x than 95 — I have tried to be clear on what target environment I'm writing about.

You will find that the 16-bit code in this chapter works fine in Windows 95.

**Structure of
this chapter**

The structure of this chapter is in two halves: the first focuses on interrupt handlers for DOS and Windows, and the second focuses on the transition between DOS and Windows, the smooth transfer of control, and communication between TSRs, WinApps, and VxDs.

Interrupt Handlers

Chapters 10 and 11 give the elements required for interrupt handlers, and I have put various example programs on the Companion Disk. This section develops the topic further.

An interrupt handler that must work regardless of whether the computer is running a DOSApp or a WinApp requires a number of special considerations.

Rather than list complete example programs that go on for many pages, I have given only partial listings here and focused on discussion of the various issues.

Chapter 10 shows a Protected mode ISR invoked from a WinApp running in Protected mode. That is, the software or hardware interrupt occurred while the CPU was in Protected mode. This is the easiest case.

If the CPU is in Real mode at the time of the hardware or software interrupt, and you want to pass control up to a Protected mode handler, beware of various constraints. Chapter 11 introduced this topic.

Example Protected Mode ISR Code

The structure of the Protected mode ISR in each case is somewhat different:

```
; This is the same example ISR from Chapter 10 ...
runtime:          ;isr for prot mode interrupts, via idt.
    int 60h       ;call old vector
                  ;(it was saved in int-60 for convenience)
    pusha
    push ds
    push es
    mov  ax,cs:hwndcs          ;post message to window
    push ax                    ;    /
    push WM_USER               ;    /
    push 0                     ;    /
    push 0                     ;    /
    push 0                     ;    /
    call POSTMESSAGE           ;    /
    mov  es,cs:dsselector  ;for writing to data in code seg.
    ...
    pop  es
    pop  ds
    popa
    iret
;
; The ISR for interrupts reflected up from Real mode has
; a different structure ... (refer Chapter 11) ...
runtime2:
;isr for Real mode ints via ivt, reflected up to
;prot-mode. entered with ds:si = Real mode ss:ip,
;es:di = Real mode call structure,
;and interrupts disabled...
;should exit with es:di still pointing to Real mode call
;structure...
    pusha
```

```
      push          ds
      push          es
;get addressability of data in code seg...
      mov   ax,cs:hwndcs              ;post message to window
      push ax                         ;    /
      push WM_USER                    ;    /
      push 0                          ;    /
      push 0                          ;    /
      push 0                          ;    /
      call POSTMESSAGE                ;    /
      mov   es,cs:dsselector ;for writing to data in code seg.
      ...
      pop   es
      pop   ds
      popa
;for returning to Real mode prog prior to interrupt...
      cld                             ;    (described in Chapter 11)
      lodsw                           ;    /
      mov   es:[di].ip1,ax            ;    /
      lodsw                           ;    /
      mov   es:[di].cs1,ax            ;    /
      lodsw                           ;    /
      mov   es:[di].flags1,ax         ;    /
      add   es:[di].sp1,6             ;    /
;
;can chain to original vector by putting it into callback
;data structure...
;   mov   ax,cs:segmentrealint
;   mov   es:[di].cs1,ax
;   mov   ax,cs:offsetrealint
;   mov   es:[di].ip1,ax
;
 iret
installint ENDP
;...........................................................
      END
```

Separate ISRs for IVT and IDT

Note that there are two ISRs, one each for interrupts that come via the IDT and those that get reflected up from Real mode via the IVT. With regard to the installation of these ISRs, note that I did not hook the vectors as soon as the WinApp received the WM_CREATE message, as this can, under certain circumstances, impair the display of the window. Instead, I posted a message, WM_USER+1, which at a later stage calls the install code (see the complete program on the Companion Disk, in \WIN2REAL and further development in \REAL2WIN).

With regard to exiting from the program, I did of course unhook the vectors upon receipt of a WM_DESTROY message.

Problems/Issues with the Protected Mode ISRs

VMs have preemptive time-slicing by the DPMI host

POSTMESSAGE() will work for both ISRs when Windows is loaded in Enhanced mode. Even when running a DOS application, POSTMESSAGE() will send the WM_USER message to the window immediately. In this example code, the DPMICALLBACK() function acknowledges receipt of the WM_USER by beeping the loudspeaker. Note that this beep occurs as soon as you press a key — how can this be, since you're in a DOSApp? The answer is that the DPMI host, as the real Windows kernel, switches VMs on a time-sliced basis and so flips over to the system VM periodically to do housekeeping, including sending the waiting WM_USER message to the callback function for the window.

ISR reentrancy

Another issue with the Protected mode ISRs is reentrancy. This is especially a problem with hardware interrupts that can come in at any time. Upon entry to the ISR, hardware interrupts are disabled, but once you put in the STI instruction, they can occur. Note that you would also send an End Of Interrupt (EOI) signal to the interrupt controller chip to tell it that it is now allowed to send more interrupts (this is done by the default handler, if you chain to it). You could argue to avoid the problem by leaving the interrupt flag clear — but this should not be done for too long. The same point applies to the EOI signal — I did it by calling the original handler (via INT-60 in the ISR reached via the IDT).

If you put in an STI (and an EOI has been sent in the case of hardware interrupts), think about reentrancy. You may have to organize the data used by the ISR to be dynamic (on the stack): I'm thinking in particular of the data register structure, in which DPMI passes the Real mode registers to and from the Protected mode ISR.

Reference source

The "DPMI Toolkit", available from Qualitas (see http://www.qualitas.com/), has mechanisms for this.

The case of the missing code segment

In your .DEF file, FIX the code segment in place, and do not mark it as DISCARDABLE. This will **not** stop Windows from removing the segment from memory, but whenever your program needs to access the segment it will be reloaded into the same place — well nearly always!

If you get a selector alias to store data into the code segment, such as a window handle to be used by the ISR, or even the alias itself, for writing data to the code segment within the ISR, it will work. The alias will not require updating, because the code segment marked as FIXED in the .DEF file will remain at the same place in memory (even though Windows may temporarily remove it). The potential problem here is that Windows does not think that you should be writing to the code segment, so will never "swap it out". Instead, it is just dumped, and when needed again it is copied from the original on disk— so you lose your data.

Bolting the segments down

If you look at the above listing on the Companion Disk, you'll see that I used GLOBALHANDLE() and GLOBALFIX(). The first returns a handle for a selector or segment address, while the second Windows function locks the segment into that linear address. This is the only sure way to stop Windows from moving the segment, and it works in both Standard and Enhanced modes. However, in Enhanced mode you can use GLOBALPAGELOCK() to prevent paging, and guarantee that the segment is locked into physical memory. What these functions will do for you is speed up operation as the ISR's will be kept in memory (and you won't lose what you write to the code segment). They are not essential, however.

... and the wayward data segment

What about getting at data in the data segment from inside the ISR? No problem, because you can store the value of DS in the code segment. The data segment doesn't even have to be FIXED, because its descriptor will be automatically updated, unlike an alias.

None of this will work under Standard mode. Why am I even bothering to discuss Standard mode — it's dead, dead, dead. Maybe in some remote parts of the world there are still people running Windows in Standard mode. I promise not to mention it again.

The Real Mode Handler

Some philosophic points about DOS drivers

Ok, now for the DOS TSR interrupt handler. Actually, this is the most fascinating part of the exercise. There is a bit of a myth that you shouldn't develop Windows-aware DOS TSRs and device drivers, but should instead be going for virtual device drivers. The DOS driver has a lot going for it.

"Using up" the first 1M of physical memory

The fact that it takes up "valuable RAM real estate" in the first 1M is always brought up as a negative factor. However, this is not such a big issue as it was in the DOS-only days. The same thing goes for locking segments in place: the Windows textbooks make a noise about how this is undesirable, yet in reality it isn't if you don't lock too many bytes — this is assembly language, remember (super compact). Lock as many segments as you want, and even lock them in the first 1M if you want. Note that Windows has functions for this (see above) and so does DPMI, apart from the specifications in the .DEF file.

Put those TSRs in that first 1M and don't worry about it!

My little DOS TSR hardly impinges on the "valuable" 1M anyway: it's under 300 bytes. It hooks INT-9, which is a special case hardware interrupt. Here it is, somewhat abridged:

```
;DOSTSR.ASM  Hardware interrupt keyboard handler for
;Windows.
.286
int9 SEGMENT BYTE PUBLIC 'CODE'
   ASSUME        cs:int9,ds:int9
   ORG  100h
install:
   jmp  start

oldoffivt2F      DW     0     ;save old int-2F vector here.
oldsegivt2F      DW     0     ;    /
winloaded        DB 0 ;set when Windows is loaded, & viceversa.
winmode          DB 0 ;bit-0=1 if Standard, =0 if Enhanced.
oldoffivt9       DW     0     ;save old vector here.
oldsegivt9       DW     0     ;    /
oldss            DW     0     ;host stack
oldsp            DW     0     ;    /
tsrpspseg        DW     0     ;seg. addr. of psp
isrbusy          DB     0     ;set to prevent reentrance.
;..............................................................
start:
   mov  tsrpspseg,es        ;save psp seg. addr.

; Test if this TSR already installed.  If so, get out.

; Code for synchronizing and co-existing with DOS (save
; segment address of this PSP, get address of "inDOS"
; flag, hook IVT vectors 28h, and maybe 1Ch)

;hook int-2Fh vector in ivt.  Windows calls this with
;AX=1605h when it loads, with regs telling useful info,
;such as if loading in Standard or Enhanced mode....
   mov  ax,352Fh            ;get int-2F vector in ivt.
   int  21h                 ;    /
   mov  oldoffivt2F,bx      ;save it
   mov  oldsegivt2F,es      ;    /
   mov  ax,252Fh            ;hook int-2F
   lea  dx,runtime2F        ;set ivt vector.
   int  21h                 ;    /
;
doit:
   mov  ax, 2561h    ;hook INT 61h so signaller can find
   lea  dx, callback  ;forwarder in system VM, from
   int  21h          ;     another VM.
;
;hook keypresses/releases ...
   mov  ax,3509h            ;get int-9 vector in ivt.
   int  21h                 ;    /
```

```
      mov  oldoffivt9,bx      ;save it
      mov  oldsegivt9,es      ;    /
      mov  ax,2509h           ;hook int-9
      lea  dx,runtime9        ;set ivt vector.
      int  21h                ;    /
;
;initialise int-60h in ivt, as used to test if WinApp has
;hooked it...
      push ds                 ;actually, this will be 0:0 anyway,
                              ; but make sure. (tough luck if some
      mov  ax,0               ; other App has hooked it!)
      mov  ds,ax
      mov  dx,0
      mov  ax,2560h           ;put 0 into int-60.
      int  21h                ;(this hook will be in all Vm's).
      pop  ds
;
      lea  dx,endprog+17  ;point past all code in this
         ;module (Note that an optimum program would relocate
         ;the install portion of code at the end, so it can
         ;be discarded).
      shr  dx,4               ;compute # paragraphs to keep.
      mov  ax,3100h           ;terminate and stay resident.
      int  21h                ;    /
;.............................................................
runtime9:
;this is now the "signaller".  it is entered at every key
;press/release...
;but only when in Real mode...
;First, I only want this ISR to work when Windows is
;loaded, so test winloaded flag...
      cmp  cs:winloaded,0 ;note cs override, since ds not set.
      jne  firsthurdle
chain:
      jmp  DWORD PTR cs:oldoffivt9   ;chain to old int-9.
firsthurdle:
      cmp  cs:isrbusy,0
      jne  chain
      mov  cs:isrbusy,1       ;prevent reentrance.
secondhurdle:
;we're in, but call old int-9 first, which will take care
;of EOI...
      pushf
      call DWORD PTR cs:oldoffivt9
;now setup registers....
      push es                 ;save working registers
      push ds                 ;    /
      pusha                   ;    /
      push cs                 ;set ds == cs
      pop  ds                 ;    /

; Code to co-exist and synchronise with DOS, if you want
; to call DOS INT-21h functions (only allowed above 0Dh)
; ... test the "inDOS" flag, wise to switch to a local
; stack, change to PSP of TSR, save "break" setting and
; turn off, redirect INT's 1Bh, 23h and 24h, save
; extended error  checking (whew!).
```

```
    sti  ; EOI already achieved by call to old vector.
;
;next hurdle is to find out if Windows is in Standard or
;Enhanced mode.
;One way is to test INT-60 to see if it is hooked -- if
;not then we must be in Enhanced mode, as WinApp only
;hooks IVT in system VM. However all we will do is test
;winmode flag...
    mov  al,winmode
    and  al,1
    jz Enhanced  ;bit-0 =0 if Enhanced.
;
Standard:
    jmp  SHORT exit4


Enhanced:
;I will be a little bit fussy here. In theory, this ISR
;could be entered when the CPU is in the system VM, hence
;we will not want to do the transfer from another VM, as
;performed by 2F/1685... though it appears that this will
;still function.  Instead I have used 2F/1683 to query
;the current VM...¹
    mov  ax,1683h
    int  2Fh                ;returns VM id in BX
    cmp  bx,1               ;1=system VM
    je Standard
;
;switch to the system virtual machine and call the forwarder
program...
    mov ax, 3561h      ; get int 61 vector address
    int 21h            ;    ..            -->ES:BX
    mov ax, 1685h      ; fcn 1685: switch VM's and callback
    mov di, bx  ; ES:DI = callback address (int 61 hdlr)
    mov bx,1           ; BX = VM to switch to (system VM)
    mov cx, 3
    xor dx, dx         ; DX:SI = priority boost (zero)
    xor si, si         ;    ..
    int 2Fh            ;switch to system VM and do INT 60
;
exit4:

; Restore host PSP, restore old break setting and IVT
; vectors 1Bh, 23h and 24h.  Restore host stack.

    popa                    ;restore registers.
    pop  ds                 ;    /
    pop  es                 ;    /
    popf                    ;    /
```

---

[1]   Windows provides various extensions to INT-2Fh, as introduced in Chapter 9. Int-2F/AX = 1683h queries the current VM. No parameters are supplied to it, and it returns only one value: the VM ID number in BX. We expect the system VM to be number 1; however, it is possible to confirm this: after Windows has initialized all virtual device drivers, it then calls INT-2F/1608h, to inform the DOS device drivers (or TSRs). Windows supplies the system VM ID number with this call.

```
        mov   cs:isrbusy,0 ;allow reentrance.
        iret                      ;return from interrupt.
;. . . . . . . . . . . . . . . . . . . . . . . . . . . . . . . . . . . . . . . . . . . . . . . . .
```

callback:
```
;this is the forwarder, entered from the signaller in
;another VM, via the int-2F/1685h mechanism...
        push es                   ; save working registers
        pushds                    ;    ..
        pusha                     ;    ..
        sti                       ;essential.
        mov ax, cs                ; set DS == CS
        mov ds, ax                ;    ..
        mov ax, 3560h             ; get current INT 60 vector address
        int 21h                   ;    ..
        mov ax, es                ; is there a WinApp handler?
        or  ax, bx                ;    ..
        jz  done60                ; if not, don't call it!
        int 60h                   ; call WinApp
done60:
        popa                      ; restore registers
        pop ds
        pop es                    ;    ..
        iret                      ; return to other VM.
;. . . . . . . . . . . . . . . . . . . . . . . . . . . . . . . . . . . . . . . . . . . . . . . . .
```

runtime2F:
```
;entered when Windows loads, with AX=1605h, and when
;Windows unloads, with AX=1606h....
;detect when Windows loads, and set a flag so that
;runtime9 will be activated...
        sti                       ;documentation says this req'd.
        cmp   ax,1605h            ;test if Win is loading.¹
        jne   notload
        cmp   cx,0       ;this must always be 0, else error.
        jne   notunload
        mov   cs:winloaded,1
        mov   cs:winmode,dl
notload: cmp   ax,1606h   ;test if Win is unloading.²
        jne   notunload
        mov   cs:winloaded,0
notunload:
        jmp  DWORD PTR cs:oldoffivt2F   ;old int-2F.
;. . . . . . . . . . . . . . . . . . . . . . . . . . . . . . . . . . . . . . . . . . . . . . . . .
```

---

[1]   This is a very interesting extension to INT-2Fh. Function 1605h is called by Windows when it first loads. This enables DOS device drivers and TSRs to perform any necessary initialization. It is important to follow the rules here, by first enabling interrupts and then calling the old INT-2Fh vector. The latter is because other drivers/TSRs may have hooked the vector. CX must have zero. If you for any reason decide that Windows should not go ahead and load, then put a non-zero value into CX and IRET: other drivers have the option of doing this also, which is why we called the old vector. Windows also supplies ES:BX = 0:0 in Standard mode, DS:SI = 0:0; DX bit-0 = 1 if Windows in Standard mode, =0 if Enhanced mode; and DI contains the version number = 030Ah for version 3.1.

[2]   This is the opposite of 1605h, called by Windows when it unloads. Windows supplies DX bit-0 = 1 for Standard mode and = 0 for Enhanced mode.

```
endprog:
int9 ENDS
     END       install
```

The entire program was too much to print, hence the sections in italics. The complete program is on the Companion Disk; however, if you know much about TSRs there is sufficient information in this listing for you to construct it.

Some coding issues

One interesting little point about this code is that we obviously don't want to try jumping VMs if Windows isn't even loaded, and indeed the INT-2Fh extensions are not even available until Windows is loaded. At one stage in the program's development, I did call INT-2F/1685h to switch VMs before Windows had loaded, and before I had put in the "winloaded" checking — and it worked! Or rather the switch didn't take place, so nothing happened. However, on another computer it crashed. The only difference I can see is that the one that worked was running DRDOS version 6, and the one that failed was running MS-DOS version 5.

Outline of DOSTSR operation

At this stage of the game you should be able to follow through the logic of DOSTSR.ASM. Note that INT-60h in the IVT is hooked by the WinApp and is where the Protected mode ISR is located. INT-61h is hooked by DOSTSR itself, merely to pass its own forwarder address to the same DOSTSR in another VM. Paradoxically, there is only the one TSR, and they only appear to be in different VMs — all virtual addresses map back to the same physical addresses. However, the DOSTSR, while executing in another VM, does not necessarily know the segment:offset address of the forwarder in the system VM.

Data contention

Something else you should pay some attention to when developing robust code is the possible contention if more than one VM wants a piece of the action at the same time. That is, programs in two different VMs enter the TSR and work on the same data. Crash! Anticipate this and either design the data to be reentrant or force instantiation by an entry in SYSTEM.INI, or use INT-2Fh/1605h to create instantiation of specific data areas (see *Writing Windows Device Drivers* by Daniel Norton, Addison Wesley, 1992, page 170). Or prevent reentrancy, as I did with my demo program. See earlier notes on the problem of reentrancy on page 323.

DOS-to-Win Device Driver/TSR

This book, so far, has dealt with various issues of how a DOS program (and TSR) can communicate with Windows and Windows programs. Now the picture is to be filled in a little more.

Device drivers for DOS and Windows

One thing that you may have noticed with some hardware products, such as network cards, is that they come with *two* sets of device drivers: one for DOS and one for Windows. Actually, in most cases, the DOS driver *will* work under Windows, but less efficiently than one written specifically for Windows. The reason for this is that Windows has to pass control down to V86 mode, to access the DOS driver, which means time overhead.

To be more correct, there are three different types of device drivers, because the old Standard mode (sorry I'm mentioning it again!) cannot use virtual device drivers. However, I won't worry about Standard mode drivers.

Therefore, a logical question arises: if you install a DOS device driver via the CONFIG.SYS file, that will be fine for DOS, but if you then load Windows, which will load device drivers specified in the SYSTEM.INI file, how do you avoid the two drivers clashing?

DOS TSRs and Windows

Now apply this line of thinking to TSRs. You have a DOS TSR, which, as described earlier, you can make Windows-aware. However, the same problem exists — a TSR is written for Real mode. Therefore, maybe you would like an automatic transition to take place from the TSR to a Windows application.

For both the device driver case and the TSR case, you want a smooth and transparent mechanism for transferring to an equivalent Windows program. Microsoft has catered to this need.

Automatic Loading of Windows Programs/Drivers

TSR2WIN example code

The example code for this section is on the Companion Disk in directory \TSR2WIN. It contains a TSR, called TSR2WIN.ASM, that assembles and links to TSR2WIN.EXE.

What this TSR does is detect when Windows is loading, automatically loads a virtual device driver (VxD), and also loads a Windows application. Note that you could just as easily have loaded the virtual device driver from a DOS device driver: the principle is the same.

Global data

Another bonus of this TSR is that it establishes a global data area and provides a FLAT 32-bit linear address for it that the Windows application and the device driver can access.

The invisible VxD

An interesting aspect to how this TSR works is that neither the VxD nor the WinApp need to be specified in a .INI file.

It is usual to put a "DEVICE= " line inside SYSTEM.INI, to cause a VxD to load, but the TSR will load the VxD without such a line. Furthermore, although Microsoft recommends that all VxDs "should" be in \WINDOWS\SYSTEM directory, the documentation does not say that they "have" to be. Thus, you can put your VxD anywhere.

TSR Installation

TSR2WIN TSR

Essential portions of TSR2WIN are reproduced here, and this is sufficient for you, without having the original source files from the Companion Disk, *if* you are familiar with basic TSRs.

Reproduced below is the portion of the installation code that sets up the data structures required for auto-loading of the VxD and WinApp.

The TSR is "inside" the VxD

A vital point must be brought out now. I chose to put the TSR, TSR2WIN.EXE, "inside" the VxD as a *DOS stub*.

All Windows programs have a DOS stub, which is a DOS program that resides inside the Windows program. Should the user execute the Windows program from the DOS prompt, only the stub will execute. It is usual for the DOS stub to display a simple message that you need Windows to run this program, then it terminates.

Putting the TSR inside a VxD is easy. I have placed a typical .DEF file, used for linking a VxD, in directory \TSR2WIN, and this file is called VDEMOD.DEF. It shows how easy it is to specify the TSR as a stub to the VxD.

It is *not* essential to do this, but it offers a simplification: for the TSR to automatically load the VxD, the TSR needs to know the path of the VxD. If the path is fixed, then you can specify it in the TSR, or maybe you could pass it to the TSR on the DOS command line-tail (when loading the TSR). *Or*, by having the TSR inside the VxD, the TSR need only look at its own path to determine where the VxD is!

It was an arbitrary choice, but I chose to put the WinApp in another directory and have specified the path in the TSR, but I could have also put the WinApp in the same path as the VxD.

Start of TSR2WIN TSR

```
;TSR2WIN.ASM  -->    TSR2WIN.EXE
;this is a windows-aware tsr, that is loaded before
;windows.
;When windows loads, this tsr will automatically cause a
;windows application to start, *and* will automatically
;load a virtual device driver.

;This tsr must *not* be a separate file --
;it is specified as the "dos stub" for the virtual device
;driver (VxD) that is to be automatically loaded.
; -- if your only requirement is to auto-load a windows
;app, then you can have this tsr stand-alone (or as stub
;for the WinApp).

;the windows application is called WINAPP.EXE
;the virtual device driver is called VDEMOD.EXE
; -- WINAPP must be in the root directory (or path spec'd
;below) -- VDEMOD.EXE can be anywhere.
;An interesting aspect of this tsr, is that it creates a
;global data structure, and passes a FLAT 32-bit pointer
;via the IVT -- WINAPP and VDEMOD can access this
; pointer.

;Note that this is a .EXE file, but data is in the code
;segment -- makes it easier to make into a tsr.

.286
.MODEL SMALL
.STACK
.DATA
;..............................................
.CODE
start:   jmp        installhooks

winloaded       DB       0
dpmiloaded      DB       0
winmode         DB       0
oldoffivt2F     Dw       0            ;2F saved vector.
oldsegivt2F     DW       0            ;        /
oldoffivt1C     DW       0            ;1C saved vector.
oldsegivt1C     DW       0            ;        /
oldoffivt9      DW       0
oldsegivt9      Dw       0
oldoffivt28     DW       0
oldsegivt28     DW       0
bypass1C        DB       0            ;fix reentrancy problems.
bypass28        DB       0            ;        /
dosbusyoff      DW       0
dosbusyseg      Dw       0

WIN386_STARTUP_INFO_STRUC       STRUC
   SIS_VERSION                  DB       3,0 ;3,0Ah for Win3.1.
   SIS_NEXT_PTR                 DD       0
   SIS_VIRT_DEV_FILE_PTR        DD       0
```

```
      SIS_REFERENCE_DATA            DD      0
      SIS_INSTANCE_DATA_PTR         DD      0
      SIS_Opt_Instance_Data_Ptr     DD 0 ;extra field Win95 only.
                                 ;i.e., if put 4,0 into first field.
WIN386_STARTUP_INFO_STRUC           ENDS

InstData        Win386_Startup_Info_Struc <>

TSR_Info_Struc  STRUC
      TSR_Next                     dd   ?
      TSR_PSP_Segment              dw   ?
      TSR_API_Ver_ID               dw   100h
      TSR_Exec_Flags               dw   0
      TSR_Exec_Cmd_Show            dw   0
      TSR_Exec_Cmd                 dd   0
      TSR_Reserved                 db   4 dup (0)
      TSR_ID_Block                 dd   0
      TSR_Data_Block               dd   0
TSR_Info_Struc ENDS

tsr_info        TSR_INFO_STRUC  <>

Exec_Path_Name db  "C:\WINAPP.EXE",0,0
                            ;path & filename of windows app.

psp_size        DW    0

My_ID_Block     dw    ?
My_Name         db   'TSR autoload WinApp & VxD',0
My_Name_End     LABEL BYTE

;this ptr must get put into INT-60h....
INCLUDE GLOBL.INC           ;global data, accessed by
                            ; WINAPP/VDEMOD.
globaldata  GLOBALSTRUCTURE <>  ;instanced here only, but
                  ; include file must be in other programs.
```

Data structures   Notice the two data structures above: TSR_INFO_STRUC, and WIN386_STARTUP_INFO_STRUC.

GLOBL.INC is not part of loading the VxD and WinApp: it has to do with global data between all programs.

I have left out most of the "ho hum" installed portion of this TSR and reproduced only the interesting bits; however, you first need to examine the installation code. Therefore, I have turned this listing around and shown the install code immediately below:

```
   DB  17  DUP(0)         ; Resident part is above here!
dumpme:
;****************************************************************
installhooks:
;is this tsr already installed?... i have given it a
;signature of CCh...
   push es                        ;just in case
   mov  ax,0CC00h             ;AL=0 is install-test code for
```

```
                                    ; my 2F handler.
    int  2Fh      ;multiplex interrupt (that we will hook)
    pop  es
    or   al,al                      ;AL=non-0 means abort.
    jz   abba
    jmp  abortload
abba:
;***********************************************************
    push cs
    pop  ds        ;note cs: overrides thus not really reqd.
;get a pointer to the name of the load file in the
;environment seg. entered with es=psp...
    mov  ax,es
    mov  bx,cs
    mov  WORD PTR cs:[TSR_info.TSR_PSP_Segment], ax
    sub  bx, ax                     ; size (in paras) of PSP
    mov  WORD PTR cs:[PSP_Size], bx
    mov  bx,2ch                     ;environment segment
    mov  es,es:[bx]
    xor  di,di
    mov  cx,-1                      ;big number
    xor  al,al                      ;search for a null
    cld
qq:
    repne scasb                     ;get past one null and stop
    cmp  byte ptr es:[di],0   ;another null
    jnz  qq                         ;no.
    add  di,3              ;skip the word before the name.
```

Setting up the VxD data structure   Continuing from above, look now at setting up the data structure for the VxD.

```
; prepare part of the instance data list. Stuff in
; pointer to the file name
; and reference data
    mov  word ptr CS:[instdata.SIS_Version],0A03h
    mov  word ptr CS:[instdata.SIS_Virt_Dev_File_Ptr],di
    mov  word ptr CS:[instdata.SIS_Virt_Dev_File_Ptr][2],es
    mov  word ptr cs:[instdata.SIS_Instance_Data_Ptr],0
    mov  word ptr cs:[instdata.SIS_Instance_Data_Ptr][2],0

;notes: above code searches the environment-block,
;looking for fully-specified path/filename of this file,
;then, inserts this address (es:di) into instdata.
```

Path of the VxD   A point of clarification about the above code is needed. A data structure has been filled in that Windows requires for loading the device driver. We need to provide its path, so the code looks into this TSRs PSP, where the path/filename is kept (we can reuse the filename for the VxD, since the TSR is embedded in the VxD).

Segment structure of TSR This code should be easy to read, but do note that this TSR is .EXE format, which means that the PSP is a separate segment from the code segment. I haven't used the data segment. For the SMALL model, the code segment and PSP get loaded into memory contiguously; that is, the code immediately follows the PSP.

That is why, to get the size of the PSP, I merely subtracted ES from CS (as ES initially points to the PSP segment).

```
;next problem, is we need to force WINAPP, our windows
;application, to load...

;Initialise length of ID string...
    mov   WORD PTR cs:[My_ID_Block],OFFSET My_Name_End - \
                                    OFFSET My_Name
```

Setting up the WinApp data structure The following code, which is another data structure required for launching the Windows application, continues from above

---

```
    mov   WORD PTR cs:[TSR_Info.TSR_Exec_Cmd], \
                                    OFFSET Exec_Path_Name
    mov   WORD PTR cs:[TSR_Info.TSR_Exec_Cmd+2], cs
    mov   WORD PTR cs:[TSR_Info.TSR_Exec_Flags], 1
                                    ;=TSR_WINEXEC
    mov   WORD PTR cs:[TSR_Info.TSR_Exec_Cmd_Show], 4
                                    ;=SW_SHOWNOACTIVATE
    mov   WORD PTR cs:[TSR_Info.TSR_ID_Block], \
                                    OFFSET My_ID_Block
    mov   WORD PTR cs:[TSR_Info.TSR_ID_Block+2], cs
    mov   WORD PTR cs:[TSR_Info.TSR_Data_Block], 0
    mov   WORD PTR cs:[TSR_Info.TSR_Data_Block+2], 0
```

Path of the WinApp As mentioned earlier, my choice to put WINAPP.EXE into a different directory is done here purely to illustrate something different. In practise, you would most likely have it in the same directory as the VxD. However, that option would introduce another small complication: you would need to extract the path from the environment segment and append the filename WINAPP.EXE (or whatever).

To see more detail on a couple of those fields, refer to Appendix D.

Hooking the vectors follows:

```
;get the addr of the dos-busy flag...
    mov   ah,34h
    int   21h                       ;-->es:bx
    mov   dosbusyoff,bx
    mov   dosbusyseg,es
```

```
    mov   ax,352Fh                    ;get int-2F vector in ivt.
    int   21h                         ;            /
    mov   oldoffivt2F,bx              ;save it
    mov   oldsegivt2F,es              ;        /
    mov   ax,252Fh                    ;hook int-2F
    lea   dx,runtime2F                ;set ivt vector. ds:dx
    int   21h                         ;            /

; ... hook any other vectors required ...
```

FLAT pointer   This is the nice final touch.
to global data

```
;finally, pass the address of our global data...
    lea   dx,globaldata
    mov   ax,ds
    shl   ax,4                        ;convert para. to offset.
    jc    over64k
    add   dx,ax             ;get FLAT linear 32-bit address
    xor   ax,ax                       ;            /
    mov   ds,ax                       ;            / -->ds:dx
bb2:
    mov   ax,2560h                    ;hook int-60h
    int   21h
    push  cs
    pop   ds                          ;restore ds
    jmp   SHORT bb3
;i.e., real address is segment=0, offset=dx (works only
;if in 1st 64K).
;WINAPP can check the hi 2 bytes of int-60h ivt, to
;confirm that they
;are zero, and that no other program has overwritten.
;no, be careful...take care of over 64K...
over64k:
    add   dx,ax   ;as above (not likely to produce carry)
    mov   ax,1                 ;this is the carry.
    mov   ds,ax
    jmp   bb2
bb3:

;terminate, leave resident....
    lea   dx,dumpme     ;point past all code in this module.
    shr   dx,4                 ;compute # paragraphs to keep.
    add   dx,psp_size          ;            /
    mov   ax,3100h             ;terminate and stay resident.
    int   21h                  ;            /

abortload:
    call  longbeep
    mov   ax,4C00h             ;don't make resident.
    int   21h
    END   start
```

Universal global pointer

The hooking of IVT vectors in the above code is very ordinary, but you will find the creation of the global data and FLAT linear address to be interesting.

A slightly negative point about this global data is that it is in the first 1M, i.e., conventional memory.

The FLAT address is simple to calculate, because it is just an addition:

DS*16 + OFFSET globaldata

It is actually only 16 or 17 bits in size, as this TSR is close to the start of linear memory, so the higher bits are zero. I stuck this pointer into the IVT, at entry-60h, which is a convenient place from which the VxD and WinApp can retrieve it. **IVT-60h is thus not in conventional segment:offset form!**

How a WinApp & VxD can access global data

This FLAT pointer is immediately usable by the VxD. All that the VxD needs to do is get it out of the IVT.

Ordinary 16-bit or 32-bit Windows applications can also use the FLAT pointer, but they do need to obtain a ring-3 FLAT data selector. The global data can then be accessed in this manner:

```
mov es,flatr3selector    ;FLAT data selector, ring-3.
mov ebx,lpglobaldata ;FLAT linear pointer, from ivt-60h.
mov es:[ebx].GLOBAL,1 ;accessing a field of global struc
```

The above code sample is *not* from the TSR — it shows how a WinApp can access the global data, where GLOBAL is a field of the global structure.

There are various ways to obtain a FLAT data selector. One thing that you could do is make an alias of DS and then modify the base address in the descriptor (if you can locate the LDT! — see Chapter 12).

Another way is for the WinApp to go to ring 0 and call a VMM service to create a FLAT ring-3 data selector — the service to call is GETAPPFLATDSALIAS(), described in the DDK. If you have a VxD as part of your system, you can get it to call this function and pass the slector back to the WinApp.

A great advantage to using GETAPPFLATDSALIAS() is that the returned selector is in the GDT and will thus continue to work across VMs. This is not such an issue with Windows 3.x and 95, because all WinApps run under the same LDT, but beware Windows NT.

The word of warning here is that if you want your selectors to be global across 32-bit Windows NT applications, which will run with private LDTs, then put your selector into the GDT

Now for the part that actually loads the VxD and WinApp.

TSR Resident Code

INT-2Fh
handler

The resident code monitors IVT-2Fh and detects when Windows is loading, as follows:

```
runtime2F:
;entered when Windows loads, with AX=1605h, and when
;Windows unloads, with AX=1606h....
;detect when Windows loads, and set a flag ...
  sti                      ;documentation says this req'd.

  cmp  ax,1608h            ;Enhanced mode loaded.
  jne  nexttry
  mov  cs:winloaded,1
  jmp  SHORT go2F

nexttry:
  cmp  ax,1605h            ;test if Win is loading
  jne  notload
  cmp  cx,0       ;this must always be 0, else error.
  jne  goerror2F
  mov  cs:winmode,dl       ;bit-0=0 if Enhanced mode.
  test dl,1                ;test bit-0
  jnz  standardload

;...............................................
;inserts our vxd into vxd chain (see my book, Appendix D)
  mov  word ptr cs:[instdata.SIS_Next_Ptr],bx
  mov  word ptr cs:[instdata.SIS_Next_Ptr][2],es
  push cs
  pop  es         ;chain, with es:bx ptg to our instdata
  lea  bx,InstData  ;structure (our VxD data structure).
;...............
standardload:
  jmp  SHORT go2F
;..........
notload:
  cmp  ax,1606h            ;test if Win is unloading.
  jne  notunload
  mov  cs:winloaded,0
  mov  cs:dpmiloaded,0
  jmp  SHORT go2F
;..........
notunload:
  cmp  ax,160Bh ;used for tsr registration with windows.
  jne  giveitanothergo
  jmp  dorego
;.....
giveitanothergo:
  cmp  ax,1687h
  je   go2F             ;otherwise will get in endless loop!
  cmp  cs:dpmiloaded,0
  jne  go2F                ;for all other cases, exit.
;.....
```

```
;after Windows has loaded, if we want to hook the IDT,
;need to test if ok...
    cmp  cs:winloaded,0
    je   go2F
;a problem exits... what if come here before the idt
;properly setup?...
    pusha
    mov  ax,1687h                 ;test for dpmi host
    int  2Fh
    cmp  ax,0                     ;ax=0 if dpmi host present.
    jne  exit5
    mov  cs:dpmiloaded,1
exit5:  popa
;.....
goerror2F:
go2f:
    jmp DWORD PTR cs:oldoffivt2F
;................................................
dorego:
; return a pointer to the TSR structure...
    mov  WORD PTR cs:[TSR_Info.TSR_Next], di
    mov  WORD PTR cs:[TSR_Info.TSR_Next+2], es
    push cs
    pop  es
    mov  di, OFFSET TSR_Info      ;this chains the tsr data
    jmp  go2f                     ;structures.
;................................................
```

Reference sources
The TSR data structure is particularly interesting, because it allows you to pass various information about your TSR to Windows (see Appendix D). Also, for Microsoft Developer Network (MSDN) members, the January 1997 set of CD-ROMs has further information on this topic in the *Archive Library* CD-ROM.

You can control whether your TSR is instantiated in each VM or not. The default is not, which means that every VM maps back to the same TSR. In most cases, this is highly desirable, including our case of a global data area in the TSR.

Getting it Together

Testing the programs
\TSR2WIN directory on your version of the Companion Disk may contain the executables, in which case you can run them immediately. You will need to place WINAPP.EXE into C:\, i.e., the root directory of the C: drive. VDEMOD.EXE can be anywhere. From the DOS prompt, not a DOS box inside Windows, go to the directory that has VDEMOD.EXE and run it by typing "VDEMOD". Then type "WIN" to load Windows.

The loudspeaker should start a continuous tone, letting you know that the VxD has loaded. Then, the WinApp will load. When I first did this on Windows 3.1, up came the Program manager, but not my WinApp — where was it? This is interesting — WINAPP.EXE loads before Program Manager, so it is hidden underneath.

This is not a problem with Windows 95, and you should see the WinApp appear on the desktop. Mind you, it's a stupid WinApp, as you'll see.

Assembling and linking

To assemble and link everything, type this:

```
C:> nmake /A winapp.mak
C:> copy winapp.exe c:\
C:> nmake /A tsr2win.mak
C:> nmake /A vdemod.mak
```

You must assemble and link TSR2WIN before VDEMOD.

VxD-Lite

To assemble and link VDEMOD, the VxD, you need special tools and Include files from the Device Development Kit (DDK), which normally means that you have to join the MSDN (which means considerable expenditure!). There is, however, *VxD-Lite*, which Microsoft did provide free a couple of years ago. It is now withdrawn and I was unable to obtain permission to place it on the Companion Disk. Although VxD-Lite targeted Windows 3.1 systems, the VxDs will run fine in Windows 95 — a VxD is a VxD.

You can find VxD-Lite bundled with *Writing Windows Virtual Device Drivers* by David Thielen and Bryan Woodruff, Addison-Wesley, 1994. Although this book is still in print, the publicity does not mention VxD-Lite — as one of the authors works for Microsoft, I fear the worst. Locate an older copy somewhere!

15

Advanced Systems Programming

Preamble

Other chapters

Chapter 12 shows how a 16-bit Windows application (WinApp) can access 32-bit ring-0 code. Chapter 13 shows how to construct a "pure" 32-bit ring-3 WinApp. Chapter 14 shows how DOS applications (DOSApps), Winapps, and Virtual Device Drivers (VxDs) can communicate. In Chapter 14, the communication is established by the DOSApp while Windows loads.

Much of this book has described how BIOS and DOS interrupts, plus the interrupt extensions (31h and 2Fh in particular) provided by Windows, can be used. However, this is all from the viewpoint of a 16-bit DOSApp or WinApp.

Execution of an interrupt causes a processor exception, and the Windows handler is in most cases in a 16-bit segment. Therefore, a 32-bit WinApp cannot use the interrupt services, even though they are there and likely to stay there for future versions of Windows.

Just as Chapter 12 shows an application going from a 16-bit segment to a 32-bit segment (called *thunking*), it is necessary for a native 32-bit WinApp to thunk down to 16-bits to use the interrupts! Thunking is introduced in Chapter 8.

32-bit WinApp access to ring 0

Actually, Chapter 12 describes how a 16-bit application can go to 32-bit ring 0. It is interesting that Matt Pietrek, in *Windows 95 Systems Programming Secrets* (IDG Books, 1995), describes a 32-bit WinApp that goes to 32-bit ring 0, but he does it by thunking down to 16 bits and using the same techniques as Chapter 12. It is amusing to see that Matt often has to resort to assembler, due to the awkwardness of C

Thinking further about Chapter 14

Chapter 14 showed how a DOSApp can get its hooks into a 16-bit WinApp and a VxD, but this could also apply to a 32-bit WinApp. It is quite feasible for a 16-bit WinApp to load, do all the low-level dirty work, such as getting selectors to forbidden areas, then call WINEXEC() to load a 32-bit WinApp. It is possible to pass parameters to a 32-bit WinApp via the command line and really easy to insert them into the argument passed to WINEXEC().

You could have a 16-bit WinApp and a 32-bit WinApp, there is no need for a DOS TSR, and run the 16-bit WinApp first.

Demo code for 16- and 32-bit WinApp communication

I don't have the room to put a detailed description into this chapter, but there is demonstration code on the Companion Disk, in directory \SHAREALL, that shows how a 16-bit WinApp and a 32-bit WinApp can communicate and share data.

This chapter

This chapter is a mixed bag. I have likened the learning process to climbing a ladder, as illustrated in Figure 15.1.

This book is intended to satisfy all the alternative needs of the pondering man, sitting with hand to mouth. It is the nuts and bolts, not the latest high-level techniques such as programming using visual OO components in Borland's C++Builder. Move onto those tools if you want, and you may well do, to produce major applications. You may also move deep into VxDs, and again, I've led you to the point where you can jump in — to the "brink" so to speak.

I also make no apologies for focusing strongly on 16-bit programming and the software interrupt services, as the needs of our pensive man in Figure 15.1 will remain valid for many years to come.

What I do need to do in this last chapter is fill in a few gaps and make some suggestions. After that, go where you will ...

Figure 15.1: The ladder of learning.

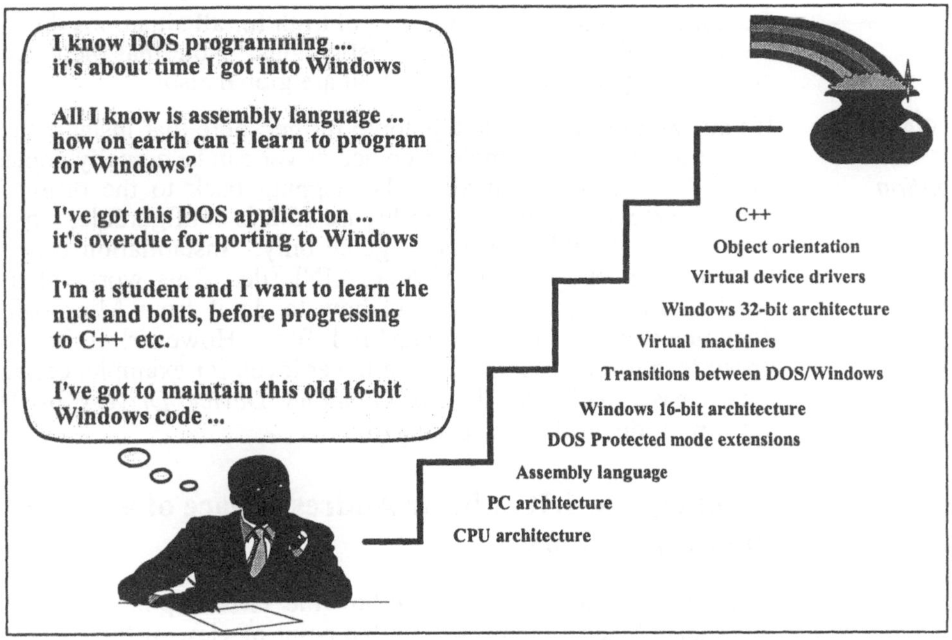

I know DOS programming ...
it's about time I got into Windows

All I know is assembly language ...
how on earth can I learn to program
for Windows?

I've got this DOS application ...
it's overdue for porting to Windows

I'm a student and I want to learn the
nuts and bolts, before progressing
to C++ etc.

I've got to maintain this old 16-bit
Windows code ...

C++
Object orientation
Virtual device drivers
Windows 32-bit architecture
Virtual machines
Transitions between DOS/Windows
Windows 16-bit architecture
DOS Protected mode extensions
Assembly language
PC architecture
CPU architecture

Mapping Across VMs

The question arises: when a new VM is created, just what is copied and what is mapped back to the original? Figure 15.2 clarifies this.

You will observe some interesting features of this mapping.

For a start, consider the TSR loaded before Windows. Because it is a one and only instance, any data in it is global across all VMs. Think about this — it may be good from the point of sharing data, but there are no safeguards. If one program accesses the TSR and causes its data to be modified, perhaps that will clash with another program's use of the TSR.

Commercial TSRs may not be designed with this in mind, so it seems wise to minimize TSR requirements prior to loading Windows.

Global mapping   Notice something else: according to my experiments, it appears that the entire *high memory*, that is the segment starting at FFFF:010h, is mapped back to the original and thus shared across all VMs.

This experimenting was done on a particular system configuration, and I cannot guarantee it to be the same on other systems.

Do not take Figure 12.1 as the gospel truth for all situations, but as a starting point for your own investigations. Note that there may be some "upper memory blocks" that are global also.

Forced instan- tiation

If you want to investigate this topic further, note that instantiation can be forced. For example, a device driver can be actually copied to all VMs, rather than all VMs mapping back to the original. Thus the memory illustrated in Figure 15.2 is configurable, which is why you should take it as a guide only. Instantiation can be forced by an entry in the SYSTEM.INI file. This works at the level of files, and you can find how to do it from Microsoft's documentation on the SYSTEM.INI file. However, it is also possible to force instantiation at a lower level, for example, certain data areas. Refer to *Writing Windows Device Drivers* by D. Norton, Addison Wesley, page 170.

Mapping the 4.3G Linear Address Space of a VM to Physical Memory

I have introduced the question of how the VMs map between each other and physical memory from the point of view of the first 1M region, but what of the entire 4.3G of linear memory?

Figure 15.2 shows the mapping below 1M, that is common between VMs. However, each VM, including the system VM, can have its own Protected mode so each can address a linear address space of 4.3G. To give a complete picture, I have elaborated below on the question of common mapping of physical memory.

Linear address

In each VM, if the CPU is in Protected mode (not V86 mode), the selector:offset address is translated via descriptor tables to a linear address. This is described in Chapter 1.

In the case of 286-based Windows Standard mode, the linear address is also the physical address, and there are no virtual machines (though there **can** be). The linear address in this case is 24 bits and can address 2^{24} physical locations.

Windows Enhanced 386-based mode uses the extra step of paging, thus the linear address no longer corresponds with the physical address. However, this 32-bit linear address gives the 4.3G linear address space we are talking about. Paging will map it into a much smaller physical memory space.

System VM & physical memory

First, a note on the first 1M of a VM. The system VM's V86 portion, being the first created, is mostly below the physical 1M. The mappings of Figure 15.2 on page 343 apply, but the virtual addresses, such as the "free" RAM within the 640K, map directly to the actual physical conventional memory. The "free" virtual conventional RAM of other VMs will be physically in extended memory.

Figure 15.2: Global memory below 1M.

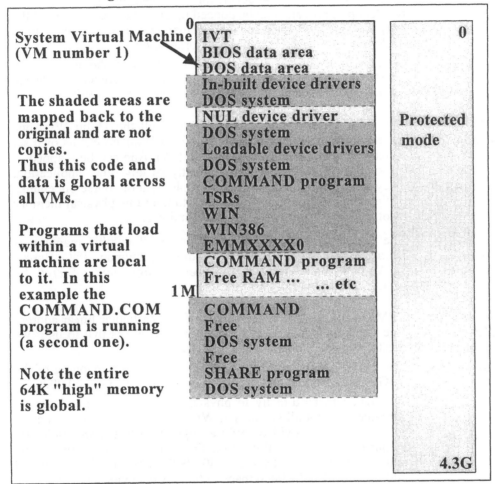

Considering the Protected mode of the system VM, Windows runs the WinApps at the linear address ranges:

0000 0000 to   000F FFFF
8050 0000 to   80FF FFFF

The first region is below 1M, wherever Windows can find some free RAM. The second region may seem like an amazingly high memory range, but remember that these are linear addresses, and paging will map them into the available physical memory.

Reference sources

You may well wonder where I got these addresses from — the source is the *Soft-ICE/W User's Manual*. The address ranges quoted here are what Windows 3.x currently uses, or so I'm led to believe, but they are not guaranteed. Incidentally, Soft-ICE/W is a specialised Windows debugger, for Enhanced mode only, that is a resident program and can be popped up at any time. It is about the only tool available for debugging virtual device drivers and similar tricky code. It is sold by NuMega Corporation.

Specific information on Windows 95 addressing should be in the latest Soft-ICE/W manual, and another reference is *Inside Windows 95* by Adrian King, Microsoft Press, USA, 1994.

32-bit WinApps

The system VM has just one LDT, and all the 16-bit WinApps have one set of pages tables. Each 32-bit WinApp has its own set of page tables. Therefore, each 32-bit WinApp can be mapped to physical memory totally independently of any other application. They sit in linear address range 2G to 4G, but of course big chunks of the linear address range map back to the same physical memory as other WinApps, DOSApps, and Windows.

Reference books

The best places to look for extreme detail on this mapping is *Unauthorized Windows 95* by Andrew Schulman, IDG Books, USA, 1994, and *Windows 95 Systems Programming Secrets* by Matt Pietrek, IDG Books, 1995.

Accessing physical memory from a VM

DPMI allows us to obtain a selector for a particular linear address (see functions 0000h and 0007h), but what use is that to us if we don't know what the linear address represents?

One extremely interesting aspect of Windows 3.x mapping of the 4.3G virtual space is the linear address starting from 0040 0000.

This range maps directly to physical memory. Again, I cannot guarantee this for all versions of Windows. Thus, if you wanted to access the physical video buffer at segment:offset A000:0000, you would convert it to a full 32-bit linear address, 000A 0000, and add it to 0040 0000. That is, 0040 0000 maps directly to physical address 0, and 004A 0000 maps to the physical video-RAM.

32-bit WinApp selectors

Chapter 12 shows that the VxDs use 32-bit ring-0 selectors 28h and 30h that are in the GDT. These are FLAT selectors, having a base address in the descriptors of zero. Thus the linear address range from 0 to 4.3G is addressable.

Ring-3 32-bit WinApps use a code segment selector of 13Fh and data segment selector of — well, I don't want to gurantee that it will always be the same. You can look at the segments using a debugger.

The same selector value is loaded into DS, ES, FS, and GS, and it is extremely interesting that the type of selector has the *expand down limit* of 1M, that is, must be greater than 1M. This prevents data accesses into the first 1M of linear memory, that maps in the DOS VM.

... and a DPMI service

The above description of linear addresses that map to physical are not the only method for getting at physical memory from Protected mode.

There is a DPMI function that performs mapping between a linear address and a physical address: function **0800**h[1] (Physical Address Mapping). You supply it with a (32-bit) physical address and it will give you a (32-bit) linear address. You could then use function **0007**h (Set Segment Base Address) to put the linear address into a descriptor. Of course, the descriptor would have to have been previously created, for example, by function **0000**h. Curiously, function 0800h is not recommended for addressing below physical 1M, I presume because there are other DPMI functions for that purpose.

Mapping between system VM and DOS VMs

Windows does set aside other parts of the linear address space for special purposes. For example, the DOS VMs are located at linear address range:

8100 0000 to FFFF FFFF

If you've been following this with an attentive mind, you may see a problem here — won't each DOS VM be in its own completely isolated virtual address space? Yes, but this is an example of where Windows maps different virtual addresses to the same physical place. Thus, from the system VM you can use the above address range to access the DOS VMs, just as though they exist within the system VM.

Another reserved area in the VM is the range:

8000 0000 to 803F FFFF

where the virtual device drivers are kept.

We lump a DOS V86 VM and its Protected mode together as one VM, so each VM has its own attached Protected mode and hence

---

[1] Intel's DPMI specification has various warnings about function 0800h:

This function should only be used by clients that absolutely require direct access to a memory mapped device at physical addresses above 1M. Clients should not use this function to access memory below the 1M boundary.

Programs and device drivers that need to perform DMA I/O to physical addresses in a virtualised hardware environment should use the Virtual DMA Services.

its own 4.3G space.   However, there needs to be a further clarification — what about overlapping of the addressing while in Protected mode or in V86 mode?   Yes, the two do overlap.   V86 addresses linear address range:

> 0000 0000 to   000F FFFF

But Protected mode in the same VM, when addressing this same range of linear addresses, also maps to the same physical addresses.

This may be a somewhat esoteric point and so may not be something to be concerned about.   However, I have thrown in this clarification in case the conceptual conflict has arisen in your mind.   In fact, you can also access that same physical address range at Protected mode linear addresses "somewhere" above 8100 0000 (mentioned above).

The Windows/DOS/DPMI Relationship

This section develops further the relationship between DOS and windows.   What are the extensions to DOS provided by Windows?   Just what is the relationship between the Windows kernel and the DPMI host?

Windows Extensions to INT-2Fh

I would like to complete the coverage of the Windows extensions to INT-2Fh, by discussing some more that can be very useful for DOS drivers and TSRs.   The others are covered at various points through the book, namely Chapters 9, 10, and 11, and can be located via the index.   I introduced INT-2Fh in Chapter 8 and have provided a detailed summary of all INT-2Fh functions in Appendix D.

- **INT-2F, AX = 1600h**
  This function queries the version number of Enhanced mode Windows.   It returns AL, and if the value is less than 3 or greater than 127 then Enhanced mode isn't running.   Otherwise AL = 3 and AH = 0Ah for version 3.10.

- **INT-2F, AX = 1680h**
  Function 1680h yields the current VM's time slice.   You would call this if your current VM isn't doing anything and you want to try and improve efficiency by releasing it.

- **INT-2F, AX = 1681h, 1682h**
  Function 1681h tells Windows not to switch VMs, whereas 1682h says go ahead. If your code is going into a critical section and must not be interrupted, then use these two. Note however that hardware interrupts still do happen.
- **INT-2F, AX = 1684h**
  Function 1684h gets an address for calling a VDD and getting information from it. The VDD ID is passed via BX and the function returns the Real mode address in ES:DI. If it returns NULL, the VDD does not provide Real mode services.

The Windows/INT-2Fh Relationship

To get a deeper feel for INT-2Fh and its relationship with Windows, I have placed a program on the Companion Disk that monitors all INT-2Fh Real mode calls via the IVT and logs them to either the printer or screen. The traffic over this interface is fascinating. INT-2Fh is Windows' main method for letting DOS device drivers know what is going on, and vice versa.

Windows can let drivers know when and how it is loading and unloading, and DOS drivers can tell Windows the address of their Protected mode code. Also, virtual drivers communicate with DOS drivers over this highway.

What I suggest is, rather than me printing out the results of my INT-2Fh monitor program, try it for yourself. It is a TSR that you load from the DOS prompt, and you will be able to see exactly the traffic on INT-2Fh at all times.

What I found most fascinating is that once Windows has loaded, the traffic is continuous. The implication here is that you have a mechanism for Windows to continuously "wake up" a TSR.

Writing Windows-Aware DOSApps

Issues involved

If you are writing a DOS application that is going to make use of extended memory, there are plenty of tools available, and some compilers automatically take care of this for you. Thus the old 1M limit is history. The extended memory that your program will use is still within the same VM as the V86 VM. Switching into Protected mode from Real mode, to run code in Protected mode and hence above 1M, in no way conflicts with the WinApps, since they are in the system VM. Also, your VM has its own LDT and IVT, so you can hook vectors to your heart's content.

There is a lot to be said for writing DOS programs that are designed to run under Windows, and quite likely this area of development will remain alive.

INT-2Fh versus INT-31h

You should remember that the DPMI INT-31h functions are only available when the CPU is in Protected mode, not while it is in V86 mode. However, the INT-2Fh extensions are available in V86 (Enhanced) Real mode and Standard Real mode. Also don't forget that V86 and Protected mode overlap, so you can readily address all of the first 1M of the VM from Protected mode (though to execute Real mode code you must perform the necessary DPMI function to transfer the CPU to Real mode, and vice versa).

Sharing memory between 16- and 32-bit WinApps

All 16- and 32-bit Winapps reside in the System VM and use the same LDT. Also, the pages tables map the linear addresses below 1M to the same physical memory. These two facts hold for Windows 3.x and 95, which means that a memory block below 1M is global.

Windows 95 32-bit applications have mechanisms for sharing memory, such as memory-mapped files (see Chapter 8), but memory global to all 16- and 32-bit WinApps requires a less unofficial solution.

A solution based upon the above two facts is on the Companion Disk in directory \SHAREALL.

A 16-bit WinApp establishes the global area, which can be freely accessed by other WinApps and can also be shared between 32-bit WinApps. Note however, that the WinApps must supply their own protocols for mutual exclusion.

GOOFEE.EXE

I have used this technique to export a data selector to the data segment of GOOFEE.EXE, my CASE tool bundled with another book written by me (see page 359). This allows other WinApps to directly control GOOFEE.EXE. There are protocols to govern the communication.

A description of GOOFEE.EXE and the mechanism for external control is to be found at:

`http://www.goofee.com/`

Anatomy of a VxD

Changing the subject somewhat, VxDs are Microsoft's preferred direction for access to the hardware. Chapter 14 introduces the VxD, and gives references. I recommend that if you want to delve deeply into VxDs, you purchase a specialised book. However, it is appropriate for me to explain a little more about how the example VxD in directory \TSR2WIN on the Companion Disk works.

The VxD is VDEMOD.ASM, and it assembles and links to VDEMOD.EXE. Note the extension .EXE, rather than .386 (that most VxDs use). This relates back to how the VxD is used in Chapter 14; that is, it is invoked from the DOS prompt prior to Windows loading, which executes the stub TSR2WIN.EXE. Note that TSR2WIN.EXE was inserted into VDEMOD.EXE by the linker, so it is *not* a separate program. Note that I have created a Make file to rebuild everything, called BUILDALL.MAK. This is designed for NMAKE.EXE and uses the /A switch.

For now, however, I am interested only in the VxD.

VxD tools   To assemble and link the example VxD requires the DDK or VxD-Lite. More specifically, the following files are required:

DEBUG.INC,   SHELL.INC,   VMM.INC,   LINK386.EXE, LINK4.EXE, MAPSYM32.EXE, MASM5.EXE, ADDHDR.EXE

These are not the standard LINK and MASM version 5 — they are special versions.

Make file   I'll start by looking at the Make file:

```
# VDEMOD.MAK
# This Make file build VDEMOD.ASM into VDEMOD.EXE, and puts
# in the dos stub TSR2WIN.EXE....
# Note that VDEMOD.EXE does not have to be in the windows
# SYSTEM directory.
# The dos stub is to be executed from the autoexec.bat file.

# Note that the Include files are in the path shown below...
# change if necessary.  masm5.exe, link386.exe, addhdr.exe,
# mapsym32.exe must
# all be in the search path ... or put in current directory.
# to run:  NMAKE /A VDEMOD.MAK
# (there is also a BUILDALL.MAK)

# I have put a path of c:\vxd for the .INC files, but
# replace as necessary.

# comment this definition out with a "#", if building a
# non-debugging version
# Debug=-DDEBUG
```

```
all : VDEMOD.exe

.asm.obj:
        masm5 -p -w2 -Mx $(Debug) -Ic:\vxd $*;

.asm.lst:
        masm5 -l -p -w2 -Mx $(Debug) -Ic:\vxd $*;

VDEMOD.obj : VDEMOD.asm c:\vxd\debug.inc c:\vxd\vmm.inc

OBJS = VDEMOD.obj

VDEMOD.exe: VDEMOD.def $(OBJS)
  link386   @VDEMOD.lnk
  addhdr    VDEMOD.exe
  mapsym32  VDEMOD
```

VxD tools   Ok, now for the .DEF file:

```
LIBRARY       VDEMOD
DESCRIPTION   'Barry Kauler VxD for Microsoft Windows'
STUB          'TSR2WIN.EXE'
EXETYPE       DEV386

SEGMENTS
            _LTEXT PRELOAD NONDISCARDABLE
            _LDATA PRELOAD NONDISCARDABLE
            _ITEXT CLASS 'ICODE' DISCARDABLE
            _IDATA CLASS 'ICODE' DISCARDABLE
            _TEXT  CLASS 'PCODE' NONDISCARDABLE
            _DATA  CLASS 'PCODE' NONDISCARDABLE

EXPORTS
  VDEMOD_DDB @1
```

The above two files are the standard red tape, adaptable to other VxDs. However, one special difference is the TSR2WIN.EXE stub. For other VxDs you can replace this with a dummy do-nothing stub, and also you may prefer to generate a VxD with a .386 extension.

VxD listing starts here   The listing of VDEMOD.ASM follows, broken up with comments:

```
TITLE VDEMOD -
;example skeleton VxD, adapted from a skeleton in
;Microsoft's DDK.
;It is possible to monitor any I/O port, and allow or
;disallow it's use.  Install_IO_Handler achieves this.

.386p

.XLIST
  INCLUDE VMM.Inc    ;supplied with DDK (or VxD-Lite)
```

```
   INCLUDE Debug.Inc ;/
   INCLUDE Shell.inc ;/
.LIST
;..................................................
Declare_Virtual_Device VDEMOD, 3, 0, VDEMOD_Control, \
     Undefined_Device_ID, VMM_INIT_ORDER,,
;..................................................
```

Parameters of Device Descriptor Block

Declare_Virtual_Device is a macro that generates a data structure with parameters as listed in the code above. The first parameter is the name of the VxD and the macro creates VDEMOD_DDB, which marks the beginning of the data structure — note that it is exported in the .DEF file.

The next two parameters are major and minor version numbers.

VDEMOD_Control is a routine called by Windows to notify the VxD of system events. This skeleton only handles initialisation and destruction — look below to see how VDEMOD_Control handles these cases.

A VxD developer is supposed to obtain a unique ID number from Microsoft. I don't know why Microsoft doesn't assign a number with each DDK, but I guess that's too simple. Undefined_Device_ID is an equate to a special number for VxDs that don't have an ID.

I could have put Undefined_Init_Order equate for the last parameter, but I wanted to make sure the printer ports got hooked by my VxD before any other. The lower the number, the earlier the initialisation, and VMM_Init_Order equates to zero, so it gets in first.

DOSApp and WinApp access to the VxD

Notice that there are two commas after the last parameter. This is because there can be two more parameters. These can specify routines for V86 and Protected mode access.

Sixteen-bit applications can call INT-2Fh, function 1684h to get access to these routines.

However, INT-2Fh assumes a 16-bit program is running, so 32-bit applications cannot access these functions. It's awful, I know, but such is the legacy of backwards compatibility. Therefore Microsoft was forced to provide a different mechanism for calling services in a VxD from a 32-bit WinApp — see the DeviceIoControl() function introduced in Chapter 8.

The alternative is that a 32-bit application can thunk down to 16 bits. Also look at the technique described on page 350.

```
;local data...

VxD_LOCKED_DATA_SEG
Device_Name   DB      "VDEMOD  ",0
```

```
VDEMOD_Owner  DD      ?
VxD_LOCKED_DATA_ENDS

;............................................................
;initilisation code...

VxD_ICODE_SEG

BeginProc VDEMOD_Device_Init

beep:
  push    eax
  mov     al,0B6h            ;turn on loudspeaker
  out     43h,al
  mov     bx,05C5h
  mov     al,bl
  out     42h,al
  mov     al,bh
  out     42h,al
  in      al,61h
  or      al,3
  out     61h,al
  pop     eax

;firstly, let's hook anything (port 200h) and show a
;generalised handler...
  mov edx, 200h
  mov esi, OFFSET32 My_VDEMOD_Hook
  VMMCall Install_IO_Handler

;now, let's prevent printing...
  mov     edx, 3BDh                      ;lpt status
  mov     esi, OFFSET32 My_VDEMOD_lpt_Hook
  VMMCall Install_IO_Handler

  mov     edx, 379h                      ;lpt status
  mov esi, OFFSET32 My_VDEMOD_lpt_Hook
  VMMCall Install_IO_Handler

  mov     edx, 279h                      ;lpt status
  mov esi, OFFSET32 My_VDEMOD_lpt_Hook
  VMMCall Install_IO_Handler
;.....
  mov     edx, 3BEh                      ;lpt control
  mov esi, OFFSET32 ctrl_VDEMOD_lpt_Hook
  VMMCall Install_IO_Handler

  mov     edx, 37Ah                      ;lpt control
  mov esi, OFFSET32 ctrl_VDEMOD_lpt_Hook
  VMMCall Install_IO_Handler

  mov     edx, 27Ah                      ;lpt control
  mov esi, OFFSET32 ctrl_VDEMOD_lpt_Hook
  VMMCall Install_IO_Handler

  xor     eax, eax
  mov VDEMOD_Owner, eax         ; no current owner
```

```
IFDEF DEBUG
  Trace_Out "VDEMOD installed"
ENDIF

        clc
        ret

EndProc VDEMOD_Device_Init

VxD_ICODE_ENDS
```

In the above code I've used a VMMCALL macro rather than inserting INT-20h directly. Install_IO_Handler is one of those low-level VMM services. This hooks the actual physical port, so if the requirement is to block printing, this will do it.

VDEMOD_Device_Init is the initialisation routine, and VDEMOD_Control tells Windows where it is (see below).

I have started the beeper (above), and there is no restriction to performing direct I/O — we are in ring 0 and IOPL = 0, remember!

```
;main code segment...

VxD_CODE_SEG

BeginProc VDEMOD_Destroy_VM

   cmp   ebx, VDEMOD_Owner         ; Destroying device owner?
   jnz   short VDM_Exit

   xor   eax, eax
   mov   VDEMOD_Owner, eax         ; no current owner

VDM_Exit:
   clc
   ret
EndProc VDEMOD_Destroy_VM

VxD_CODE_ENDS

;....................................................
;locked code... the callbacks...

VxD_LOCKED_CODE_SEG

BeginProc VDEMOD_Control

   Control_Dispatch Device_Init, VDEMOD_Device_Init
   Control_Dispatch Destroy_VM, VDEMOD_Destroy_VM
   clc
   ret

EndProc VDEMOD_Control
```

```
VxD_LOCKED_CODE_ENDS

;............................................
;the hooked ports get redirected here...

VxD_CODE_SEG

BeginProc My_VDEMOD_Hook

;firstly, resolve contention...
    pushad                              ; save regs
    mov   eax, VDEMOD_Owner          ; get previous owner
    cmp   eax, ebx                   ; same as current owner?
    jz    short process_io           ; yes, just handle it
    or    eax, eax                   ; was there an old owner?
    jz    short new_owner            ; no

    mov   esi, OFFSET32 Device_Name
    VxDCall Shell_Resolve_Contention
    jc    short dont_process         ; hmmm, couldn't resolve

    cmp   ebx, VDEMOD_Owner          ; if contention winner is
                                     ; the current owner,
    je    short dont_process         ; then we shouldn't process

new_owner:

IFDEF DEBUG
    Trace_Out "VDEMOD: New Owner #EBX"
ENDIF

    mov   edx, 200h      ; our arbitrary port address
    VMMCall Disable_Local_Trapping    ; give winner free access
    xchg  ebx, VDEMOD_Owner          ; save new owner, get old
    or    ebx, ebx                   ; no old owner?
    jz    short process_io           ; no, just process
    VMMCall Enable_Local_Trapping    ; old owner now locked out

process_io:
    popad

;secondly, handle I/O...
    Dispatch_Byte_IO Fall_Through, <SHORT VDEMOD_Out>
    in    al,dx                      ; do real in
    ret

VDEMOD_Out:
    out   dx,al                      ; do real out
    ret

dont_process:
    popad                            ; restore regs
    mov   al, 0Eh                    ; indicate error to sample
                                     ; apps
IFDEF DEBUG
    Trace_Out "VDEMOD: I/O inhibited for VM #EBX"
```

```
        ENDIF

    ret
EndProc My_VDEMOD_Hook
;..............................................................

BeginProc My_VDEMOD_lpt_Hook
;we come here if the printer status ports are trapped...
;set bit-3,7, clear bit-4...
    pushad

    popad

    Dispatch_Byte_IO Fall_Through, <SHORT VDEMOD_lpt_Out>
    in      al,dx                           ; do real in
    mov     al,10101000b     ;this should stuff-up printing!
    ret                      ; (busy,out-0f-paper,offline,error)

VDEMOD_lpt_Out:
    out     dx,al                           ; do real out
    ret

EndProc My_VDEMOD_lpt_Hook
;..............................................................
BeginProc ctrl_VDEMOD_lpt_Hook
;we come here if the printer control ports are trapped...
    pushad

    popad

    Dispatch_Byte_IO Fall_Through, <SHORT ctrl_lpt_Out>
    in      al,dx                           ; do real in
;   mov     al,0     ;
    ret

ctrl_lpt_Out:
;   out     dx,al                           ; do real out
    mov al,00001000b         ;bit-3=1,printer not selected.
    out dx,al
    ret

EndProc ctrl_VDEMOD_lpt_Hook
;..............................................................
VxD_CODE_ENDS
    END
```

To take this further, you need a good book with all the details on the VMM services. Install_IO_Handler for example, hooks a port and redirects to a callback routine. The routine is entered with EBX = current VM, ECX = type of I/O, EDX = port number, and EAX = output data (if type of I/O is output). When the callback exits, if the type of I/O is input, the value placed in EAX is the input value.

The book should also explain the VMM.INC macros and data structures, such as the Dispatch_Byte_IO macro used above.

Maximum Productivity

Now for something completely different ...

The philosophy of extremes

What I would like to do is present you with an idea, and an implementation of the idea. Basically, the idea is that you should "do" only two extremes of programming — *very* low-level, or *very* high-level, and nothing in between.

The rationale for this is very simple — for programming, you want, above all else, productivity, that is, the most program for your money. Now, productivity also includes maintainability, because that is part of the cost equation. To maximize productivity, you want the most powerful programming tools.

On the other hand, if you need to do low-level work, you should use the tool best suited to the job — assembly language — rather than try to torture a high-level language into performing low-level tasks. Sure, you can use something like C to perform low-level work, but it is really just a very awkward mimicking of assembly language and is definitely hardware dependent; therefore, you should really be doing it in assembly, in-line or as separate modules.

I have included a justification for assembly language, at the end of this chapter, taken directly from the first edition of this book.

True visual programming

The main point is that the most productive programming language is *not* C or C++, nor is it some augmentation using class libraries and front-end code generators. Nor is it Pascal, Basic, Fortran, Cobol, etc.

The way to go is *visual programming*, and that does *not* mean Visual Basic or Visual C++. These two products are not visual programming languages (VPLs), as they are still text-based languages. Borland's Delphi and C++Builder move slightly further toward true visual programming, but it is still mostly the user interface only that is developed visually.

If you want a definition of VPLs, and you have Internet access, look at the frequently-asked-questions (FAQ) file on COMP.LANG.VISUAL, a newsgroup. Much to the dismay of the people who started this newsgroup, very few programmers know what a VPL is, and COMP.LANG.VISUAL is bombarded with mail about Visual C++ and Visual Basic.

VPLs potentially can increase your programming productivity ten times or more, and probably the best of all is *LabView*, developed by National Instruments Corp.

Pragmatic programming   I am a realist.   Obviously, writing entire applications, or substantial portions, in assembly language is not everyones' cup of tea.   It may be justified for developing device drivers and other low-level work, and, yes, weird people like me do put large applications together entirely in assembly.   Most of my work involves close interaction with hardware and/or operating systems, so I suppose I would be classified as a *systems programmer*.

However, even systems programmers tend to use C, depending on what operating system they are using and on just what kind of systems-level work they are doing.

Really, this chapter presents a certain philosophy and is not to be taken as the "truth engraved in stone".   Use these ideas as "food for thought".

GOOFEE diagrams   The new kid on the block is GOOFEE diagrams, which I developed.   GOOFEE is a visual analysis, design, and implementation methodology that is unique.   It is a truly unified, wholistic, approach, targeting embedded systems.

Reference sources   Checkout my book *Flow Design for Embedded Systems*, R&D Books / Miller Freeman, USA, 1997.

The R&D Books site on the Internet is:

http://www.rdbooks.com/

My research site is:

http://www.goofee.com/

One, out of many, on-line bookstores is:

http://www.amazon.com/

GOOFEE is not (yet) a 100% visual development tool, but I wanted to mention it, in case anyone is interested!   True 100% VPLs are rare indeed. What is there?

LabView Programming

LabView is at least as powerful as any text-based language and has evolved, since 1986, into a very mature and sophisticated product.   In a nutshell, LabView is an environment in which you can very rapidly develop applications with a single line of coding. LabView applications can be recompiled, unchanged, to run on Macintosh, PC with Windows, Windows NT, and Sun workstations, with more to come.

The speed with which you can put programs together has to be seen to be believed, and phenomenal productivity improvements are well documented. But, enough of that; I am not a salesperson for National Instruments, nor is this book about LabView programming. So, I'll give you all sides of the picture and get onto how I see assembly language fitting in.

Dataflow visual programming

LabView is based upon a *dataflow* model, but there are enough control-flow mechanisms built-in to get around the limitations of the pure dataflow concept. Dataflow means that you construct programs by drawing data paths between icons on the screen, and execution follows the data paths.

Figure 15.3 shows this. Look at the diagram at the bottom of the figure, and you will see how "icons" have been interconnected, to form the program, or "diagram". Control structures are also shown, such as a while-loop and a case-structure. Note that the case-structure is very efficient in its use of screen real-estate; cases are superimposed, like a deck of cards, with a simple selection-box on top, for flicking through them.

This leads to an interesting consideration — a classical problem with visual programming environments is that they tend to very rapidly run out of screen, but LabView is the most compact and "screen efficient" of any VPL that I have encountered.

Just imagine being able to rapidly wire-up a program, without having to remember any text-based *syntax*, or even how to use *pointers*.

Disadvantages of LabView

Speed & size

Ok, ok, the negative points. The power comes at a price, which is inefficiency of the generated code. LabView programs are HUGE, and they don't "run at the speed of compiled C", as National Instruments publicity would have you believe. Maybe you can contrive such a situation, but practical programs would be lucky to achieve half the speed of an equivalent C program.

However, "equivalent" is a difficult issue, because LabView applications have so much extra stuff built-in. Compared with one of LabView's main competitors, *HP-VEE* developed by Hewlett Packard, the former is much faster and, in my personal opinion, far more sophisticated and mature.

Figure 15.3: LabView icon, front panel, and diagram.

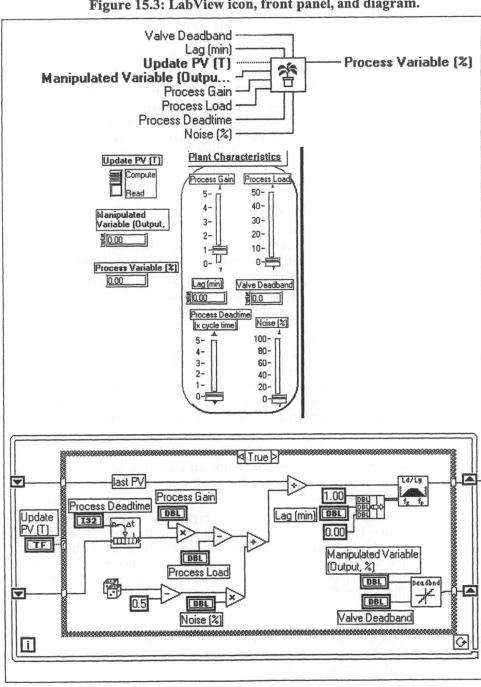

Narrow target
market
The next negative point is that LabView is optimised for use by engineers and scientists, and this is reflected throughout its design. It is absolutely superb for data acquisition, control, mathematic processing, and image work. It was never really intended to be a commercial high-volume product, so you don't use it to develop high-volume applications.

You use LabView for custom, one-off, or low-quantity jobs.

However, since people have discovered that LabView is good for just about everything and is very easy to use (once you get used to the dataflow concept), the original market domain has tended to diversify. People are now selling stand-alone applications, to a larger market — in particular, this has happened since LabView was released on the PC in 1993.

Integration with Assembly

Since LabView is optimised for engineering and scientific applications, its strengths are on the I/O side, which ties in strongly with the kind of things you normally use assembly for.

So if you use LabView to control all the machinery in your factory, and you also want to do some low-level optimisation, how can you integrate assembly into the picture?

Front panel
& top-icon
Have another look at Figure 15.3. Each diagram, or code-module, has a front panel, which is a window through which all inputs and outputs travel. Note, however, that this front panel may or may not appear at execution-time. Think of it as a handy development aid, since it gives you total control over the diagram for testing purposes and interactive monitoring while executing.

Look higher, and you will see that there is an icon, that has input and output terminals, which all go to and from the diagram, via the front panel. This icon makes the diagram into the equivalent of a subroutine or procedure and is a software object that can be reused with the greatest of ease.

An important point to note about Figure 15.3 is that you are seeing it in black and white, when in fact, it is in full color, and all wires clearly show the data types they carry. Furthermore, LabView will not allow a connection if the data types are incompatible — also note that most built-in LabView icons are polymorphic, meaning that they will accept almost any data type.

Icon
hierarchy
A LabView program can be composed of a hierarchy of icons. That is, the top-icon in Figure 15.3 is itself composed of icons wired into a diagram, which may in turn be composed of underlying diagrams, and so on.

Any one of these icons could be a program that has been written in another language.

Note also, that a small help window automatically shows you the meaning of each input and output terminal on an icon as you wire it up.

Code
interface
nodes

An icon that has been written in another language is known as a *code interface node* (CIN), and because LabView (I'll call it LV from now on!) was originally developed for the Macintosh, that platform has the best language support. For the PC, only Watcom C and Microsoft C/C++ are supported.

Dynamic
link
libraries

What if you want use your own brand of assembler, or any assembler for that matter? There is a way. LabView can call any dynamic link library (DLL) function — Sun workstations and Macintoshes also have the equivalent to the DLL mechanism.

This is great, as you can put all your hardware-dependent low-level code into a DLL and write a version for each platform — the rest of LabView will automatically work on any of the platforms.

A standard skeleton DLL, written in assembly language, is on the Companion disk in directory \LV-DLL.

Reference
source

To find out more about LabView, National Instruments has a Web page at:

http://www.natinst.com/

Layout Programming

There is another, easy, way to generate a DLL — its name is *Layout*.

Layout is another VPL. Comparing Layout with LabView is like comparing chalk and cheese, but there is a method in my madness.

Layout had its origins on DOS, and Layout for DOS still exists. Layout for Windows made its appearance in the world in November 1994.

Flowchart
programming

Layout uses a *flowchart* model of programming, in which the programmer interconnects modules representing flow of control, as shown in Figure 15.4. The representation is different from conventional flowcharts, as you can see in the way the repeat-loop is constructed. The little arrows going off to the left point to subroutines or procedures, and clicking on one of those little boxes brings up the subroutine flowchart.

The flowchart concept is ancient, and its popularity comes and goes. Modern flowcharts do need constraints, unlike the old representations that we oldies used with Fortran, that allowed spaghetti-code. Layout also forces some concurrency mechanisms, such as being able to send a message to a procedure, and have the procedure execute concurrently. Layout also allows flowcharts to be attached to specific objects on a window, so, for example, if you press a button on the screen, the attached code will execute, asynchronously to the rest of the application.

However, the way in which the flowchart paradigm has been forced into these avenues is, to me, very odd.

Comparison with LabView

Visual efficiency

Layout's on-screen graphical program construction representation is very crude and cumbersome, compared with LabView. The flowchart just about immediately fills the screen, and you have to scroll a long way down to see everything. Of course, a tight hierarchy helps a little bit.

Also, I think that some of the linkages need to be shown. For example, it has to be clearer what variables belong to what code, and if a code module is attached to an on-screen object, the relationship should be clearly shown. That is, the structure of everything in the program, links, and relationships, should all be clearly and visually viewable.

It has some rough edges, but it is a VPL, and quite a good one. It belongs to the very exclusive club of VPLs that completely avoid the necessity of having to write lines of code.

However, like LabView, you can drop down to code, if required.

Target market

Layout is marketed as a mass-market full-cycle CASE tool for Windows and DOS program development. This is in contrast to LabView. Layout is not just a front-end or back-end CASE tool — it is full cycle, and directly generates very efficient .EXE files or C or C++ output. I suppose LabView can be used as a full-cycle CASE tool, but it is not a mass-market tool.

Layout is what you would use to develop a product that is optimised to work with Windows, utilises Windows features to the full extent (such as OLE, DDE, and networking, though note that LabView also supports DDE and TCP/IP), and generates very fast and small code. Even though Layout was developed first for DOS, it followed the Windows event-driven style. In fact, it mimicked Windows in just about every way, so it was a natural contender for conversion to Windows.

I began to wonder when the developers became very quiet for a couple of years, but they were working frantically on the Windows version.

Figure 15.4: Layout.

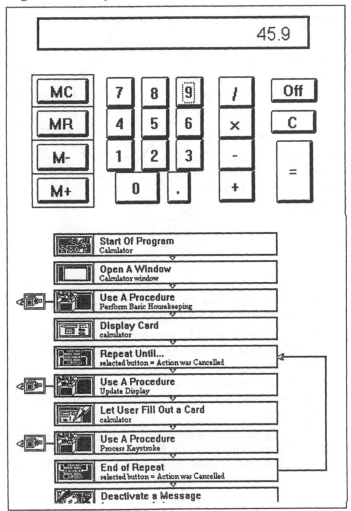

Integrating with Assembly

In-line assembly

In many respects, Layout gives you the best programming environment. My philosophy of the two extremes is applicable here — you can merrily construct a flowchart, then at any point in the flowchart, you can stick in in-line C, C++, or assembly code. It is a super-high-level language that allows in-line assembly!

Figure 15.5: Integration with assembly.

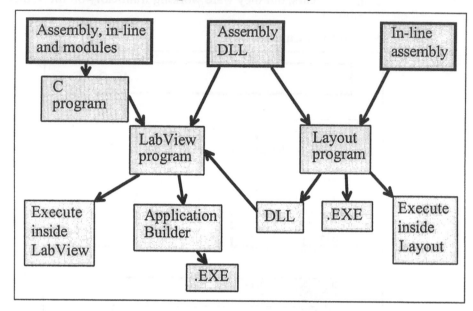

Assembly DLL

Each of the boxes in Figure 15.4 is called a *blackbox*, and, as with LabView, they can be developed in another language. In fact, Layout blackboxes are simply DLLs!

That's it: write a DLL, register it with Layout, and it becomes a blackbox that you can put into your flowchart, just like any other blackbox.

Figure 15.5 gives an overall picture. Look at the figure, and you can see how assembly language fits in with both Layout and LabView.

One very interesting possibility that you can see in Figure 15.5 is that Layout can generate DLL output. That is, any program that you construct inside Layout can, simply by selecting "Make blackbox" from the menu, generate a DLL.

DLLs are mighty handy, because they can readily be used by various development environments, including LabView. Because Layout is a totally open environment (with no run-time distribution licencing required, unlike LabView), you might find it useful as an adjunct for use with LabView.

Reference source

Layout was developed in the USA by VTools Inc. (formerly Objects Inc.), and their Web page is:

http://www.vtools.com/

Assembly versus High-Level Languages

My vision of the viability of programming at two extremes is an interesting diversion, but for many people the argument is more focused. The current fight, that has been going on for some years, is between languages such as C and assembly. C (and C++) is winning, but I would like to put in my two-cents worth.

Gloves off...

Always the debate rages about the applicability of assembly language in our modern programming world. An excellent and balanced argument is to be found in the article "Assembly Language Lives!" by Michael Abrash, *Dr Dobb's Journal*, March 1990. It is several years old but still relevant!

Faster, smaller, etc.

The simple bottom-line truth is that assembly language produces far faster and smaller code than compilers can produce. This is true to the present day, no matter how optimised compilers have become. For example, Quantasm Corp. sell a floating point emulation library to replace the standard ones. It's written in assembly language and is typically 4% to 10% of the size and 5 to 10 times as fast as most C floating-point libraries!

Apart from size and speed, there are other major issues. The contention that assembly language is harder to write in than a high-level language is, I hesitate to say, "nonsense". I'll just say "maybe in some cases". I can throw assembly code together as fast as with C. My assembler does not perform the kind of checking that a compiler will, but Microsoft MASM version 6.x and Borland TASM version 5 have significant C-like checking.

Portability

Transportability is a major consideration. If code is to be transported to other platforms, that is, other CPUs running Windows, shouldn't I avoid assembly language? Yes and no. For a start, the issue only applies to Windows NT. If you examine any high-level program such as a C program, you will find that it is riddled with hardware dependency. Hardware-dependent statements are written in a high-level language throughout typical programs. Whenever I see low-level functionality being coded in a high-level language, I find it to be cumbersome, inefficient, and in many cases unreadable. Why force a square peg into a round hole? Why not just bite the bullet and code those portions directly in assembly? Why not use your compiler's in-line assembly capability? What you gain is code that does what it was developed for and is efficient, compact, and highly readable. Assembly language code is far more readable than some C code I've seen! In fact, what you end up with is code that is **more portable** than if the whole thing had been done in the high-level language. The reason for this is that the boundary between strictly non-hardware-dependent code and hardware-dependent code is clarifed.

In fact, you can bet that some programs written exclusively for Windows NT will have assembly code. To transport the program to another platform, the developers will simply recode the easily distinguishable hardware-dependent portions.

Binary compatibility is a fascinating aspect of Windows NT. This means that assembly language code will work on all hardware platforms. The x86 instruction set, and hardware dependency, will be emulated, to the extent that most Windows programs will run without recompiling, but it remains to be seen how far we can push this. This is another exciting area to research!

The bottom line

There will always be a demand for assembly language programmers, due mostly to competition. If word processor A runs twice as fast as word processor B, which one has the edge? And if word processor A runs in half the memory, again the choice is obvious. We have not by any means reached the point where we can ignore these issues, despite faster CPUs and more memory. In fact, our programs run slower than ever before and 16M is no longer enough!

Let me finish this argument by quoting Karl Wright and Rick Schell, from "Mixed-Language Programming with ASM", *Dr Dobb's Journal*, March 1990:

> "It is not only practical but advisable to mix languages and models in order to achieve the best results. Modern assembly language is a vital part of this mix, and will continue to be important in the future, because space and performance are always important for competitive software, no matter how powerful the hardware becomes."

As a final note, I wrote the GOOFEE Diagrammer, bundled with my book *Flow Design for Embedded Systems* (R&D Books, 1997), entirely in assembly language. It is a 16-bit WinApp, a very sophisticated graphics drawing program, that is ridiculously small at only 100K. By using the high-level features of modern assemblers, my assessment is that coding time would not have been any faster if done in C, nor would it have been any more readable (maybe *less* readable). I did use Layout for some of the dialog boxes and mathematics, but I found the automatic code generation was doing things I didn't want. Hence, I rewrote those portions in assembly language also!

A

Instruction Summary

Abbr.	NAME
CF	Carry flag
PF	Parity flag
AF	Auxiliary carry flag
ZF	Zero flag
SF	Sign flag
TF	Trap flag
IF	Interrupt enable
DF	Direction flag
OF	Overflow
IOPL	I/O privilege level
NT	Nested tank flag
RF	Resume flag
VM	Virtual 8086 mode

#	=	flags affected by instruction
$	=	ignore
cnt	=	count
Reg	=	register
Mem	=	memory
Accum	=	AL, AX, EAX
Imm	=	immediate
Deststr	=	destination string
Sourcestr	=	source string
Segreg	=	segment register

Number preceding item
indicates number of bits

Note that some coprocessor flags are also indicated in the tables.
The legend is:

IE	Invalid
UE	Underflow
PE	Precision
DE	Denormalised
ZE	Divide by zero
OE	Overflow

Format

Instruction
Description
Flags affected   **86/286/386**

AAA
ASCII adjust AL after add
AF, CF   ...

AAD
ASCII adjust before divide
SF, ZF, PF   ...

AAM
ASCII adjust after multiply
SF, ZF, PF   ...

AAS
ASCII adjust after subtract
AF, CF   ...

ADC accum,imm
ADC mem,imm
ADC mem,reg
ADC reg,imm
ADC reg,mem
ADC reg, reg
Add with CF
OF,SF,ZF,AF,PF,CF   ...

ADD accum, imm
ADD mem,imm
ADD mem,reg
ADD reg,imm
ADD reg,mem
ADD reg,reg
Add
OF,SF,ZF,AF,PF,CF   ...

AND accum, imm
AND mem, imm
AND mem, reg
AND reg, imm
AND reg, reg
AND reg, mem
Logical AND
OF=0,SF,ZF,PF,CF=0   ...

ARPL
Adjust requested privilege
level
ZF   ..

BOUND reg, source

Detect array index out of range
None   ..

BSF reg,mem or reg,reg
Bit scan forward
ZF   .

BSR reg,mem or reg,reg
Bit scan reverse
ZF   .

BT
Test bit
CF   .

BTC
Test bit and complement
CF   .

BTR
Test bit and reset
CF   .

BTS
Test bit and scan
CF   .

CALL 16 memptr
CALL 16 regptr
CALL 32 memptr
CALL farproc
CALL nearproc
Call (control transfer)
None except if task sw.   ...

CBW/CWDE
Convert byte to word/word to
double word
None   ...

CLC
Clear CF
CF = 0   ...

CLD
Clear DF
DF = 0   ...

CLI
Clear IF
IF = 0   ...

CLTS
Clear task switched flag
TS = 0 in CRO reg.   .

CMC
Complement CF
CF   ...

CMP accum, imm
CMP mem, imm
CMP mem, reg

CMP reg, imm
CMP reg, mem
CMP reg, reg
Compare
OF,SF,ZF,AF,PF,CF   ...

CMPS (rep)deststr,sourc
CMPS deststr,sourcestr
Compare byte, word or double
word string
OF,SF,ZF,AF,PF,CF   ...

CWD/CDQ
Convert word to doubleword /dw
to qw
None   ...

DAA/DAS
Decimal adjust after
add/subtract
SF, ZF, AF, PF, CF   ...

DEC mem
Decrement
OF, SF, ZF, AF, PF   ...

DEC reg
Decrement
OF, SF, ZF, AF, PF   ...

DIV 16 mem
DIV 16 reg
DIV 8 mem
DIV 8 reg
Divide
None   ...

ENTER 16 imm, 0
ENTER 16 imm, 1
ENTER 16 imm, level
Enter procedure
None   ..

ESC imm, mem
ESC imm, reg
Escape to external device
None   ...

HLT
Halt
None   ...

IDIV 16 mem
IDIV 16 reg
IDIV 8 mem
IDIV 8 reg
Signed integer divide
None   ...

IMUL 16 mem
IMUL 16 reg
IMUL 8 mem
IMUL 8 reg

Signed integer multiply
OF, CF   ...

IMUL destreg,16,reg,imm
IMUL destreg, mem, imm
Signed integer multiply
OF, CF   ..

IN accum, 8 imm
IN Accum, DX
Input from port/DX-port
None   ...

INC mem
INC reg
Increment
OF, SF, ZF, AF, PF   ...

INS (rep) deststr, DX
INS deststr, DX
Input string
None   ..

INT 8 imm
Interrupt
IF = 0, trap = 0   ...

INTO
Interrupt on overflow
IF = 0, trap = 0   ...

IRET
Interrupt return
All   ...

JMP 16 memptr
JMP 16 regptr
JMP 32 memptr
JMP far
JMP near
JMP short
Unconditional jump
None   ...

Jxxx short
Conditional jump
None   ...

LAHF
Load AH with flags(LO byte
flags)
None   ...

LAR
Load access rights byte
ZF   ..

LDS 16 reg, 32 mem
Load pointer to DS:reg
None   ...

Format

Instruction
Description
Flags affected   **86/286/386**

LEA 16 reg, 16 mem
Load effective address to
register
None   ...

LEAVE
Leave procedure
None   ..

LES 16 reg, 32 mem
Load pointer to ES:reg
None   ...

LFS
Load pointer to FS:reg
None   .

LGDT
Load global descriptor table
None   ..

LGS
Load pointer to GS
None   .

LIDT
Load IF descriptor table
None   ..

LLDT
Load local descriptor table
None   ..

LMSW
Load machine status word
None   ..

LOCK
Bus lock prefix
None   ...

LODS (rep) sourcestr
Repeat load byte/word/dword
None   ...

LODS sourcestr
Load byte/word/dword
None   ...

LOOP short

Loop
None   ...

LOOPE/LOOPZ short
Loop equal/zoom
None   ...

LOOPNE/LOOPNZ short
Loop while not equal/not zero
None   ...

LSL
Load segment limit
ZF   ..

LSS
Load pointer to SS
None   .

LTR
Load task register
None   ..

MOV 16 reg, segreg
MOV accum,mem
MOV mem,accum
MOV mem, imm
MOV mem, reg
MOV mem, segreg
MOV reg, imm
MOV reg, mem
MOV reg, reg
MOV segreg, 16 mem
MOV segreg, 16 reg
Move
None   ...

MOVS (repeat)deststr,sourcstr
MOVS deststr,sourcestr
Move string
None   ...

MOVSX
Move with sign extension
None   .

MOVZX
Move with zero extension
None   .

MUL 16 mem
MUL 16 reg
MUL 8 mem
MUL 8 reg
Multiply
OF, CF   ...

NEG mem
NEG reg
Change sign
OF,SF,ZF,AF,PF,CF   ...

NOP

No operation
None   ...

NOT mem
NOT reg
Invert
None   ...

OR accum, imm
OR mem, imm
OR mem, reg
OR reg, imm
OR reg, mem
OR reg, reg
Logical OR
OF=0,SF,ZF,PF,CF=0   ...

OUT 8 immed, accum
OUT DX, accum
Output to port/DX-port
None   ...

OUTS (rep) DX,sourcestr
Output string
None   ...

OUTS DX, sourcestr
Output string
None   ..

POP mem
POP reg
Restore from stack
None   ...

POPA
Restore general reg's
fromstack
All   ..

PUSH imm
Save to stack
None   ..

PUSH mem
PUSH reg
Save to stack
None   ...

PUSHA
Save all to stack
None   ..

PUSHF
Save flags to stack
None   ...

RCL/RCR/ROL/ROR mem,1
RCL/RCR/ROL/ROR reg, 1
rotate thru CF left/CF
right/rot. left/right
OF, CF   ...

RCL/RCR/ROL/ROR mem,CL
RCL/RCR/ROL/ROR reg,CL
rotate thru CF left/CF right/
rot. left/right
CF   ...

RCL/RCR/ROL/ROR mem,cnt
RCL/RCR/ROL/ROR reg,cnt
rotate thru CF left/CF right/
rot. left/right
CF   ..

REP
Repeat
None   ...

REPE/REPZ
Repeat equal/zero
None   ...

REPNE/REPNZ
Repeat not equal/not zero
None   ...

RET immed (far with pop)
RET (far, no pop)
RET immed (near with pop)
RET (near, no pop)
Return
None   ...

SAHF
Store AH into flags
SF, ZF, AF, PF, CF   ...

SAL/SHL/SAR/SHR mem,1
SAL/SHL/SAR/SHR reg,1
Shft arith'tic-left/left/shift
arith.-right/right
OF=0,SF,ZF,PF,CF   ...

SAL/SHL/SAR/SHR mem,CL
SAL/SHL/SAR/SHR mem,cnt
Shft arith'tic-left/left/shift
arith.-right/right
SF,(SHR=0)ZF,PF,CF   ...

SAL/SHL/SAR/SHR reg,CL
Shft arith'tic-left/left/shift
arith.-right/right
SF, ZF, PF, CF   ...

SAL/SHL/SAR/SHR reg,cnt
Shft arith'tic-left/left/shift
arith.-right/right
SF,(SHR=0)ZF,PF,CF   ..

```
Format

Instruction
Description
Flags affected        86/286/386
```

SBB accum, imm
SBB mem, imm
SBB mem, reg
SBB reg, imm
SBB reg, mem
SBB reg, reg
Subtract with borrow
OF,SF,ZF,AF,PF,CF   ...

SCAS (repeat) deststr
SCAS deststr
Scan byte/word
OF,SF,ZF,AF,PF,CF   ...

SETxxx
Conditional byte set
None   .

SGDT
Store global descriptor table
None   ..

SIDT
Store Int. descriptor table
None   ..

SLDT
Store local descriptor table
None   ..

SMSW
Store machine status word
None   ..

STC/STD/STI
Set CF/DF/IF
CF = 1/DF = 1/IF = 1   ...

STOS (repeat) deststr
STOS deststr
Store byte/word
None   ...

STR
Store task register
None   ..

SUB accum, imm
SUB mem, imm
SUB mem, reg

SUB reg, imm
SUB reg, mem
SUB reg, reg
Subtract
OF,SF,ZF,AF,PF,CF   ...

TEST accum,imm
TEST mem, imm
TEST reg, imm
TEST reg, mem
TEST reg, reg
AND function to flags
OF=0,SF,ZF,PF,CF=0   ...

VERR
Verify read access
ZF   ..

VERW
Verify write access
ZF   ..

WAIT
Wait for 80X87
None   ...

XCHG accum, 16 reg
XCHG mem, reg
XCHG reg, reg
Exchange
None   ...

XLAT sourcetable
Translate byte
None   ...

XOR accum, imm
XOR mem, imm
XOR mem, reg
XOR mem, reg
XOR reg, imm
XOR reg, mem
XOR reg, reg
Exclusive OR
OF=0,SF,ZF,PF,CF=0   ...

**Instruction Set of the 80x87
Mathematics Coprocessor**

Refer beginning of Appendix A
for legend.

Format

Instruction
Description
Flags affected   **87/287/387**

F2XM1
$2^{ST(0)}-1$
UE,PE   . . .

FABS
Absolute value
IE   . . .

FADD dest,source
Add real
IE,DE,OE,UE,PE   . . .

FADDP dest,source
Add real & pop
IE,DE,OE,UE,PE   . . .

FBLD source
Packed decimal (BCD) load
IE   . . .

FBSTP
Packed dec. (BCD) store & pop
IE   . . .

FCHS
Change sign
IE   . . .

FCLEX/FNCLEX
Clear exceptions
None   . . .

FCOM source
compare real
IE,DE   . .

FCOMP source
compare real & pop
IE,DE   . .

FCOMPP
compare real & pop twice
IE,DE   . .

FCOS
Cosine

FDECSTP
Decrement stack pointer
None   . . .

FDISI/FNDISI
Disable interrupts
None   ·$$

FDIV dest,source
Divide real
IE,DE,ZE,OE,UE,PE   . . .

FDIVP dest,source
Divide real & pop
IE,DE,ZE,OE,UE,PE   . . .

FDIVR dest,source
Divide real reversed
IE,DE,ZE,OE,UE,PE   . . .

FDIVRP dest,source
Divide real reversed & pop
IE,DE,ZE,OE,UE,PE   . . .

FENI/FNENI
Enable interrupts
None   ·$$

FFREE dest
Free register
None   . . .

FIADD source
Integer add
IE,DE,OE,PE   . . .

FICOM source
Integer compare
IE,DE   . . .

FICOMP source
Integer compare & pop
IE,DE   . . .

FIDIV source
Integer divide
IE,DE,ZE,OE,UE,PE   . . .

FIDIVR source
Integer divide reversed
IE,DE,ZE,OE,UE,PE   . . .

FILD source
Integer load
IE   . . .

FIMUL source
Integer multiply
IE,DE,OE,PE   . . .

Format

Instruction
Description
Flags affected   **87/287/387**

FINCSTP
Increment stack pointer
None   . . .

FINIT/FNINIT
Initialise processor
None   . . .

FIST dest
Integer store
IE,PE   . . .

FISTP dest
Integer store & pop
IE,PE   . . .

FISUB source
Integer subtract
IE,DE,OE,PE   . . .

FISUBR source
Integer subtract reversed
IE,DE,OE,PE   . . .

FLD source
Load real
IE,DE   . . .

FLD1
Load +1.0
IE   . . .

FLDCW source
Load control word
None   . . .

FLDENV source
Load environment
None   . . .

FLDL2E
Load log (2^e)
IE   . . .

FLDL2T
Load log (2^10)
IE   . . .

FLDLG2
Load log (10^2)
IE   . . .

FLDLN2
Load log (e^2)
IE   . . .

FLDPI
Load pi
IE   . . .

FLDZ
Load +0.0
IE   . . .

FMUL dest,source
Multiply real
IE,DE,ZE,OE,UE,PE   . . .

FMULP dest,source
Multiply real & pop
IE,DE,ZE,OE,UE,PE   . . .

FNOP
No operation
None   . . .

FPATAN
Partial arctangent
UE,PE   . . .

FPREM
Partial remainder
IE,DE,UE   . . .

FPREM1
Partial remainder (IEEE)
                             .

FPTAN
Partial tangent
IE,PE   . . .

FRNDINT
Round to integer
IE,PE   . . .

FRSTOR source
Restore saved state
None   . . .

FSAVE/FNSAVE dest
Save state
None   . . .

FSCALE
Scale
IE,OE,UE   . . .

FSETPM
Enter Protected mode
None   . .

FSIN
Sine

.

FSINCOS
Sine & cosine

.

FSQRT
Square root
IE,DE,PE   . . .

FST dest
Store real
IE,OE,UE,PE   . . .

FSTCW/FNSTCW dest
Store control word
None   . . .

FSTENV/FNSTENV dest
Store environment
None   . . .

FSTP dest
Store real & pop
IE,OE,UE,PE   . . .

FSTSW/FNSTSW dest
Store status word
None   . . .

FSUB dest,source
Subtract real
IE,DE,OE,UE,PE   . . .

FSUBP dest,source
Subtract real & pop
IE,DE,OE,UE,PE   . . .

FSUBR dest,source
Subtract real reversed
IE,DE,OE,UE,PE   . . .

FSUBRP dest,source
Subtract real reversed & pop
IE,DE,OE,UE,PE   . . .

FTST
Test stack top against +0.0
IE,DE   . .

FUCOM
Unordered compare
.

FUCOMP
Unordered compare & pop
.

FUCOMPP
Unordered comp. & pop twice
.

FWAIT
Wait while 8087 is busy
None   . . .

FXAM
Examine stack top
None   . . .

FXCH dest
Exchange registers
IE   . . .

FXTRACT
Extract exponent & Significand
IE   . . .

FYL2X
$ST(1)*\log_2(ST(0))$
PE   . . .

FYL2XP1
$ST(1)*\log_2(ST(0)+1)$
PE   . . .

B

Keyboard Tables

Virtual Keys

NAME	HEX	KEY
VK_LBUTTON	01	Left mouse button
VK_RBUTTON	02	Right mouse button
VK_CANCEL	03	Cancel key
VK_MBUTTON	04	Middle mouse button
	05-07	Unassigned
VK_BACK	08	BACKSPACE key
VK_TAB	09	TAB key
	0A-0B	Unassigned
VK_CLEAR	0C	CLEAR key
VK_RETURN	0D	ENTER key
	0E-0F	Unassigned
VK_SHIFT	10	SHIFT key
VK_CONTROL	11	CONTROL key
VK_MENU	12	ALT key
VK_PAUSE	13	PAUSE key
VK_CAPITAL	14	CAPS LOCK key
	15-19	Reserved for Kanji

NAME	HEX	KEY
	1A	Unassigned
VK_ESCAPE	1B	ESCAPE key
	1C-1F	Reserved for Kanji
VK_SPACE	20	SPACEBAR
VK_PRIOR	21	PAGE UP key
VK_NEXT	22	PAGE DOWN key
VK_END	23	END key
VK_HOME	24	HOME key
VK_LEFT	25	LEFT ARROW key
VK_UP	26	UP ARROW key
VK_RIGHT	27	RIGHT ARROW key
VK_DOWN	28	DOWN ARROW key
VK_SELECT	29	SELECT key
VK_PRINT	2A	OEM specific
VK_EXECUTE	2B	EXECUTE key
VK_SNAPSHOT	2C	PRINTSCREEN key
VK_INSERT	2D	INSERT key
VK_DELETE	2E	DELETE key
VK_HELP	2F	HELP key
VK_0 to VK_9	30-39	0-9 (same as ASCII)
	3A-40	Unassigned
VK_A to VK_Z	41-5A	A-Z (same as ASCII)
	5B-5F	Unassigned
VK_NUMPAD0 to VK_NUMPAD9	60-69	Numeric keypad
VK_MULTIPLY	6A	Multiply key
VK_ADD	6B	Add key
VK_SEPARATER	6C	Separater key
VK_SUBTRACT	6D	Subtract key
VK_DECIMAL	6E	Decimal point key (.)
VK_DIVIDE	6F	Divide key
VK_F1 to VK_16	70-7F	Function keys 0-16
	80-87	OEM specific
	88-8F	Unassigned
VK_NUMLOCK	90	NUM LOCK key

NAME	HEX	KEY
VK_OEM_SCROLL	91	SCROLL LOCK key
	92-B9	Unassigned
VK_OEM_1	BA	Keyboard specific
VK_OEM_PLUS	BB	Plus key
VK_OEM_COMMA	BC	Comma key
VK_OEM_MINUS	BD	Minus key
VK_OEM_PERIOD	BE	Period key
VK_OEM_2	BF	Keyboard specific
VK_OEM_3	C0	Keyboard specific
	C1-DA	Unassigned
VK_OEM_4 to VK_OEM_8	DB-DF	Keyboard specific
	E0-E1	OEM specific
VK_OEM_102	E2	<> or \ (non-USA kbrd)
	E3-E4	OEM specific
	E5	Unassigned
	E6	OEM specific
	E7-E8	Unassigned
	E9-F5	OEM specific
	F6-FE	Unassigned

Extended ANSI

Dec.[1]	0	1	2	3	4	5	6	7	8	9
30			sp	!	"	#	$	%	&	'
40	(	)	*	+	,	-	per	/	0	1
50	2	3	4	5	6	7	8	9	:	;
60	<	=	>	?	@	A	B	C	D	E
70	F	G	H	I	J	K	L	M	N	O
80	P	Q	R	S	T	U	V	W	X	Y
90	Z	[	\	]	^	_	`	a	b	c
100	d	e	f	g	h	i	j	k	l	m
110	n	o	p	q	r	s	t	u	v	w
120	x	y	z	{	|	}	~			
130										
140						'	'	"	"	•
150	—	—	~	™	š	›	œ			Ÿ
160	sp.	¡	¢	£	¤	¥	¦	§	¨	©
170	ª	«	¬	-	®	¯	°	±	²	³
180	´	µ	¶	·	¸	¹	º	»	¼	½
190	¾	¿	À	Á	Â	Ã	Ä	Å	Æ	Ç
200	È	É	Ê	Ë	Ì	Í	Î	Ï	Ð	Ñ
210	Ò	Ó	Ó	Ô	Õ	Ö	×	Ø	Ù	Ú
220	Ü	Ý	Þ	ß	à	á	â	ã	ä	å
230	æ	ç	è	é	ê	ë	ì	í	î	ï
240	ð	ñ	ò	ó	ô	õ	ö	÷	ø	ù
250	ú	û	ü	ý	þ	ÿ				

[1] Codes are in decimal. Example: 162 = ¢. sp = space. per = period.

IBM ASCII

Dec.[1]	0	1	2	3	4	5	6	7	8	9
30			sp	!	"	#	$	%	&	'
40	(	)	*	+	,	-	per	/	0	1
50	2	3	4	5	6	7	8	9	:	;
60	<	=	>	?	@	A	B	C	D	E
70	F	G	H	I	J	K	L	M	N	O
80	P	Q	R	S	T	U	V	W	X	Y
90	Z	[	\	]	^	_	`	a	b	c
100	d	e	f	g	h	i	j	k	l	m
110	n	o	p	q	r	s	t	u	v	w
120	x	y	z	{	|	}	~			
130										
140						æ	Æ	ô	ö	ò
150	û	ù	ÿ	Ö	Ü	¢	£			ƒ
160	sp.	í	ó	ú	ñ	Ñ	ª	º	¿	⌐
170	¬	½	¼	¡	«	»	▓	▓	■	|
180	┤	╡	╢	□	╕	╣	║	╗	╝	╜
190	╛	┐	└	┴	┬	├	─	┼	╞	╟
200	╚	╔	╩	╦	╠	═	╬	╧	╨	╤
210	╥	╙	╘	╒	╓	╫	╪	┘	┌	
220	█	▌	▐	▄	α	ß	Γ	π	Σ	σ
230	µ	τ	Φ	Θ	Ω	δ	∞	φ	ε	∩
240	≡	±	≥	≤	⌠	⌡	÷	≈	°	·
250	·	√	η	²	■	□				

[1] Codes are in decimal. Example: 162 = ó. sp = space. per = period.

Windows documentation refers to IBM ASCII as the OEM, or Terminal, font.   It is the standard 8-bit ASCII character set supported by DOS.   Some other fonts also support this character set, such as LotusLineDraw.

C

DPMI Services

This material is extracted with permission from the *DOS Protected Mode Interface (DPMI) Specification*, Version 1.0, by the DPMI Committee, hosted by Intel Corporation.

Detailed information is provided here for version 0.9 because Windows 3.x and 95 only support v0.9. For further information, refer to the above source, the body of this book, and the Companion Disk. The complete specification is also on-line at:

```
http://www.delorie.com/djgpp/doc/
```

This Appendix is in two portions: first a listing of all services grouped functionally, and second a detail description of all v0.9 functions.

DPMI INT-31h Functions Listed by Functional Group

Function Number	Function Name	DPMI 0.9	1.0
LDT MANAGEMENT SERVICES			
0000h	Allocate LDT Descriptor	*	*
0001h	Free LDT Descriptor	*	*
0002h	Map Real mode Segment to Descriptor	*	*
0003h	Get Selector Increment Value	*	*
0006h	Get Segment Base Address	*	*

Function Number	Function Name	DPMI 0.9	1.0
0007h	Set Segment Base Address	*	*
0008h	Set Segment Limit	*	*
0009h	Set Descriptor Access Rights	*	*
000Ah	Create Alias Descriptor	*	*
000Bh	Get Descriptor	*	*
000Ch	Set Descriptor	*	*
000Dh	Allocate Specific LDT Descriptor	*	*
000Eh	Get Multiple Descriptors		*
000Fh	Set Multiple Descriptors		*

EXTENDED MEMORY MANAGEMENT SERVICES

Function Number	Function Name	DPMI 0.9	1.0
0500h	Get Free Memory Information	*	*
0501h	Allocate Memory Block	*	*
0502h	Free Memory Block	*	*
0503h	Resize Memory Block	*	*
0504h	Allocate Linear Memory Block		*
0505h	Resize Linear Memory Block		*
0506h	Get Page Attributes		*
0507h	Set Page Attributes		*
0508h	Map Device in Memory Block		*
0509h	Map Conventional Memory in Memory Block		*
050Ah	Get Memory Block Size and Base		*
050Bh	Get Memory Information		*
0800h	Physical Address Mapping	*	*
0801h	Free Physical Address Mapping		*
0D00h	Allocate Shared Memory		*
0D01h	Free Shared Memory		*
0D02h	Serialize on Shared Memory		*
0D03h	Free Serialization on Shared Memory		*

DOS MEMORY MANAGEMENT SERVICES

Function Number	Function Name	DPMI 0.9	1.0
0100h	Allocate DOS Memory Block	*	*
0101h	Free DOS Memory Block	*	*
0102h	Resize DOS Memory Block	*	*

INTERRUPT MANAGEMENT SERVICES

Function Number	Function Name	DPMI 0.9	1.0
0200h	Get Real Mode Interrupt Vector	*	*
0201h	Set Real Mode Interrupt Vector	*	*
0202h	Get Processor Exception Handler Vector	*	*

Function Number	Function Name	DPMI 0.9	1.0
0203h	Set Processor Exception Handler Vector	*	*
0204h	Get Protected Mode Interrupt Vector	*	*
0205h	Set Protected Mode Interrupt Vector	*	*
0210h	Get Extended Processor Exception Handler Vector in Protected Mode		*
0211h	Get Extended Processor Exception Handler Vector in Real Mode		*
0212h	Set Extended Processor Exception Handler Vector in Protected Mode		*
0213h	Set Extended Process Exception Handler Vector in Real Mode		*
0900h	Get and Disable Virtual Interrupt State	*	*
0901h	Get and Enable Virtual Interrupt State	*	*
0902h	Get Virtual Interrupt State	*	*

TRANSLATION SERVICES

0300h	Simulate Real Mode Interrupt	*	*
0301h	Call Real Mode Procedure with Far Return Frame	*	*
0302h	Call Real Mode Procedure with Int. Return Frame	*	*
0303h	Allocate Real Mode Callback Address	*	*
0304h	Free Real Mode Callback Address	*	*
0305h	Get State Save/Restore Addresses	*	*
0306h	Get Raw CPU Mode Switch Addresses	*	*

PAGE MANAGEMENT SERVICES

0600h	Lock Linear Region	*	*
0601h	Unlock Linear Region	*	*
0602h	Mark Real Mode Region as Pageable	*	*
0603h	Relock Real Mode Region	*	*
0604h	Get Page Size	*	*
0702h	Mark Page as Demand Paging Candidate	*	*
0703h	Discard Page Contents	*	*

DEBUG SUPPORT SERVICES

0B00h	Set Debug Watchpoint	*	*
0B01h	Clear Debug Watchpoint	*	*
0B02h	Get State of Debug Watchpoint	*	*
0B03h	Reset Debug Watchpoint	*	*

Function Number	Function Name	DPMI 0.9	1.0
MISCELLANEOUS SERVICES			
0400h	Get DPMI Version	*	*
0401h	Get DPMI Capabilities		*
0A00h	Get Vendor-Specific API Entry Point	*	*
0C00h	Install Resident Service Provider Callback		*
0C01h	Terminate and Stay Resident		*
0E00h	Get Coprocessor Status		*
0E01h	Set Coprocessor Emulation		*

DPMI Services Detail

Int 2Fh Function 1686h
Get CPU Mode
Returns information about the current CPU mode. Programs which only execute in Protected mode do not need to call this function.
Call with:
AX   =   1686h
Returns:
if executing in Protected mode
AX   =   0
if executing in Real mode or Virtual 86 mode
AX   =   nonzero

Int 2Fh Function 1687h
Obtain Real-to-Protected Mode Switch Entry Point.
This function can be called in Real mode only to test for the presence of a DPMI host, and to obtain an address of a mode switch routine that can be called to begin execution in Protected mode.
Call with:
AX   =   1687h
Returns:
if function successful
AX   =   0
BX   =   flags
            Bit Significance
            0   0 = 32 bit programs are
                not supported
                1 = 32 bit programs are
                supported
            1-15 not used
CL   =   processor type
            02h = 80286
            03h = 80386
            04h = 80486
            05h-FFh Preserved for future
            Intel processors
DH   =   DPMI major version as a
            decimal number (represented
            in binary)
DL   =   DPMI minor version as a
            decimal number (represented
            in binary)
SI   =   number of paragraphs
            required for DPMI host
            private data (may be 0)
ES:DI   =   segment:offset of procedure to
            call to enter Protected mode
            if function unsuccessful

(no DPMI host present)
AX   =   nonzero

Int 31h Function 0000h
Allocate LDT Descriptors
Allocates one or more descriptors in the task's Local Descriptor Table (LDT). The descriptor(s) allocated must be initialized by the application with other function calls.
Call with:
AX   =   0000h
CX   =   number of descriptors to
            allocate
Returns:
if function successful
CF   =   clear
AX   =   base selector
if function unsuccessful
CF   =   set
AX   =   error code
            8011h descriptor unavailable

Int 31h Function 0001h
Free LDT Descriptor
Frees an LDT descriptor.
Call with:
AX   =   0001h
BX   =   selector for the descriptor to
            free
Returns:
if function successful
CF   =   clear
AX   =   error code
            8022h invalid selector

Int 31h Function 0002h
Segment to Descriptor
Maps a Real mode segment (paragraph) address onto an LDT descriptor that can be used by a Protected mode program to access the same memory.
Call with:
AX   =   0002h
BX   =   Real mode segment address
Returns:
if function successful
CF   =   clear

> **FORMAT:**
>
> *Heading*
> *Description*
> *Call with*
> *Returns*

AX   =   selector for Real mode
            segment
if function unsuccessful
CF   =   set
AX   =   error code
            8011h descriptor unavailable

Int 31h Function 0003h
Get Selector Increment Value
The DPMI functions Allocate LDT Descriptors
(Int 31h Function 0000h) and Allocate DOS
Memory Block (Int 31h Function 0100h) can
allocate an array of contiguous descriptors, but
only return a selector for the first descriptor.
The value returned by this function can be used
to calculate the selectors for subsequent
descriptors in the array.
Call with:
AX   =   0003h
Returns:
CF   =   clear (this function always
            succeeds)
AX   =   selector increment value

Int 31h Function 0006h
Get Segment Base Address
Returns the 32-bit linear base address from the
LDT descriptor for the specified segment.
Call with:
AX   =   0006h
BX   =   selector
Returns:
if function successful
CF   =   clear
CX:DX   =   32-bit linear base
            address of segment
if function unsuccessful
CF   =   set
AX   =   error code
            8022h invalid selector

Int 31h Function 0007h
Set Segment Base Address
Sets the 32-bit linear base address field in the
LDT descriptor for the specified segment.
Call with:
AX   =   0007h
BX   =   selector

CX:DX   =   32-bit linear base
            address of segment
Returns:
if function successful
CF   =   clear
if function unsuccessful
CF   =   set
AX   =   error code
            8022h invalid selector
            8025h invalid linear address
(changing the base would cause the descriptor
to reference a linear address range outside that
allowed for DPMI clients)

Int 31h Function 0008h
Set Segment Limit
Sets the limit field in the LDT descriptor for the
specified segment.
Call with:
AX   =   0008h
BX   =   selector
CX:DX   =   32-bit segment limit
Returns:
if function seccessful
CF   =   clear
if function unsuccessful
CF   =   set
AX   =   error code
            8021h invalid value (CX<>0on
            a 16-bit DPMI host; or the
            limit is greater than 1MB, but
            the low 12 bits are not set)
            8022h invalid selector
            8025h invalid linear address
(changing the limit would cause the descriptor
to   reference a linear address range outside
that allowed for DPMI clients.)

Int 31h Function 0009h
Set Descriptor Access Rights
Modifies the access rights and type fields in the
LDT descriptor for the specified segment.
Call with:
AX   =   0009h
BX   =   selector
CL   =   access rights/type byte
CH   =   80386 extended access
            rights/type byte
Returns:
if function successful
CF   =   clear
if function unsuccessful
CF   =   set
AX   =   error code
            8021h invalid value (access
            rights/type bytes invalid)
            8022h invalid selector
            8025h invalid linear   address
(changing the access rights/type bytes would

cause the descriptor to reference a linear address range outside that allowed for DPMI clients.)

Int 31h Function 000Ah
Create Alias Descriptor
Creates a new LDT data descriptor that has the same base and limit as the specified descriptor.
Call with:
AX = 000Ah
BX = selector
Returns:
if function successul
CF = clear
AX = date selector (alias)
if function unsuccessful
CF = set
AX = error code
8011h descriptor unavailable
8022h invalid selector

Int 31h Function 000Bh
Get Descriptor
Copies the local descriptor table (LDT) entry for the specified selector into an 8-byte buffer.
Call with:
AX = 000Bh
BX = selector
ES:(E)DI = selector:offset of 8 byte buffer
Returns:
if function successful
CF = clear
and buffer pointed to by ES:(E)DI contains descriptor if function unsuccessful
CF = set
AX = error code
8022h invalid selector

Int 31h Function 000Ch
Set Descriptor
Copies the contents of an 8-byte buffer into the LDT descriptor for the specified selector.
Call with:
AX = 000Ch
BX = selector
EX:(E)DI = selector:offset of
8-byte buffer containing
descriptor
Returns:
if function successful
CF = clear
if function unsuccessful
CF = set
AX = error code
8021h invalid value (access rights/types byte invalid)
8022h invalid selector
8025h invalid linear address (descriptor references a linear

address range outside that allowed for DPMI clients)

Int 31h Function 000Dh
Allocate specific LDT Descriptor
Allocates a specific LDT descriptor.
Call with:
AX = 000Dh
BX = selector
Returns:
if function successful
CF = clear
and descriptor has been allocated
if function unsuccessful
CF = set
AX = error code
8011h descriptor unavailable (descriptor is in use)
8012h invalid selector (references GDT or beyond the LDT limit)

Int 31h Function 0100h
Allocate DOS Memory Block
Allocates a block of memory from the DOS memory pool, i.e. memory below the 1 MB boundary that is controlled by DOS. Such memory blocks are typically used to exchange data with Real mode programs, TSRs, or device drivers. The function returns both the Real mode segment base address of the block and one or more descriptors that can be used by Protected mode applications to access the block.
Call with:
AX = 0100h
BX = number of (16-byte)
paragraphs desired
Returns:
if function successful
CF = clear
AX = Real mode segment base
address of allocated block
DX = selector for allocated block
if function unsuccessul
CF = set
AX = error code
0007h memory control blocks damaged (also returned by DPMI 0.9 hosts)
0008h insufficient memory (also returned by DPMI 0.9 hosts).
BX = size of largest available block
in paragraphs

> **FORMAT:**
>
> *Heading*
> *Description*
> *Call with*
> *Returns*

Int31h Function 0101h
Free DOS Memory Block
Frees a memory block that was previously allocated with the Allocate DOS Memory Block function (Int 31h Function 0100h).
Call with:
AX   =   0101h
DX   =   selector of block to be freed
Returns:
if function successful
CF   =   clear
if function unsuccessful
CF   =   set
AX   =   error code
            0007h memory control blocks damaged (also returned by DPMI 0.9 hosts).
            0009h incorrect memory segment specified (also returned by DPMI 0.9 hosts).
            8022h invalid selector

Int 31h Function 0102h
Resize DOS Memory Block
Changes the size of a memory block that was previously allocated with the Allocate DOS Memory Block function (int 31h Function 0100h).
Call with:
AX   =   0102h
BX   =   new block size in (16-byte) paragraphs
DX   =   selector of block to modify
Returns:
if functions successful
CF   =   clear
if function unsuccessful
CF   =   set
AX   =   error code
            0007h memory control blocks damaged (also returned by DPMI 0.9 hosts).
            0008h insufficient memory (also returned by DPMI 0.9 hosts).

            0009h incorrect memory segment specified (also returned by DPMI 0.9 hosts).
            8011h descriptor unavailable
            8022h invalid selector
BX   =   maximum possible block size (paragraphs)

Int 31h Function 0200h
Get Real Mode Interrupt Vector
Returns the contents of the current virtual machine's Real mode interrupt vector for the specified interrupt.
Call with:
AX   =   0200h
BL   =   interrupt number
Returns:
CF   =   clear (this function always succeeds)
CX:DX =   segment:offset of real mode interrupt handler

Int 31h Function 0201h
Set Real Mode Interrupt Vector
Sets the current virtual machine's Real mode interrupt vector for the specified interrupt.
Call with:
AX   =   0201h
BL   =   interrupt number
CX:DX =   segment:offset of real mode interrupt handler
Returns:
CF   =   clear (this function always succeeds)

Int 31h Function 0202h
Get Processor Exception Handler Vector
Returns the address of the current client's Protected mode exception handler for the specified exception number. This function should be avoided by DPMI 1.0 clients.
Call with:
AX   =   0202h
BL   =   exception number (00h-1Fh)
Returns:
if function successful
CF   =   clear
CX:(E)DX = selector:offset of exception handler
if function unsuccessful
CF   =   set
AX   =   error code
            8021h invalid value (BL not in range 0-1Fh)

Int31h Function 0203h
Set Processor Exception Handler Vector
Sets the address of a handler for a CPU exception or fault, allowing a Protected mode application to intercept processor exceptions (such as segment not present faults) that are not handled by the DPMI host and would otherwise generate a fatal error. This function should be avoided by DPMI 1.0 clients.
Call with:
AX   = 0203h
BL   = exception/fault number
         (00h-1Fh)
CX:(E)DX = selector:offset of
         exception handler
Returns:
if function successful
CF   = clear
if function unsuccessful
CF   = set
AX   = error code
         8022h invalid value (BL not in
         range 0-1Fh)
         8022h invalid selector

Int 31h Function 0204h
Get Protected Mode Interrupt Vector
Returns the address of the current Protected mode interrupt handler for the specified interrupt.
Call with:
AX   = 0204h
BL   = interrupt number
Returns:
CF   = clear (this function always
         succeeds)
CX:(E)DX = selector:offset of
         exception handler

Int 31h Function 0205h
Set Protected Mode Interrupt Vector
Sets the address of Protected mode handler for the specified interrupt into the interrupt vector.
Call with:
AX   = 0205h
BL   = interrupt number
CX:(E)DX = selector:offset of
         exception handler
Returns:
if function successful
CF   = clear
if function unsuccessful
CF   = set
AX   = error code
         8022h invalid selector

Int 31h Function 0300h
Simulate Real Mode Interrupt
Simulates an interrrupt in Real mode. The function transfers control to the address specified by the Real mode interrupt vector. The Real mode handler must return by executing an IRET.
Call with:
AX   = 0300h
BL   = interrupt number
BH   = flags
        Bit Significance
        0   reserved for historical
            reason, must be zero
        1-7 reserved, must be zero
CX   = number of words to copy
        from Protected mode to real
        mode stack
ES:(E)DI = selector:offset of real
        mode register data structure in
        the following format:

Offset	Length	Contents
00h	4	DI or EDI
04h	4	SI of ESI
08h	4	BP or EBP
0Ch	4	reserved, should be 0
10h	4	BX or EBX
14h	4	DX or EDX
18h	4	CX or ECX
1Ch	4	AX or EAX
20h	2	CPU status flags
22h	2	ES
24	2	DS
26h	2	FS
28h	2	GS
2Ah	2	IP (reserved, ignored)
2Ch	2	CS (reserved,ignored)
2Eh	2	SP
30h	2	SS

Returns:
if function successful
CF   = clear
ES:(E)DI = selector:offset of
        modified Real mode register
        data structure
if function unsuccessful
CF   = set
AX   = error code
        8012h linear memory
        unavailable (stack)
        8013h physical memory
        unavailable (stack)
        8014h backing store
        unavailable (stack)
        8021h invalid value (CX too
        large)

> **FORMAT:**
>
> *Heading*
> *Description*
> *Call with*
> *Returns*

Int 31h Function 0301h
Call Real Mode Procedure With Far Return Frame
Simulates a FAR CALL to a real Mode procedure. The called procedure must return by executing a RETF (far return) instruction.
Call with:

AX	=	0301h
BH	=	flags

Bit Significance
0   reserved for historical reason, must be zero
1-7 reserved, must be zero

CX	=	number of words to copy from Protected mode to real mode stack
ES:(E)DI	=	selector:offset of real mode register data structure in the following format:

Offset	Length	Contents
00h	4	DI or EDI
04h	4	SI or ESI
08h	4	BP or EBP
0Ch	4	reserved, ignored
10h	4	BX or EBX
14h	4	DX or EDX
18h	4	CX or ECX
1Ch	4	AX or EAX
20h	2	CPU status flags
22h	2	ES
24h	2	DS
26h	2	FS
28h	2	GS
2Ah	2	IP
2Ch	2	CS
2Eh	2	SP
30h	2	SS

Returns:
if function successful

CF	=	clear
ES:(E)DI	=	selector:offset of modified Real mode register data structure

if function unsucessful

CF	=	set
AX	=	error

8012h linear memory unavailable (stack)
8013h physical memory unavailable (stack)
8014h backing store unavailable (stack)
8021h invalid value (CX too large)

Int 31h Function 0302h
Call Real Mode Procedure With IRET Frame
Simulates a FAR CALL with flags pushed on the stack to a Real mode procedure. The Real mode routine must return by executing an IRET instruction.
Call with:

AX	=	0302h
BH	=	flags

Bit Significance
0   reserved for historical reason, must be zero
1-7 reserved, must be zero

CX	=	number of words to copy from Protected mode to real mode stack
ES:(E)DI	=	selector:offset of real mode register data structure in the following format:

Offset	Length	Contents
00h	4	DI or EDI
04h	4	SI or ESI
08h	4	BP or EBP
0Ch	4	reserved, ignored
10h	4	BX or EBX
14h	4	DX or EDX
18h	4	CX or ECX
1Ch	4	AX or EAX
20h	2	CPU status flags
22h	2	ES
24h	2	DS
26h	2	FS
28h	2	GS
2Ah	2	IP
2Ch	2	CS
2Eh	2	SP
30h	2	SS

Returns:
if function successful

CF	=	clear
ES:(E)DI	=	selector:offset of modified Real mode register data structure

if function unsuccessful

CF	=	set
AX	=	error code

8012h linear memory unavailable (stack)
8013h physical memory unavailable (stack)

8014h backing store
unavailable (stack)
8021h invalid value (CX too
large)

Int 31h Function 0303h
Allocate Real Mode Callback Address
Returns a unique Real mode segment:offset,
known as a "Real mode callback," that will
transfer control from Real mode to a Protected
mode procedure. Callback addresses obtained
with this function can be passed by a Protected
mode program to a Real mode application,
interrupt handler, device driver, or TSR, so that
the Real mode program can call procedures
within the Protected mode program or notify the
Protected mode program of an event.
Call with:
AX   =   0303h
DS:(E)SI=   selector:offset of
            Protected mode procedure to
            call
ES:(E)DI=   selector:offset of
            32h-byte buffer for Real mode
            register data structure to be
            used when calling callback
            routine.
Returns:
if function successful
CF   =   clear
CX:DX   =   segment:offset of real
             mode callback
if function unsuccessful
CF   =   set
AX   =   error code
             8015h callback unavailable

Int 31h Function 0304h
Free Real Mode Callback Address
Releases a Real mode callback address that was
previously allocated with the Allocate Real
Mode Callback Address function (Int 31h
Function 0303h).
Call with:
AX   =   0304h
CX:DX   =   Real mode callback
             address to be freed
Returns:
if function successful
CF   =   clear
if function unsuccessful
CF   =   set
AX   =   error code
             8024h invalid callback address

Int 31h Function 0305h
Get State Save/Restore Addresses
Returns the addresses of two procedures used to
save and restore the state of the current task's

registers in the mode which is not currently
executing.
Call with:
AX   =   0305h
Returns:
CF   =   clear (this function always
             succeeds)
AX   =   size of buffer in bytes
             required to save state
BX:CX   =   Real mode address of
             routine used to save/restore
             state
SI:(E)DI =   Protected mode
             address of routine used to
             save/restore state

Int 31h Function 0306h
Get Raw Mode Switch Addresses
Returns addresses that can be called for
low-level mode switching
Call with:
AX   =   0306h
Returns:
CF   =   clear (this function
             always succeeds)
BX:CX   =   real to Protected mode switch
             address
SI:(E)DI =   protected to-Real mode
             switch address

Int 31h Function 0400h
Get Version
Returns the version number of the DPMI
Specification implemented by the DPMI host.
Clients can use this information to determine
which function calls are supported in the current
environment.
Call with:
AX   =   0400h
Returns:
CF   =   clear (this function always
             succeeds)
AH   =   DPMI major version as a
             binary number
AL   =   DPMI minor version as a
             binary number
BX   =   flags
             Bits Significance
             0   0=host is 16-bit DPMI
                   implementation
                   1=host is 32-bit (80386)
                   DPMI implementation
             1   0=CPU returned to
                   Virtual 86 mode for
                   reflected interrupts
                   1=CPU returned to real
                   mode for reflected
                   interrupts

> **FORMAT:**
>
> *Heading*
> *Description*
> *Call with*
> *Returns*

2		0=virtual memory not supported 1=virtual memory supported
3		reserved, for historical reasons
4-15		reserved for later use

CL	=	processor type 02h = 80286 03h = 80386 04h = 80486 05h-FFh reserved for future Intel processors
DH	=	current value of virtual master PIC2 base interrupt
DL	=	current value of virtual slave PIC base interrupt

Int 31h Function 0500h
Get Free Memory Information
Returns information about the amount of available physical memory, linear address space, and disk space for page swapping. Since DPMI clients will often run in multitasking environments, the information returned by this function should only be considered as advisory. DPMI 1.0 clients should avoid use of this function.
Call with:

AX	=	0500h
ES:(E)DI=		selector:offset of 48-byte buffer

Returns:

CF	=	clear (this function always succeeds)

and the buffer is filled with the following information:

Offset	Length	Contents
00h	4	Largest available free block in bytes
04h	4	Maximum unlocked page allocation in pages
08h	4	Maximum locked page allocation in pages
0Ch	4	Linear address space size in pages
10h	4	Total number of unlocked pages
14h	4	Total number of free pages
18h	4	Total number of physical pages
1Ch	4	Free linear address space in pages
20h	4	Size of paging file/partition in pages
24h	0Ch	Reserved, all bytes set to 0FFh

Int 31h Function 0501h
Allocate Memory Block
Allocates and commits a block of linear memory.
Call with:

AX	=	0501h
BX:CX	=	size of block (bytes, must be nonzero)

Returns:
if function successful

CF	=	clear
BX:CX	=	linear address of allocated memory block
SI:DI	=	memory block handle (used to resize and free block)

if function unsuccessful

CF	=	set
AX	=	error code 8012h linear mem. unavailable 8013h physical mem. unavailable 8014h backing store unavailable 8016h handle unavailable 8021h invalid value (BX:CX = 0)

Int 31h Function 0502h
Free Memory Block
Frees a memory block that was previously allocated with either the Allocate Memory Block function (Int 31h Function 0501h) or the Allocate Linear Memory Block function (Int 31h Function 0504h).
Call with:

AX	=	0502h
SI:DI	=	memory block handle

Returns:
if function successful

CF	=	clear

if function unsuccessful

CF	=	set
AX	=	error code 8023h invalid handle

Int 31h Function 0503h

Resize Memory Block

Changes the size of a memory block that was previously allocated with either the Allocate Memory Block function (Int 31h Function 0501h) or the Allocate Linear Memory Block function (Int 31h Function 0504h).

Call with:

AX = 0503h
BX:CX = new size of block
(bytes, must be nonzero)
SI:DI = memory block handle

Returns:

if function successful
CF = clear
BX:CX = new linear address of
memory block
SI:DI = new handle for memory
block

if function unsuccessful
CF = set
AX = error code
8012h linear memory
unavailable
8013h physical memory
unavailable
8014h backing store
unavailable
8016h handle unavailable
8021h invalid value (BX:CX =
0)
8023h invalid handle (In
SI:DI)

Int 31h Function 0600h

Lock Linear Region

Locks the specified linear address range.

Call with:

AX = 0600h
BX:CX = starting linear address
of memory to lock
SI:DI = size of region to lock (bytes)

Returns:

if function successful
CF = clear

if function unsuccessful
CF = set
AX = error code
8013h physical memory
unavailable
8017h lock count exceeded
8025h invalid linear address
(unallocated pages)

Int 31h Function 0601h

Unlock Linear Region

Unlocks a linear address range that was previously locked using the Lock Linear Region function (Int 31h Function 0600h).

Call with:

AX = 0601h
BX:CX = starting linear address
of memory to unlock
SI:DI = size of region to unlock
(bytes)

Returns:

if function successful
CF = clear

if function unsuccessful
CF = set
AX = error code
8002h invalid state (page not
locked)
8025h invalid linear address
(unallocated pages)

Int 31h Function 0602h

Mark Real Mode Region as Pageable

Advises the DPMI host that the specified memory below the 1 MB boundary may be paged to disk.

Call with:

AX = 0602h
BX:CX = starting linear address
of memory to mark as
pageable
SI:DI = size of region to be marked
(bytes)

Returns:

if function successful
CF = clear

if function unsuccessful
CF = set
AX = error code
8002h invalid state (region
already marked as pageable)
8025h invalid linear address
(region is above 1MB
boundary)

Int 31h Function 0603h

Relock Real Mode Region

Relocks a memory region that was previously declared as pageable with the Mark Real Mode Region as Pageable function (Int 31h Function 0602h).

Call with:

AX = 0603h
BX:CX = starting linear address
of memory to relock
SI:DI = size of region to relock
(bytes)

Returns:

if function successful
CF = clear

if function unsuccessful
CF = set
AX = error code

```
FORMAT:

Heading
Description
Call with
Returns
```

8002h invalid state (region not marked as pageable)
8013h physical memory unavailable
8025h invalid linear address (region is above 1 MB boundary)

Int 31h Function 0604h
Get Page Size
Returns the size of a single memory page in bytes.
Call with:
AX   =   0604h
Returns:
if function successful
CF   =   clear
BX:CX =   page size in bytes
if function unsuccessful
CF   =   set
AX   =   error code
8001h unsupported function (16-bit host)

Int 31h Function 0702h
Mark Page as Demand Paging Candidate
Notifies the DPMI host that a range of pages may be placed at the head of the page-out candidate list, forcing these pages to be replaced ahead of other pages even if the memory has been accessed recently. The contents of the pages will be preserved.
Call with:
AX   =   0702h
BX:CX =   starting linear address of pages to mark as paging candidates
SI:DI   =   size of region to mark (bytes)
Returns:
if function successful
CF   =   clear
if function unsuccessful
CF   =   set
AX   =   error code
8025h invalid linear addreses (range unallocated)

Int 31h Function 0703h
Discard Page Contents
Discards the entire contents of a given linear memory range. This function is used when a memory object (such as a data structure) that occupies a given area of memory is no longer needed, so that the area will not be paged to disk unnecessarily.
The contents of the discarded region will be undefined.
Call with:
AX   =   0703h
BX:CX =   starting linear address of pages to discard
SI:DI   =   size of region to discard (bytes)
Returns:
if function successful
CF   =   clear
if function unsuccessful
CF   =   set
AX   =   error code
8025h invalid linear address (range unallocated)

Int 31h Function 0800h
Physical Address Mapping
Converts a physical address into a linear address. This function allows device drivers running under DPMI hosts which use paging to reach physical memory that is associated with their devices above the 1 MB boundary Examples of such devices are the Weitek numeric coprocessor (usually mapped at 3 GB), buffers that hold scanner bit maps, and high-end displays that can be configured to make display memory appear in extended memory.
Call with:
AX   =   0800h
BX:CX =   physical address of memory
SI:DI   =   size of region to map (bytes)
Returns:
if function successful
CF   =   clear
BX:CX =   linear address that can be used to access the physical memory
if function unsuccessful
CF   =   set
AX   =   error code
8003h system integrity (DPMI host memory region)
8021h invalid value (address is below 1 MB boundry)

Int 31h Function 0900h
Get and Disable Virtual Interrupt State
Disables the virtual interrupt flag and returns the previous state of the virtual interrupt flag.

Call with:
AX   =   0900h
Returns:
Virtual interrupts disabled
CF   =   clear (this function always
           succeeds)
AL   =   0 if virtual interrupts were
           previously disabled
      =   1 if virtual interrupts were
           previously enabled

Int 31h Function 0901h
Get and Enable Virtual Interrupt State
Enables the virtual interrupt flag and returns the
previous state of the virtual interrupt flag.
Call with:
AX   =   0901h
Returns:
Virtual interrupts enabled
CF   =   clear (this function always
           succeeds)
AL   =   0 if virtual interrupts were
           previously disabled
      =   1 if virtual interrupts were
           previously enabled

Int 31h Function 0902h
Get Virtual Interrupt State
Returns the current state of the virtual interrupt
flag.
Call with:
AX   =   0902h
Returns:
CF   =   clear (this function always
           succeeds)
AL   =   0 if virtual interrupts are
           disabled
      =   1 if virtual interrupts are
           enabled

Int 31h Function 0A00h
Get Vendor-Specific API Entry Point
Returns an address which can be called to use
host-specific extensions to the standard set of
DPMI functions. DPMI 1.0 clients should avoid
use of this function.
Call with:
AX   =   0A00h
DS:(E)SI= selector:offset of
           ASC11Z (null-terminated
           string which identifies the
           DPMI host vendor
Returns:
if function successful
CF   =   clear
ES:(E)DI= selector:offset of
           extended API entry point, and
           DX, GS,EAX,ECX,EDX,ESI,
           and EBP may be modified.

if function unsuccessful
CF   =   set
AX   =   error code
           8001h unsupported function
           (extension not found)

Int 31h Function 0B000h
Set Debug Watchpoint
Sets a debug watchpoint at the specified linear
address.
Call with:
AX   =   0B00h
BX:CX =   linear address of
           watchpoint
DL   =   size of watchpoint
           (1,2, or 4 bytes)
DH   =   type of watchpoint
           0 = execute
           1 = write
           2 = read/write
Returns:
if function successful
CF   =   clear
BX   =   watchpoint handle
if function unsuccessful
CF   =   set
AX   =   error code
           8016h to many breakpoints
           8021h invalid value (in DL or
           DH)
           8035h invalid linear address
           (linear address not mapped or
           alignment error)

Int 31h Function 0B01h
Clear Debug Watchpoint
Clears a debug watchpoint that was previously
set using the Set Debug Watchpoint function
(Int 31h Function 0B00h), and releases the
watchpoint handle.
Call with:
AX   =   0B01h
BX   =   watchpoint handle
Returns:
if function successful
CF   =   clear
if function unsuccessful
CF   =   set
AX   =   error code
           8023h invalid handle

Int 31h Function 0B02h
Get State of Debug Watchpoint
Returns the state of a debug watchpoint that was
previously set using the Set Debug Watchpoint
function (Int 31h Function 0B00h).
Call with:
AX   =   0B02h
BX   =   watchpoint handle

FORMAT:

Heading
Description
Call with
Returns

Returns:
if function successful
CF   =   clear
AX   =   watchpoint status
            Bit Significance
            0   0 = watchpoint has not
                 been encountered
                 1 = watchpoint has been
                 encountered
            1-15 reserved
if function unsuccessful
CF   =   set
AX   =   error code
            8023h invalid handle

Int 31h Function 0B03h
Reset Debug Watchpoint
Resets the state of a previously defined debug
watchpoint; i.e. a subsequent call to Int 31h
Function 0B02h will indicate that the debug
watchpoint has not been encountered.
Call with:
AX   =   0B03h
BX   =   watchpoint handle
Returns:
if function successful
CF   =   clear
if function unsuccessful
CF   =   set
AX   =   error code
            8023h invalid handle

D

INT-2F Extensions

These extensions to INT-2Fh are provided by Windows. Two others, functions 1686h and 1687h, are part of the DPMI and are described in Appendix C.

The source of this reference information is Microsoft's *Device Development Kit*, and more material is to be found there, with many example programs.[1]

Function	Description
1600h	Get Enhanced-Mode Windows Installed State
1602h	Get Enhanced-Mode Windows Entry-Point Address
1605h	Windows Initialization Notification
1606h	Windows Termination Notification
1607h	Device Call Out
1608h	Windows Initialization Complete Notification
1609h	Windows Begin Exit
1680h	Release Current VM Time-Slice
1681h	Begin Critical Section
1682h	End Critical Section
1683h	Get Current Virtual Machine ID
1684h	Get Device Entry Point Address
1685h	Switch VMs and CallBack

---

[1]   The specific reference for Appendix D is the *Device Driver Adaptation Guide*, Microsoft Device Development Kit, version 3.1, © 1992 Microsoft Corporation. All rights reserved. Reprinted with permission from Microsoft Corporation.

Function	Description
4000h	Enable VM-Assisted Save/Restore
4001h	Notify Background Switch
4002h	Notify Foreground Switch
4003h	Enter Critical Section
4004h	Exit Critical Section
4005h	Save Video Register State
4006h	Restore Video Register State
4007h	Disable VM-Assisted Save/Restore

INT-2F Extensions Detail

Interrupt 2Fh Function 1600h
Get Enhanced-Mode Windows Installed State
Determines whether 386 Enhanced-mode Windows is running. If a program intends to use a 386 Enhanced-mode Windows function, it must first use this function to make sure that 386 Enhanced-mode Windows is running.
This function is valid under all versions of 386 Enhanced-mode Windows.
Call with:
AX   =   1600h
Return value:
The return value is 00h or 80h in the AL register if 386 Enhanced-mode Windows is not running. If 386 Enhanced-mode Windows is running, the return value depends on the version of Windows. Windows/386 version 2.x returns 01h or 0FFh. Windows version 3.x returns the major version number in the AL register, and the minor version number in the AH register.

Interrupt 2Fh Function 1602h
Get Enhanced-Mode Windows Entry Point
Returns the address of the 386 Enhanced-mode Windows entry-point procedure. Applications can call this procedure to direct Windows/386 version 2.x to carry out specific actions.
This function applies to Windows/386 version 2.x only. It is provided under Windows version 3.x for compatibility reasons.
Call with:
AX   =   1602h
Return value:
The return value contains the Windows entry-point address in the ES:DI registers.
Comments
Although the Windows entry-point address is the same for every virtual machine, an application can call this function any number of times.
To direct Windows to carry out a specific action, the application sets the AX register to 0000h. This function retrieves the current VM identifier and copies the identifier to the BX register.
Additionally, the application must place a return address in the ES:DI register pair and use the jmp instruction to transfer control to the Windows entry point.

FORMAT:

Heading
Description
Call with
Returns
[Comments]

Interrupt 2Fh Function 1605h
Windows Initialization Notification
Notifies MS-DOS device drivers and TSRs that standard- or 386 Enhanced-mode Windows is starting. Windows calls this function as it starts allowing MS-DOS device drivers and TSRs that monitor Interrupt 2Fh the opportunity to prepare for running in the Windows environment.
Call with:

AX	=	1605h
ES:BX	=	0:0
DS:SI	=	0:0
CX	=	0 (normally)
DX	=	Specifies whether standard- or 386 Enhanced-mode Windows is initializing. 386 Enhanced-mode Windows sets bit 0 to 0; standard-mode Windows sets bit 0 to 1. Only bit 0 is used; all other bits reserved and undefined.
DI	=	Specifies the version number of Windows. The major version number is in the high-order byte, the minor version number in low-order byte.

Return value:
The return value is 0 in the CX register if all MS-DOS device drivers and TSRs monitoring Interrupt 2Fh can run in the Windows environment, and Windows can proceed with initialization. Otherwise, the CX register is nonzero and Windows must terminate.
Comments
Any MS-DOS device driver or TSR that either cannot run in the Windows environment or must adapt its operation when in the Windows environment should add itself to the Interrupt 2Fh chain and watch for this function.
If the device driver or TSR cannot run in the Windows environment, it should set the CX register to a nonzero value, display a message informing the user of its incompatibility with Windows, and return. Windows does not print a message of its own. Instead, it calls Windows

> **FORMAT:**
>
> *Heading*
> *Description*
> *Call with*
> *Returns*
> *[Comments]*

Termination Notification (Interrupt 2Fh Function 1606h) and returns to MS-DOS.
If the device driver or TSR can run in the Windows environment, it should do the following:

1   Call the next device driver or TSR in the Interrupt 2Fh chain to allow all device drivers and TSRs in the chain an opportunity to respond to this function.

2   Upon return from the interrupt chain, carry out the following actions:

   a Free any extended memory. The device driver or TSR takes this action only if it has previously allocated extended memory using the Extended Memory Specification (XMS) interface.

   b Switch the processor to real mode, or set the DS:SI register pair to the address of an Enable/Disable Virtual 8086 Mode callback function. The device driver or TSR takes this action only if it has previously switched the processor to virtual 8086 mode. If standard-mode Windows is starting, the device driver or TSR must switch the processor to real mode. The callback function is permitted for 386 Enhanced-mode Windows only.

   c Initialise a Win386_Startup_ Info_Struc structure, and copy the address of the structure to the ES:BX register pair. The device driver or TSR carries out this action only if 386 Enhanced-mode Windows is starting.

3   Return (using the iret instruction) but without changing the CX register.

For more information about these procedures, see the following comments:

The device driver or TSR must preserve all registers and restore the original values before returning. The only exceptions to this rule are changes made to the BX, CX, DS, ES, and SI registers as a result of following the previous procedure.

Enable/Disable Virtual 8086 Mode Callback Function
Some device drivers and TSRs, such as expanded memory emulators, switch the processor to virtual 8086 mode. Because 386 Enhanced-mode Windows cannot start successfully while the processor is in this mode, any device driver or TSR that switches to virtual 8086 mode must either switch back to Real mode or supply the address of a callback function that can switch between real and virtual 8086 modes.
Windows uses the callback function to disable virtual 8086 mode before Windows itself enters Protected mode. Windows calls the callback function again to enable virtual 8086 mode after Windows exits Protected mode. Windows calls the callback function using a far call instruction, and it specifies which action to take by setting the AX register to 0 or 1.
To disable virtual 8086 mode, Windows sets the AX register to 0, disables interrupts, and calls the callback function. The function should switch the processor to Real mode, clear the carry flag to indicate success, and return. If an error occurs, the function sets the carry flag and returns. Windows checks the carry flag and terminates if it is set.
To enable virtual 8086 mode, Windows set the AX register to 1, disables interrupts, and calls the callback function. The function should switch the processor to virtual 8086 mode, clear the carry flag, and return. If an error occurs, the function sets the carry flag and returns. However, Windows ignores the carry flag, so if an error occurs no action is taken and the processor is left in Real mode.
Whether an error occurs when enabling or disabling virtual 8086 mode, it is up to the callback function to display any error message to the user. Also, the callback function must not enable interrupts unless an error occurs, and the function sets the carry flag.
A device driver or TSR supplies a callback function by copying the address of the function to the DS:SI register pair when it processes the Windows Initialization Notification (Interrupt 2Fh Function 1605h). Windows permits only one callback function, so the device driver or TSR should first check to make sure that the DS and SI registers are both zero. If they are

nonzero, the device driver or TSR should set the CX register to a nonzero value and return, directing Windows to terminate without starting.

Initializing a Win386_Startup_Info_Struc Structure

An MS-DOS device driver or TSR initializes a Win386_Startup_Info_Struc structure to direct 386 Enhanced-mode Windows to load the virtual device and to reserve the instance data the device driver or TSR needs to operate in the Windows environment. The device driver or TSR is also responsible for establishing a chain of startup structures by copying the contents of the ES:BX register pair to the Next_Dev_Ptr member. It is assumed that any other device driver or TSR in the Interrupt 2Fh chain will have set the ES:BX register pair to the address of its own startup structure prior to returning.

Any device driver or TSR can use a Windows virtual device to help support its operation in the 386 Enhanced-mode Windows environment. To specify a virtual device, the device driver or TSR sets the SIS_Virt_Dev_File_Ptr member to the address of the virtual device's filename. The device file is assumed to be in the Windows SYSTEM directory. The device driver or TSR can also set the SIS_Reference_Data member to specify additional data to be passed to the virtual device when loaded.

Any device driver or TSR can reserve instance data for itself. Instance data is one or more blocks of memory used by the device or TSR, and managed by Windows. For device drivers or TSRs loaded before 386 Enhanced-mode Windows starts, reserving instance data allows the device driver or TSR to keep separate data for each virtual machine. Whenever Windows switches virtual machines, it saves the previous VM's instance data and loads the current VM's instance data. If a device driver or TSR does not specify instance data, the same data is used for all virtual machines.

A device driver or TSR reserves instance data by appending an array of Instance_Item_Struc structures to the Win386_Startup_Info_Struc structure. The last structure in the array must be set to zero. Each Instance_Item_Struc structure specifies the starting address and size (in bytes) of an instance data block.

The device driver or TSR must copy the address of its startup structure to the ES:BX register pair before returning.

See also at end of this appendix:
Win386_Startup_Info_Struc,
Instance_Item_Struc

Interrupt 2Fh Function 1606h
Windows Termination Notification
Notifies MS-DOS device drivers and TSRs that standard- or 386 Enhanced-mode Windows is terminating. Windows calls this function as it terminates allowing MS-DOS device drivers and TSRs that monitor Interrupt 2Fh the opportunity to prepare for leaving the Windows environment.
Call with:
AX = 1606h
DX = Specifies whether standard- or 386 Enhanced-mode Windows is terminating. 386 Enhanced-mode Windows sets bit 0 to 0; standard-mode Windows sets bit 0 to 1. Only bit 0 is used; all other bits reserved and undefined.
Return value:
This function has no return value.
Comments
Windows calls this function when the processor is in Real mode.

Interrupt 2Fh Function 1607h
Device Call Out
Directs an MS-DOS device driver or TSR to provide information to the calling virtual device. Although the BX register specifies a device identifier, other registers may be used to specify the action to take.
Called with:
AX = 1607h
BX = device identifier for a virtual device
Return value:
The return value depends on the specific action requested.
Comments
This function typically is used by a virtual device to communicate with the driver or TSR that explicitly loaded the virtual device. It is up to the virtual device to supply a correct device identifier and any other parameters that specify what action to take. It is up to the driver or TSR to monitor Interrupt 2Fh and respond to the function appropriately.

A virtual device can call this function at any time, either in Real mode or after 386 Enhanced-mode Windows has started.

Interrupt 2Fh Function 1608h
Windows Initialization Complete Notification
Notifies MS-DOS device drivers and TSRs that 386 Enhanced-mode Windows has completed its initialization. Windows calls this function after it has installed and initialized all virtual

FORMAT:

Heading
Description
Call with
Returns
[Comments]

devices allowing MS-DOS device drivers and TSRs that monitor Interrupt 2Fh the opportunity to identify instance data and perform other functions that are restricted to 386 Enhanced-mode Windows initialization.

Call with:
AX   =   1608h
Return value:
This function has no return value.

Comments
When Windows calls this function, all virtual-device initialization is complete, so a device driver or TSR can call virtual-device entry points.

Windows does not necessarily call this function immediately after calling Windows Initialization Notification (Interrupt 2Fh Function 1605h). In particular, virtual devices may call Device Call Out (Interrupt 2Fh Function 1607h) or other functions prior to Windows calling this function. In such cases, any MS-DOS device driver or TSR responding to these calls is responsible for detecting and properly handling these calls.

Interrupt 2Fh Function 1609h
Windows Begin Exit
Notifies MS-DOS device drivers and TSRs that Windows is about to terminate. Windows calls this function when it first begins termination to allow a device driver or TSR to prepare for a return to a non-Windows environment.

Call with:
AX   =1609h
Return value:
This function has no return value.

Comments
Windows calls this function at the start of the Sys_VM_Terminate device control call. All virtual devices still exist, so a device driver or TSR can call a virtual device's entry point if necessary.

Windows does not call this function in the event of a fatal system crash.

Windows may execute Real mode code after this function has been called and before 386 Enhanced-mode Windows returns to Real mode. It is the responsibility of the device driver or TSR to detect and properly handle these situations.

Interrupt 2Fh Function 160Bh
Windows TSR Identify
(This information is missing from the DDK.)
When Windows and Windows Setup start, they broadcast Interrupt 2Fh Function 160Bh to notify TSRs and MS-DOS device drivers. A Windows-aware TSR must identify itself by responding to the notification and adding itself to a linked list of Windows-aware TSRs. A Windows-aware TSR can also direct Windows to load a Windows-based application or a DLL (such as an installable driver) to allow the TSR's services to continue running properly inside the Windows environment.

Each TSR must monitor Interrupt 2Fh and respond to the startup broadcast by attaching itself to a linked list. A TSR that intercepts this broadcast must do the following:

1   Allocate space for a TSR_Info_Struc structure.
2   At the minimum, fill in the TSR_PSP_Segment, TSR_API_Ver_ID, and TSR_ID_Block fields.
3   Save the current content of the ES:DI register pair in the TSR_Next field.
4   Set the ES:DI register to point to its own TSR_Info_Struc structure.
5   Pass execution control to the next handler in the Interrupt 2Fh chain.

A TSR should never assume that the ES:DI register pair is 0:0 because it may not be the first TSR on the Interrupt 2Fh chain. A TSR must always chain to the next Interrupt 2Fh handler so that TSRs following it in the chain can add themselves to the Windows-aware linked list.

The TSR_Info_Struc structure (defined in the DDK include file INT2FAPI.INC) is a Microsoft Macro assembler structure — see end of this Appendix.

Interrupt 2Fh Function 1680h
Release Current VM Time-Slice
Directs Windows to suspend the time slice of the current VM and start a new time slice for another VM. MS-DOS programs use this function when they are idle, such as when waiting for user input, to allow 386 Enhanced-mode Windows to run other programs that are not idle.

Call with:
AX   =1680h
Return value:

The return value is 00h in the AL register if the function is supported. Otherwise, AL is unchanged (contains 80h).

Comments

Only non-Windows programs should use Release Current VM Time-Slice; Windows applications should yield by calling the WaitMessage function. A program can call this function at any time, even when running in environments other than 386 Enhanced-mode Windows environment. If the current environment does not support the function, the function returns and the program continues execution.

Windows suspends the current VM only if there is another VM scheduled to run. If no other VM is ready, the function returns to the program and execution continues. A program should call the function frequently (for example, once during each pass of the program's idle loop) to give Windows ample opportunity to check for other VMs that are ready for execution.

Before calling this function, a program should check that the Interrupt 2Fh address is not zero.

Interrupt 2Fh Function 1681h

Begin Critical Section

Prevents Windows from switching execution from the current VM to some other. MS-DOS device drivers and TSRs use this function to prevent a task-switch from occurring.

Call with:

AX   =   1680h

Return value:

This function has no return value.

Comments

When a virtual machine is in a critical section, no other task will be allowed to run except to service hardware interrupts. For this reason, the critical section should be released using End Critical Section (Interrupt 2Fh Function 1682h) as soon as possible.

Interrupt 2Fh Function 1682h

End Critical Section

Releases the critical section previously started using Begin Critical Section (Interrupt 2Fh Function 1681h). Every call to Begin Critical Section must be followed by a matching call to End Critical Section.

Call with:

AX   =   1682h

Return value:

This function has no return value.

Interrupt 2Fh Function 1683h

Get Current Virtual Machine ID

Returns the identifier for the current virtual machine. MS-DOS device drivers, TSRs, and other programs use this function to determine which virtual machine is running. This is especially important for programs that independently manage separate data or execution contexts for separate virtual machines.

Call with:

AX   =   1683h

Return value:

The return value is the current virtual-machine identifier in the BX register.

Comments

Each virtual machine has a unique, nonzero identifier. Although Windows currently runs in virtual machine 1, programs should not rely on this. Windows assigns the identifier when it creates the virtual machine, and releases the identifier when it destroys the virtual machine. Since Windows may reuse identifiers from previous, destroyed virtual machines, programs should monitor changes to virtual machines to ensure no mismatches.

Interrupt 2Fh Function 1684h

Get Device Entry Point Address

Retrieves the entry point address for a virtual device's service functions. MS-DOS device drivers or TSRs typically use this function to communicate with virtual devices they have explicitly loaded.

Call with:

AX   =   1684h

BX   =   device identifier (ID).

Return value:

The return value is the entry-point address contained in the ES:DI register pair if the function is supported. Otherwise, ES:DI contain zero.

Comments

Any virtual device can provide service functions to be used by MS-DOS programs. For example, the virtual-display device provides services that the Windows old application program uses to display MS-DOS programs in a window.

It is the responsibility of the MS-DOS program to provide the appropriate virtual-device identifier. The function returns a valid address if the virtual device supports the entry point.

MS-DOS programs call the entry point using a far call instruction. The services provided by the virtual device depend on the device. It is the responsibility of the MS-DOS program to set registers to values that are appropriate to the specific virtual device.

For versions of Windows prior to version 3.0, the program must set the ES:DI register pair to zero before calling this function.

> **FORMAT:**
>
> *Heading*
> *Description*
> *Call with*
> *Returns*
> *[Comments]*

Interrupt 2Fh Function 1685h
Switch VMs and CallBack

Directs Windows to switch to a specific virtual machine and begin execution. After Windows switches, it calls the specified callback function allowing a device driver or TSR to access the data associated with the specified virtual machine. This function is typically used by MS-DOS device drivers and TSRs that support networks, and that need to perform functions in a specific virtual machine.

Call with:

AX = 1685h
BX = Virtual machine ID
CX = Flags. Specifies when to switch. This parameter is a combination of the following bit values.

Bit Meaning
0   Set to 1 to wait until interrupts are enabled.
1   Set to 1 to wait until critical section is released.
All other bits are reserved and must be 0.

DX:SI = Priority. Specifies the priority boost for the virtual machine. It can be one of the following values.

Value: Meaning
Critical_Section_Boost: VM priority is boosted by this value when Begin_Critical_Section is called.
Cur_Run_VM_Boost: Time-slice scheduler boosts each VM in turn by this value to force them to run for their allotted time slice.
High_Pri_Device_Boost: Time critical operations that should not circumvent the critical section boost should use this boost.
Low_Pri_Device_Boost: Used by virtual devices that need an event to be processed in a timely fashion but that are not extremely time critical.
Reserved_High_Boost: Reserved; do not use.
Reserved_Low_Boost: Reserved; do not use.
Time_Critical_Boost: Events that must be processed even when another VM is in a critical section should use this boost. For example, VPICD uses this when simulating hardware interrupts.

ES:DI = CallBack. Points to the callback function.

Return value:

The return value is a cleared carry flag if successful. Otherwise, the function sets the carry flag and sets the AX register to one of the following error values.

Value Meaning
0001h Invalid VM ID
0002h Invalid priority boost
0003h Invalid flags

Comments

Windows calls the callback function as soon as the conditions specified by the Flags parameter are met. This may be before or after Switch VMs and Callback returns.

The callback function can carry out any action, but must save and restore all registers it modifies. The function must execute an iret instruction to return to Windows. The priority for the virtual machine remains at the level specified by Priority until the callback function executes the iret instruction.

Interrupt 2Fh Function 1686h
See Appendix C

Interrupt 2Fh Function 1687h
See Appendix C

Interrupt 2Fh Function 168Bh
Set Focus

(This information is missing from the DDK.)
A TSR uses this function to force the input focus to a given VM. This service has some inherent risks. The function has the following syntax:

```
mov ax, 168Bh  ; set focus.
mov bx, VMId   ; 0 to set focus to current VM,
               ; or VM ID to set focus to a
               ; given VM.
int  2Fh
cmp  al,0       ; 0 if focus is set.
je   success
```

The VMId parameter must either specify a valid virtual machine ID or must be 0 to specify the current virtual machine. (In Windows 3.1, the VMId of the system virtual machine is 1.) The function returns 0 if it changes the focus successfully.

Interrupt 2Fh Function 4000h
Enable VM-Assisted Save/Restore
Directs the virtual-display device (VDD) to notify the virtual machine (VM) application whenever the VDD needs to access the video hardware registers. The VDD returns a value specifying the number and type of video modes the VDD supports when the VM application is in the background.
A VM application calls this function during its initialization.
Call with:
AX   =   4000h
Return value:
The return value is one of the following values, if successful.

Value	Meaning
01h	No modes virtualized in background.
02h	Only text modes virtualized in background.
03h	Only text and single-plane graphics modes virtualized.
04h	Only text, single-plane, and VGA multiplane graphics modes virtualized.
0FFh	All supported video modes virtualized.

Otherwise, the function returns zero in the AL register if virtualization is not supported.
Comments
When a VM application calls this function, the VDD disables I/O trapping of unreadable registers. Thereafter, the VDD calls Save Video Register State and Restore Video Register State (Interrupt 2Fh Functions 4005h and 4006h) if it needs to access the video registers. The VM application must provide an appropriate interrupt handler to process these functions.
When an VM application calls this function, the VDD saves the current state of the video registers. The VDD uses this saved state later to restore video registers before it calls Notify Foreground Switch and Restore Video Register State (Interrupt 2Fh Functions 4002h and 4006h).
After a VM application calls Enable VM-Assisted Save/Restore, the VDD no longer saves video memory across screen switches; it

becomes the application's responsibility to completely reinitialize video memory after a Notify Foreground Switch request.

Interrupt 2Fh Function 4001h
Notify Background Switch
Notifies a VM application that it is being switched to the background. The VM application can carry out any actions, but should do so within 1000ms. This is the amount of time the system waits before switching the application.
Call with:
AX   =   4001h
Return value:
This function has no return value.
Comments
After switching to the background, the application continues to run unless it attempts to access video memory. If the video adapter is in a video mode that the virtual display device (VDD) does not support in the background, the VDD freezes the application until the application can be switched back to the foreground.
VM applications that have called Enable VM-Assisted Save/Restore (Interrupt 2Fh Function 4000h) should avoid accessing video memory and registers to avoid being frozen. Applications that have not called Enable VM-Assisted Save/Restore call access video memory and registers since the VDD saves these after this function returns.

Interrupt 2Fh Function 4002h
Notify Foreground Switch
Notifies a VM application that it has been switched to the foreground and can now access the video memory and registers without being frozen.
The virtual-display device (VDD) calls this function.
Call with:
AX   =   4002h
Return value:
This function has no return value.
Comments
If the VM application has called Enable VM-Assisted Save/Restore (Interrupt 2Fh Function 4000h), VDD restores the video registers to their state prior to the call to Enable VM-Assisted Save/Restore; the application is responsible for restoring video memory. If the VM application has not called Enable VM-Assisted Save/Restore (Interrupt 2Fh Function 4000h), the VDD automatically restores both video memory and registers.

> ## FORMAT:
>
> *Heading*
> *Description*
> *Call with*
> *Returns*
> *[Comments]*

Under certain error conditions, the VDD may call this function without calling a corresponding Notify Background Switch (Interrupt 2Fh Function 4001h).

Interrupt 2Fh Function 4003h
Enter Critical Section
Notifies the virtual-display device that the VM application has entered a critical section and cannot respond to Save Video Register State (Interrupt 2Fh Function 4005h).
A VM application calls this function when it has started critical section processing.
Call with:
AX   =   4003h
Return value:
This function has no return value.
Comments
After the VM application enters the critical section, the virtual display device postpones calling Save Video Register State for up to 1000ms or until the VM application calls Exit Critical Section (Interrupt 2Fh Function 4004h), whichever comes first.
If time elapses without the VM application calling Exit Critical Section, the virtual-display device reprograms the video hardware anyway and, when its operation is complete, calls Notify Foreground Switch (Interrupt 2Fh Function 4002h) in an attempt to reinitialize the application properly.

Interrupt 2Fh Function 4004h
Exit Critical Section
Notifies the virtual-display device that a VM application has completed a critical section and can now respond to Save Video Register State (Interrupt 2Fh Function 4005h).
A VM application calls this function when it has completed critical section processing.
Call with:
AX   =   4004h
Return value:
This function has no return value.
Comments

Calls to Exit Critical Section not preceded by a corresponding call to Enter Critical Section (Interrupt 2Fh Function 4003h) are ignored.

Interrupt 2Fh Function 4005h
The Save Video Register State
Notifies VM applications that the virtual-display device (VDD) requires access to the video hardware registers. The VDD calls this function, for example, when preparing to copy the entire screen to the clipboard.
Call with:
AX   =   4005h
Return value:
This function has no return value.
Comments
The VDD calls this function only if the VM application has called Enable VM-Assisted Save/Restore (Interrupt 2Fh Function 4000h). VM applications that receive Save Video Register State must save any data necessary to restore the current video state and must return within 1000ms. If the application fails to return in time, the virtual-display device accesses the video hardware anyway. After accessing the video hardware registers, the VDD calls Restore Video Register State (Interrupt 2Fh Function 4006h) to notify the application that it can restore its video state.
The VDD calls Save Video Register State only at times when the hardware must be reprogrammed for what are essentially brief and nonvisible operations. For example, the VDD does not call this function prior to calling Notify Background Switch (Interrupt 2Fh Function 4001h).

Interrupt 2Fh Function 4006h
Restore Video Register State
Notifies a VM application that the virtual-display device (VDD) has relinquished its access to the video registers. The VM application should restore the video registers to any state necessary to continue uninterrupted foreground operation.
Call with:
AX   =   4006h
Return value:
This function has no return value.
Comments
The VDD calls this function only if the VM application has called Enable VM-Assisted Save/Restore (Interrupt 2Fh Function 4000h). Before calling this function, the VDD restores any registers it modified to the values they had when the VM application originally called Enable VM-Assisted Save/Restore. In other words, every register is guaranteed to be either unchanged or reset to a previous state; precisely

which registers may be reset is undefined, but the set is restricted to those Sequencer and Graphics Controller registers that do not affect the display.

Interrupt 2Fh Function 4007h
Disable VM-Assisted Save/Restore
Directs the virtual-display device (VDD) to discontinue notifying the VM application when it needs access to video registers.
VM applications call this function when they terminate.
Call with:
AX   =   4007h
Return value:
This function has no return value.
Comments
This function directs the VDD to restore I/O trapping of unreadable registers and to discontinue calling Save Video Register State and Restore Video Register State (Interrupt 2Fh Functions 4005h and 4006h) when it needs access to the registers. Furthermore, the VDD ignores any subsequent calls to Enter Critical Section and Exit Critical Section (Interrupt 2Fh Functions 4003h and 4004h).
This function does not disable Notify Background Switch and Notify Foreground Switch (Interrupt 2Fh Functions 4001h and 4002h).

DATA STRUCTURES

Win386_Startup_Info_Struc

```
Win386_Startup_Info_Struc STRUC
SIS_Version                db  3, 0
SIS_Next_Dev_Ptr           dd  ?
SIS_Virt_Dev_File_Ptr      dd  0
SIS_Reference_Data         dd  ?
SIS_Instance_Data_Ptr      dd  0
SIS_Opt_Instance_Data_Ptr  dd  0
Win386_Startup_Info_Struc ENDS
```

The Win386_Startup_Info_Struc structure contains information that Windows uses to prepare an MS-DOS device driver or TSR program for operation with Windows.

Member: Description
SIS_Version: Specifies the version number of the structure. 386 Enhanced-mode Windows uses this member to determine the size of the structure. This member should contain 3 in the low-order byte and 10 in the high-order byte to indicate that it is version 3.1.
SIS_Next_Dev_Ptr: Points to the next structure in the list. The address of the next structure must be supplied by the next handler in the

Interrupt 2Fh chain. A driver or TSR calls the next handler, then sets this member to the address returned by the handler in the ES:BX register pair.
SIS_Virt_Dev_File_Ptr: Points   to   a null-terminated string that contains the name of a 386 Enhanced-mode Windows virtual device file. MS-DOS devices such as networks use this to force a special 386 Enhanced-mode Windows virtual device to be loaded. If this member is zero, no device is loaded.
SIS_Reference_Data: Specifies reference data for the virtual device. This member, used only when SIS_Virt_Dev_File_Ptr is nonzero, is passed to the virtual device when it is initialized. The member can contain any value and often contains a pointer to some device-specific structure.
SIS_Instance_Data_Ptr: Points to a list of data to be instanced, that is, allocated uniquely for each new virtual machine. If the member is zero, then no data is instanced. Each entry in the list is an Instance_Item_Struc structure. The list is terminated with a 32-bit zero.
SIS_Opt_Instance_Data_Ptr: Points to a list of data to be optionally instanced. Available only if SIS_Version is 4.0.

Instance_Item_Struc

```
Instance_Item_Struc STRUC
  IIS_Ptr      dd  ?
  IIS_Size     dw  ?
Instance_Item_Struc ENDS
```

The Instance_Item_Struc structure specifies the starting address and size of a block of instance data.

Member: Description
IIS_Ptr: Points to the starting address of a block of instance data.
IIS_Size: Specifies the size (in bytes) of the block of instance data pointed to by the IIS_Ptr member.

TSR_Info_Struc

```
TSR_Info_Struc STRUC
  TSR_Next            dd  ?
  TSR_PSP_Segment     dw  ?
  TSR_API_Ver_ID      dw  100h
  TSR_Exec_Flags      dw  0
  TSR_Exec_Cmd_Show dw  0
  TSR_Exec_Cmd        dd  0
  TSR_Reserved        db  4 dup (0)
  TSR_ID_Block        dd  0
  TSR_Data_Block      dd  0
TSR_Info_Struc ENDS
```

The TSR_Info_Struc structure has the following fields:

TSR_Next: Contains the 32-bit address (segment:offset) of the next TSR_Info_Struc structure in the TSR chain.

TSR_PSP_Segment: Specifies the segment address of the TSR's program segment prefix (PSP).

TSR_API_Ver_ID: Specifies the version number of the structure.

TSR_Exec_Flags: Specifies the execution flag. This field must have one of the following values:·

>   TSR_WINEXEC: Execute a
>   Windows-based application.·
>   TSR_LOADLIBRARY: Load a DLL.·
>   TSR_OPENDRIVER: Load a driver.This
>   field is ignored if TSR_Exec_Cmd is 0:0.

TSR_Exec_Cmd_Show: Specifies the flags (for example, SW_SHOWNOACTIVE) to be passed to the WinExec function. This field is ignored if TSR_Exec_Cmd is 0:0.

TSR_Exec_Cmd: Contains a 32-bit address (segment:offset) of a command line string specifying the name and command line parameters of a Windows-based application or a DLL. This field should be 0:0 if automatic loading of a Windows-based application or a DLL is not required.

TSR_Reserved: Contains 4 bytes of reserved space. Do not use this field.

TSR_ID_Block: Contains a 32-bit address (segment:offset) of the TSR's ID block. The first word in the block specifies the number of bytes in the block and is followed by the zero-terminated ID string. Windows uses the ID string to identify the TSR in error messages.

TSR_Data_Block: Contains a 32-bit address (segment:offset) of application-specific data. This pointer is not used by Windows; it is included in the structure so that the TSR can communicate with other TSRs or with Windows-based applications.

INDEX

The INT-2Fh extensions provided by DPMI are described in Appendix C. Other extensions are described in Appendix D. Where further described in the book, see below.

NOTES

1

The DPMI Specification is available from Intel, order number 240977-001. Enquire with your local Intel office or the Intel Literature Distribution Center, P.O. Box 7641, Mt Prospect, IL 60056-7641, USA.

2

The specific reference for Appendix D is the Device Driver Adaptation Guide, Microsoft Device Development Kit, version 3.1, © 1992 Microsoft Corporation. All rights reserved. Reprinted with permission from Microsoft Corporation.

Printed and bound by CPI Group (UK) Ltd, Croydon, CR0 4YY

21/10/2024

01777098-0004